Astronomy

A VISUAL GUIDE

Astronomy

A VISUAL GUIDE

IAN RIDPATH

ADDITIONAL CONTRIBUTORS
GILES SPARROW, CAROLE STOTT

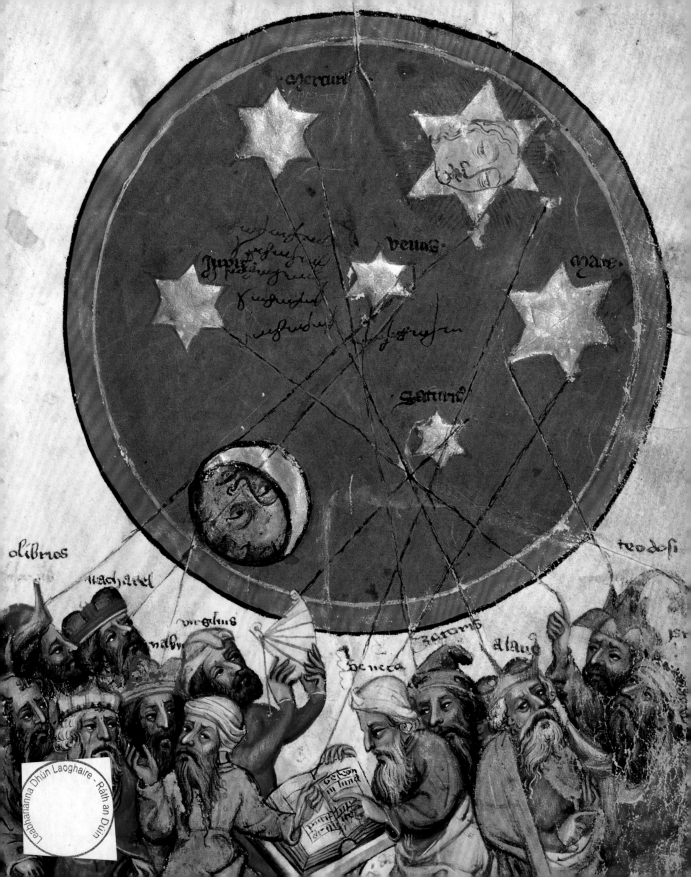

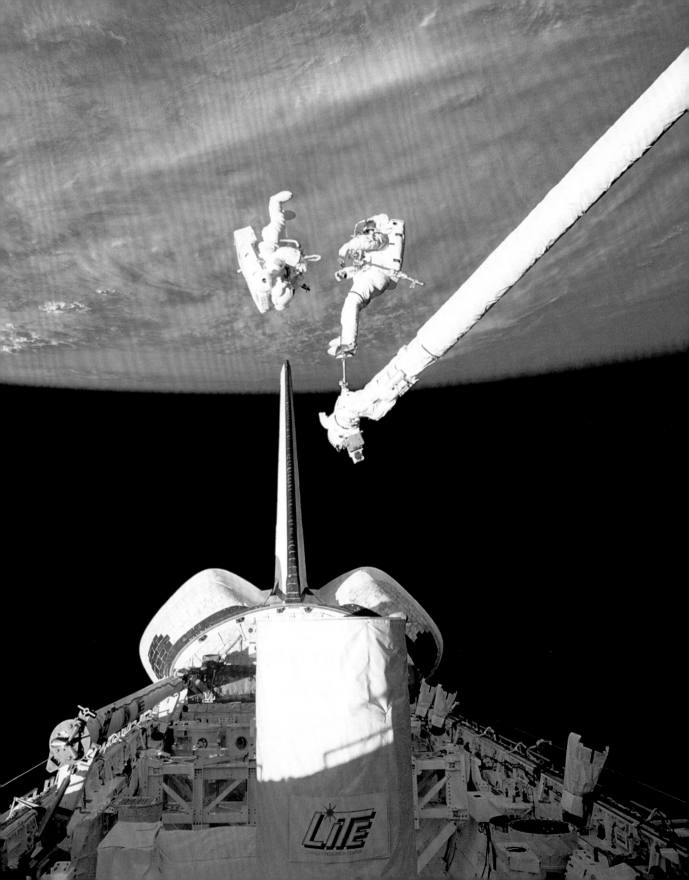

 Penguin Random House

Revised Edition

DK DELHI
Senior Editor Arani Sinha
Art Editor Priyanka Bansal
Assistant Editors Rishi Bryan, Tanya Singhal
Assistant Art Editor Kanupriya Lal
Managing Editor Soma B. Chowdhury
Managing Art Editor Govind Mittal
Assistant Picture Researcher Vishal Ghavri
Picture Research Manager Taiyaba Khatoon
Senior DTP Designer Tarun Sharma
DTP Designer Pawan Kumar
Production Manager Pankaj Sharma
Pre-Production Manager Balwant Singh
Senior Jackets DTP Designer Harish Aggarwal
Jacket Designers Suhita Dharamjit, Juhi Sheth
Jackets Editorial Coordinator Priyanka Sharma
Managing Jackets Editor Saloni Singh

DK LONDON
Editors Francesco Piscitelli, Kate Taylor
Art Editor Renata Latipova
Managing Editor Gareth Jones
Senior Managing Art Editor Lee Griffiths
Producer, Pre-Production David Almond
Senior Producer Mandy Inness
Jacket Designer Surabhi Wadhwa
Design Development Manager Sophia MTT
Jacket Editor Claire Gell
Associate Publishing Director Liz Wheeler
Art Director Karen Self
Publishing Director Jonathan Metcalf

This edition published in 2022
First published in Great Britain in 2006 as
Eyewitness Companions Astronomy by
Dorling Kindersley Limited
DK, One Embassy Gardens, 8 Viaduct Gardens,
London, SW11 7BW

The authorised representative in the EEA is
Dorling Kindersley Verlag GmbH. Arnulfstr. 124,
80636 Munich, Germany

Copyright © 2006, 2018, 2022 Dorling Kindersley Limited
A Penguin Random House Company
10 9 8 7 6 5 4 3 2 1
001–308117–Nov/2022

A CIP catalogue record for this
book is available from the British Library.
ISBN 978-0-2416-2097-7

Printed and bound in China

For the curious
www.dk.com

Contents

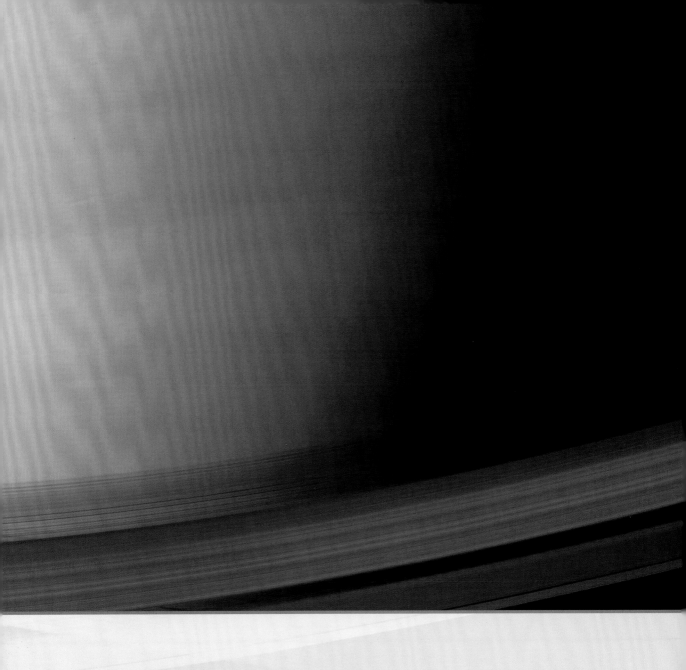

Introduction

From the study of the smallest members of the Solar System to the most distant galaxies, astronomy is a science that knows no bounds. It attempts to answer the most fundamental questions: where did we come from, and are we alone? Yet it remains a science in which amateurs can play a role.

A casual glance at the sky transports us across gulfs of space and time. What you see may appear to be unchanging points of light, but the real truth is far more complex. There is the Orion Nebula, for example, a cloud of gas nearly 1,500 light-years distant, within which are being re-enacted the processes that led to the formation of the Sun and planets, 4.6 billion years ago.

There is the Pleiades, a clutch of hot blue-white stars that emerged from a gaseous nebula like Orion's, and whose youngest members began to glow around the same time as the first humans appeared on Earth.

There is Betelgeuse, a star distended and reddened with age, destined to end its life in a blinding explosion which will spill its constituent atoms into the interstellar mix. Over subsequent aeons, those atoms will be recycled into new generations of stars, planets, and perhaps even life. When humans first began to study the heavens a few thousand years ago, no one knew what stars were. Only in the past few hundred years has it become clear that they are distant versions of our own Sun, but an understanding of what made the Sun and other stars shine awaited 20th-century advances in nuclear physics. Now we know that all stars are powered by the energy of nuclear fusion, and a combination of theory and observation has allowed us to piece together the story of how stars are born, the ways in which they may develop, and how they die. Most excitingly, astronomers have begun to discover planets around other stars, confirming that planetary systems are a natural by-product of star birth and increasing the chances that there may be life elsewhere.

central pivot

alidade for sighting stars

mater, or main body, with degree scale around limb, or edge

» Medieval astrolabe
Widely used in medieval times, an astrolabe is a disc-shaped device used for finding latitude and measuring time by sighting on stars, like a sextant.

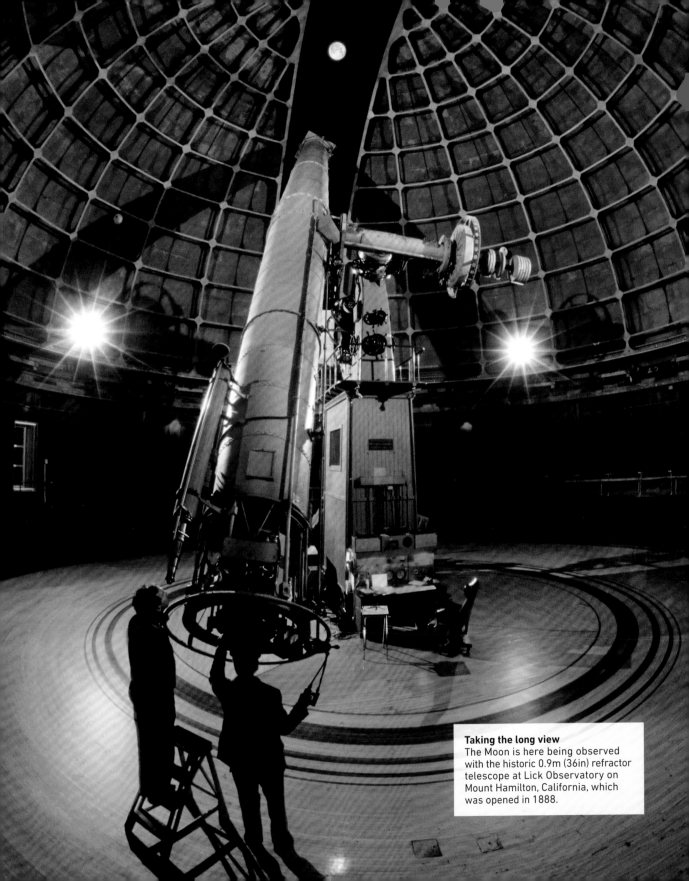

Taking the long view
The Moon is here being observed with the historic 0.9m (36in) refractor telescope at Lick Observatory on Mount Hamilton, California, which was opened in 1888.

The stars and nebulae visible to our unaided eyes all lie within our own galaxy, the Milky Way. Binoculars and small telescopes expand our horizons to other galaxies millions of light-years beyond, while modern instruments have stretched our visual limits to a few hundred thousand years from the Big Bang, the cosmic eruption that marked the birth of space and time, some 13.8 billion years ago.

Dating the Big Bang has been one of the great achievements of modern cosmology, but along with this success has come a new surprise – the discovery that the expansion of the Universe is not slowing down, as had been supposed, but is speeding up, due to a mysterious force known as dark energy. Under the impulsion of dark energy, the destiny of the Universe is to expand forever, gradually thinning out and fading into eternal darkness. Understanding the nature of dark energy is the major challenge facing cosmology at the start of the 21st century.

The contents of this book, brought together by a team of writers, editors, and designers, provide a wide-ranging introduction to the Universe and the objects within it. For those

◤ Watching the shadow
Amateur astronomers use small telescopes to follow the progress of a lunar eclipse on a cloudy evening. Forthcoming eclipses are listed in the Almanac section on pages 324–39.

" Gazing at **remote galaxies**, glowing with **subtle starlight**, we can only **wonder** if there is **someone** on a planet orbiting one of its stars, **looking back** at us. "

◀ In the Earth's shadow
Lunar eclipses are a fascinating sight and easy to observe. When the Moon enters the Earth's shadow, it takes a red tinge. The progress of an eclipse can be followed with the naked eye and with binoculars.

who wish to see for themselves, a greater array of equipment is available to the would-be observer than ever before, from humble binoculars to computer-driven telescopes fitted with the latest electronic imaging technology. Hints on the selection of suitable equipment can be found in the chapter on Observation (pp.147–65), while charts and descriptions at the end of the book will guide you to the major sights to be observed. With the knowledge provided by professional astronomers, we can better understand the different types of object that are within view, from star-forming nebulae to young clusters and dying stars shedding gas to form planetary nebulae. Further off, we can appreciate the varied forms of elliptical and spiral galaxies, and catch a glimpse of galaxies in the process of merging. Gazing at those remote galaxies, glowing with subtle starlight, we can only wonder if there is someone on a planet orbiting one of its stars looking back at us.

IAN RIDPATH

History

The beginnings of astronomy

Astronomy has been called the oldest of the sciences, and rightly so. Since the dawn of civilization, humans have struggled to make sense of the complex motions of celestial objects, and countless ancient monuments and artefacts reflect their fascination.

⌃ MUL.APIN Tablet
One of a pair, this Babylonian tablet is inscribed with lists of constellations in cuneiform script. Just 8.4cm (3.3in) high, it is a masterpiece of miniature writing.

The Babylonian tradition

Stonehenge in England and the Pyramids of Egypt, both dating from around 2500 BCE, embody astronomically significant alignments based on knowledge of the skies, but the true birthplace of astronomy was in the Middle East.

Two baked clay tablets produced around 700 BCE by the Babylonians of present-day Iraq summarize information on the motions of stars and planets.

The list of stars and constellations known to the Babylonians is clear evidence of a long-standing tradition of celestial observation. Some constellations, such as Leo and Scorpius, have come down to us virtually unchanged. The Babylonians made another lasting contribution to astronomy: having measured the length of the year as approximately 360 days, they divided the circle of the sky into 360 degrees, subdivided each degree into 60 parts, and introduced the 24-hour day, with each hour also divided into 60 parts.

The Greek view of the heavens

Knowledge of Babylonian astronomy spread to Greece around 500 BCE. Unlike the Babylonians, who were mainly concerned with divining celestial omens – what we would term astrology – the Greeks sought to understand the physical principles on which the Universe worked, thus initiating the separation of science from superstition. Eudoxus, a Greek astronomer of the 4th century BCE, developed a scheme of 27 crystalline spheres all nested within each other, rotating on different axes and at different speeds, which carried the celestial bodies around the spherical Earth. Later Greeks modified his system, but the principles of perfect circular motion and an Earth-centred (geocentric) Universe remained entrenched in astronomical thinking until the 17th century.

The greatest observational astronomer of the Greeks was Hipparchus, who compiled the first accurate catalogue of the naked-eye stars in the

◁ Greek astronomers
In Renaissance Europe, the ancient Greeks were still regarded as the ultimate authorities on scientific matters, as demonstrated by this 15th-century German painting of astronomers on Mount Athos.

〉〉 FAR EASTERN ASTRONOMY

Other cultures developed constellations quite different from those of the Greeks. The Chinese, for example, recognized a total of 283 constellations, many of them small and faint. Whereas the Greeks pictured mythological beasts and heroes in the sky, Chinese constellations represented scenes from court and social life. Far Eastern astronomers kept a particular lookout for unexpected phenomena termed "guest stars", which we now know as comets, novae, and supernovae. Among the events they chronicled was the Crab Nebula supernova in 1054 CE.

〉〉 Gyeongju observatory
This stone tower in Korea was reportedly used by astronomers on every clear night of the year. Dating from 634 CE, it is the world's oldest surviving astronomical observatory.

2nd century BCE. As well as measuring their positions, Hipparchus also classified stars into six categories of brightness, establishing the magnitude scale we use today.

In the 2nd century CE, Ptolemy presented a summary of Greek astronomical knowledge in a work usually known as the *Almagest,* meaning "greatest", a name given to it by later Arabic astronomers. This included an updated version of Hipparchus's catalogue, expanded from 850 stars to over 1,000 and arranged into 48 constellations – the foundation of our present-day constellation system.

Ptolemy also offered a new model for the motions of celestial bodies. The basic orbit of each body consisted of a large circle, called the deferent, with its centre offset from the Earth. As each object moved along the deferent, it also traced out a smaller circle, known as an epicycle.

Arabic astronomy

After the decline of Greek and Roman civilization, the centre of astronomical research moved east to Baghdad, where Ptolemy's work was translated into Arabic. Shortly before 1000 CE, an Arab astronomer named al-Sufi produced a revised version of Ptolemy's star catalogue, called the *Book of the Fixed Stars.* As well as the star catalogue, al-Sufi's book contained drawings of each constellation. Widely copied and reissued with various illustrations, this became one of the most popular Arab books of astronomy. Between the 10th and 13th centuries, the ancient Greek works were reintroduced to Europe via Arab-dominated Spain.

〈〈 Turkish astronomers
This 16th-century illustration of an observatory founded by Suleyman the Magnificent shows the great traditions of Arab astronomy being carried on by their successors, the Ottoman Turks.

The rebirth of Western astronomy

European astronomy was awoken from its dormancy in the 16th century by a Polish clergyman and astronomer, Nicolaus Copernicus (1473–1543), who revived the Sun-centred or heliocentric theory proposed by the Greek philosopher Aristarchus in the 3rd century BCE. Such an arrangement explained why Mercury and Venus never strayed far from the Sun, because their orbits were now recognized to be closer to the Sun than the Earth's. It also explained why Mars, Jupiter, and Saturn took occasional backward, or "retrograde", loops in the sky, because the Earth was overtaking them on its faster, smaller orbit.

Tycho Brahe (1546–1601), a Danish nobleman, realized the need for new and improved observations against which theories of planetary motion could be judged. Between 1576 and 1586, he built two observatories, called Uraniborg and Stjerneborg, on the island of Hven, between Denmark and Sweden, where he built up a detailed series of observations of the motions of the planets. Tycho could never bring himself to accept the heliocentric theory. Instead, he developed his own ingenious compromise in which the Earth remained stationary at the centre, orbited by the Moon and Sun, while the planets orbited the moving Sun.

⌃ **Celestial globe**
The positions of the stars on this globe from 1603, made by Willem Janzsoon Blaeu, were plotted according to the catalogue of Tycho Brahe.

The laws of planetary motion

Tycho bequeathed his observations to his assistant, a brilliant German mathematician named Johannes Kepler (1571–1630). After many years of diligent calculation, Kepler discovered

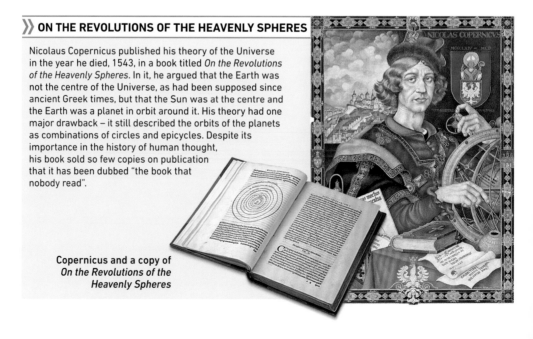

》 ON THE REVOLUTIONS OF THE HEAVENLY SPHERES

Nicolaus Copernicus published his theory of the Universe in the year he died, 1543, in a book titled *On the Revolutions of the Heavenly Spheres*. In it, he argued that the Earth was not the centre of the Universe, as had been supposed since ancient Greek times, but that the Sun was at the centre and the Earth was a planet in orbit around it. His theory had one major drawback – it still described the orbits of the planets as combinations of circles and epicycles. Despite its importance in the history of human thought, his book sold so few copies on publication that it has been dubbed "the book that nobody read".

Copernicus and a copy of
On the Revolutions of the
Heavenly Spheres

◀ **Stjerneborg**
Tycho Brahe equipped this observatory with improved instruments capable of measuring positions with an accuracy 10 times greater than before.

that the planets do indeed orbit the Sun as Copernicus had proposed, but not in complex combinations of circles and epicycles. Instead, planetary orbits are elliptical, and the orbital period of each planet is mathematically linked to its average distance from the Sun.

Galileo's discoveries

While Kepler was laying the theoretical basis for a new understanding of the cosmos, another revolution was taking place in observational astronomy. Unimagined wonders were coming into view as the first telescopes were turned towards the heavens. The greatest pioneer of telescopic astronomy was an Italian, Galileo Galilei (1564–1642). Wherever Galileo looked, he found innumerable faint stars, beyond the reach of the human eye. The Milky Way, in particular, was resolved into a mass of faint stars. Whereas planets could be magnified to discs by the telescope, the stars remained points of light, confirming that the Universe was infinitely vaster than supposed. As an

additional blow to the ancient view that the heavens were perfect, Galileo saw that the Moon's surface was not a smooth, polished sphere but was scarred by craters and mountains. Most significantly of all, he found that Jupiter was orbited by four moons, now known as the Galilean satellites. He went on to discover that Venus shows phases, proof that it orbits the Sun, and glimpsed the rings of Saturn although he did not recognize what they were.

This new view of the heavens, coming hard on the heels of Kepler's theoretical breakthrough, swept away the old Earth-centred view of the Universe for good. But a more fundamental problem remained. What was the force that made planets orbit the Sun as they do?

Galileo's experiments helped found modern physics. He dropped objects of different weights from a tall tower, reputedly the Leaning Tower of Pisa, and found that they all reached the ground at the same time, whereas Greek scientists such as Aristotle had taught that heavier objects should fall faster.

▲ **Galileo Galilei**
Condemned by the Roman Catholic Church in 1633 for declaring that the Earth moved round the Sun, Galileo was placed under house arrest for the remainder of his life.

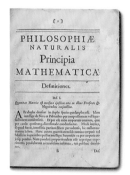

⌃ Newton's Principia
The laws of motion demonstrated by Newton in his *Principia Mathematica* of 1687 provided a sound mathematical basis for all subsequent students of physics and astronomy.

⟫ Halley's Comet
Halley correctly calculated that one elliptical orbit of the comet that bears his name took around 76 years.

What is more, Galileo discovered that the velocity of a falling object doubled for every 9.8m (32ft) that it fell, a constant figure that later became known as the acceleration due to gravity.

Newton and gravity

Half a century later, an English scientist, Isaac Newton (1642–1727), was inspired to think about gravity by another falling object, in this case an apple from a tree in his garden in Lincolnshire. He realized that the same force that made the apple fall to the ground must also be responsible for keeping the Moon in orbit around the Earth.

Newton went on from this realization to deduce his law of gravity, publishing it in 1687 in *Principia Mathematica*. According to Newton, an object's gravitational attraction depends on its mass (that is, the amount of matter it contains), and the strength of the attraction falls off with the square of the distance from the object. This law explained for the first time why the planets orbited the Sun as they did and why the Moon raised tides in the Earth's oceans. In due course, it would apply also to the motions of artificial satellites and space probes.

Using Newton's theory of gravity, the English astronomer Edmond Halley (1656–1742) calculated that comets move around the Sun on highly elliptical orbits. Convinced that the comets seen in 1531, 1607, and 1682 were one and the same, Halley predicted that the comet would return around 1758. When it duly reappeared 16 years after his death, it was named Halley's Comet.

⌃ Bayeux Tapestry
Confirmation of Halley's calculations came from the 11th-century Bayeux Tapestry, which records the appearance of the comet in 1066 shortly before King Harold of England was defeated at the Battle of Hastings by William, Duke of Normandy.

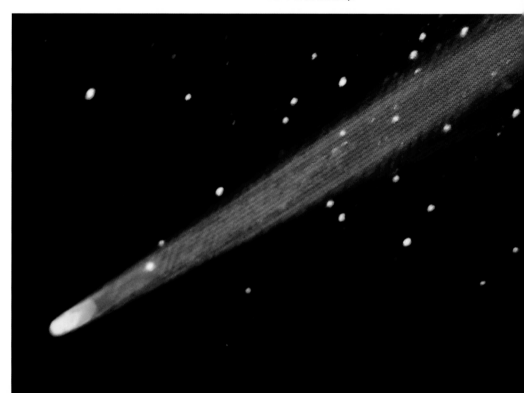

» THE DEVELOPMENT OF THE TELESCOPE

Despite references by earlier writers, including Roger Bacon in the 13th century, to the idea of combining lenses to see objects at a distance, the first person to actually make a telescope seems to have been Hans Lippershey, a Dutch spectacle-maker, in 1608. News of his invention spread fast, and Galileo heard reports of it on a visit to Venice the following year.

copper binding around wooden tube covered with paper

small objective lens, giving narrow field of view

Galileo's Telescope
Galileo immediately made a telescope himself, fitting a convex lens in one end of a tube and a concave one in the other. This basic refracting telecope enabled him to make the discoveries that amazed his fellow scientists. His most powerful instrument magnified up to 30 times. Later 17th-century astronomers, notably Huygens (see p.129), improved and refined Galileo's design.

Newton's Telescope
One disadvantage of refracting telescopes was that light was broken up into its component colours, which focused at different points. This "chromatic aberration" could be avoided by using a mirror rather than a lens to collect and focus the light. In 1672, Newton produced a reflecting telescope, just 30cm (12in) long, with a concave mirror, made of copper and tin, at the base. This focused the light back onto a plane mirror set at an angle, which directed the image to an eyepiece at the side of the tube.

Herschel's Telescope
For much of the 18th century, Newton's telescope was not widely imitated. Large curved mirrors tended to distort under their own weight, and the development of achromatic lenses made refracting telescopes a more attractive option. However, German-born English astronomer William Herschel (1738–1822) favoured huge reflectors, for which he ground and polished his own mirrors and lenses. He was duly rewarded for his labours by the chance discovery of Uranus in 1781.

eyepiece

mirror

overlapping tubes slide to change focus

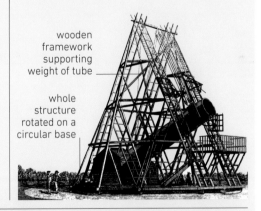

wooden framework supporting weight of tube

whole structure rotated on a circular base

Lord Rosse's Telescope
William Parsons, 3rd Earl of Rosse (1800–67) was a wealthy aristocrat, whose great ambition was to build a giant reflecting telescope. In 1845, he completed one at his Irish estate at Birr Castle with a 1.8m (6ft) solid mirror and a focal length of 16.45m (54ft). Rosse and his descendants used the huge instrument chiefly for the study of nebulae, star clusters, and galaxies.

Galileo demonstrates his telescope
In the summer of 1609, Galileo Galilei demonstrated his new telescope to the governors of Venice, who awarded him increased funding for his research. He went on to make improved telescopes with which he made many pioneering astronomical discoveries, including the four main moons of Jupiter, now known as the Galilean satellites.

The rise of astrophysics

From the late 18th century onwards, astronomers and physicists made a series of key discoveries and advances in techniques and technology. These enabled astronomers to learn for the first time about the physical properties of stars other than our own Sun.

Secrets and light

Throughout the early history of astronomy, the stars were mere pinpricks of light in the night sky – too distant to reveal discs even when studied with the highest magnifications. But as the light-gathering power of astronomical instruments improved, astronomers were able to make more sophisticated use of distant starlight. Although Isaac Newton had split sunlight through a prism in the late 17th century, it was English chemists such as William Hyde Wollaston (1766–1826) who refined the process, developing the spectroscope – a device for analysing the wavelengths of light emitted from an object. In 1821, German scientist Joseph von Fraunhofer discovered that sunlight split through a spectroscope was not a continuous spectrum of colours, but was crossed by numerous dark lines. In 1859, his fellow Germans Robert Bunsen and Gustav Kirchhoff successfully explained these

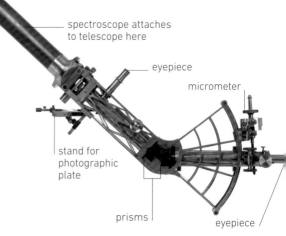

spectroscope attaches to telescope here

eyepiece

micrometer

stand for photographic plate

prisms

eyepiece

⌃ Splitting light
A spectroscope passes light from a star through a prism or "diffraction grating" (a piece of dark glass etched with very narrow transparent slits), deflecting different wavelengths and colours of light onto slightly different paths.

⌃ ⟫ The nature of nebulae
Nebulae like M33 (right) remained a mystery during the 19th century. William Parsons, Earl of Rosse, made studies of all kinds of nebulous objects, including M51, the Whirlpool Galaxy (above), using his telescope in Ireland.

lines as the result of chemicals in the Sun's atmosphere absorbing the same wavelengths of light they emitted when heated in a laboratory.

Light from the stars was at first too faint to analyse in this way, but the arrival of photography during the 19th century eventually allowed long exposures that captured the spectra of stars over time. Meanwhile, in 1838, German astronomer Friedrich Bessel made another breakthrough, using parallax to measure the distance to a star for the first time (see p.65).

Order among the stars

Armed with the new tools of spectroscopy and photography, astronomers set about the great task of cataloguing the objects in the sky. At first the multitude of different absorption lines in stellar spectra and the variety of colours and luminosities appeared chaotic, but a group of women astronomers working at Harvard College Observatory in the USA in the 1890s eventually discovered a scheme that made sense. Led by Annie Jump Cannon, the "Harvard Computers" compiled the great Henry Draper Catalogue, sifting through thousands of stellar spectra to sort the stars into "spectral types"

distinguished by their spectral lines and their colours (which by then were known to represent surface temperatures). Elsewhere astronomers were busy compiling parallax measurements for all the nearby stars, but it was not until 1906 and 1913 that Ejnar Hertzsprung and Henry Norris Russell independently hit upon the idea of comparing the two sets of data on a graph (see panel, below). The resulting Hertzsprung–Russell diagram revealed that the majority of stars obeyed a simple relationship between spectral type and brightness, with the exceptions falling in distinct regions of the graph.

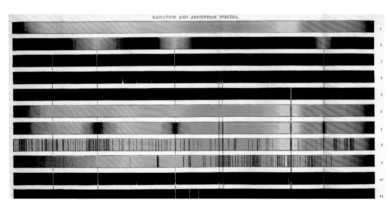

⌃ First photographs
Photography made possible a permanent record of astronomical observations for the first time. These stellar spectra are from the Henry Draper Catalogue, named after the pioneer astrophotographer.

》 HERTZSPRUNG AND RUSSELL

In 1905, Danish astronomer Ejnar Hertzsprung (1873–1967) became the first to suggest an absolute standard of brightness for stars ("absolute magnitude"). He defined it as the magnitude of a star as seen from a distance of ten parsecs (32.6 light-years). A year later, he published a paper in which he compared the absolute magnitudes of stars in the Pleiades with their colours and spectral types, plotting

them on a graph and noting the relationship between the two, as well as the existence of bright "giants" and dim "dwarfs". However, Hertzsprung published his work in an obscure German photographic journal, and it went unnoticed until 1913, when the American Henry Russell (1877–1957) presented his own work, independently developed, to the Royal Astronomical Society.

Ejnar Hertzsprung

Henry Russell

Our place in the Universe

The early 20th century saw a revolution in astronomy, as the true size and nature of the Universe became apparent. This great shift of perspective was followed by the realization that the Universe is expanding, and must have originated at a particular point in time.

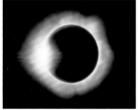

⌃ **General relativity**
Einstein's theories were confirmed in 1919 by Arthur Eddington (see p.38). His observations of a solar eclipse showed that positions of stars are distorted as their light bends in the Sun's gravitational field.

The galaxy debate

The true nature of nebulae had been the subject of heated debate since the first spectra from them were obtained in the 19th century. Most astronomers agreed that "spiral nebulae" were made up of countless stars, so small or far away that they blended into a single fuzzy object – but just how distant were they? Some thought they were relatively small and in orbit around the Milky Way, while others argued that they were huge and unimaginably distant independent galaxies.

The debate was finally settled by the work of Henrietta Leavitt (see p.59) and Edwin Hubble. Leavitt developed a method for measuring the absolute distances of stars, which Hubble then applied to prove that galaxies were millions of light-years from Earth.

⌃ **Early map of the Milky Way**
William Herschel made the first serious attempt at mapping the Universe in 1785. From the distribution of stars in the sky, he correctly deduced that the Solar System sits in the plane of a flattened cloud of stars.

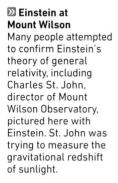

» **Einstein at Mount Wilson**
Many people attempted to confirm Einstein's theory of general relativity, including Charles St. John, director of Mount Wilson Observatory, pictured here with Einstein. St. John was trying to measure the gravitational redshift of sunlight.

⟫ EDWIN HUBBLE

Hubble (1889–1953) trained as a lawyer before turning to astronomy. In 1919 he joined the staff of Mount Wilson Observatory in California, where he specialized in the study of galaxies. In 1923 he discovered the first Cepheid variable star (see p.76) in the "Andromeda Nebula", M31, and he went on to find many more, enabling him to calculate the true scale of the Universe for the first time. By 1929 he had also proved the link between red shift and distance, known as Hubble's Law. Hubble also devised the system of galaxy classification that is still in use today.

The nature of space-time

While Hubble's discoveries expanded estimates of the size of the Universe immeasurably, Albert Einstein's theories of relativity changed our understanding of its very nature. Einstein confronted the major problems in physics head-on – in particular, the fact that light always appeared to travel at the same speed, regardless of the motion of its source. In order to accommodate this fact, he formulated a completely new concept of four-dimensional "spacetime", in which measurements of space and time could become distorted in extreme conditions, such as during travel at high speeds or in strong gravitational fields. The implications of relativity are too many to cover here, but they set the stage for the next great cosmological revolution.

Expansion and origins

Hubble's distance measurements clinched the case for an enormous Universe and led on to an even more important discovery – that the further away a galaxy lies, the faster it is receding. Hubble reached the conclusion that the entire Universe is expanding at a uniform rate.

To most astronomers, Hubble's discovery implied that the Universe had originated at a single point in space at some time in the distant past. Belgian astronomer Georges Lemaître was the first to suggest that the Universe originated in a "primordial atom", in 1927, but it was not until 1948 that George Gamow (see p.50) and his colleagues worked out the details of the Universe's explosive origin. The term "Big Bang" was actually coined by one of the theory's strongest opponents, Fred Hoyle, who believed that the Universe was in a "steady state" of continuous expansion and creation of matter.

⚠ **Clinching evidence**
The Big Bang theory predicted a faint glow of heat left over from the origin of the Universe. This Cosmic Microwave Background Radiation was discovered in 1964 by Americans Arno Penzias and Robert Wilson.

pink areas are hotter and denser

blue areas are cooler and less dense

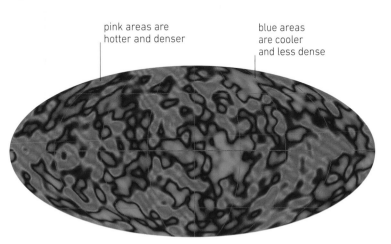

⚠ **Afterglow of creation**
The Cosmic Microwave Background Radiation causes the entire sky to glow at just 3 degrees above absolute zero, but it has minute temperature fluctuations that reveal the state of the Universe at the instant it became transparent (see p.53). These "ripples" were mapped for the first time by the COBE satellite in 1992.

Into orbit

Until the mid-20th century, space travel was a dream pursued by a few visionaries and eccentrics. The military rockets of the Second World War brought it within reach, while the Cold War between the United States and Soviet Union made it a reality.

⊼ Robert Goddard
Goddard (1882–1945) is shown here alongside his first successful liquid-fuelled rocket, launched from a Massachusetts field in 1926. Although it only reached an altitude of 12.5m (41ft), this pioneering flight paved the way for Goddard and others to develop increasingly powerful rockets throughout the 1920s.

Early ideas on space travel

Although space travel had been a popular topic for fantasy since Roman times, it was not until the 19th century that writers began to seriously consider its problems. French author Jules Verne launched his heroes to the Moon with a giant cannon (in fact, the acceleration would have killed them), while British writer H.G. Wells invented a material that shielded his lunar capsule from the effects of Earth's gravity. In reality, the only practical solution was the rocket. Long used as a military weapon, a rocket's self-propelled nature means that it can push itself forward without a medium to travel through, making it

ideal for the vacuum of space. Many of the principles of rocketry for use in space travel were worked out by Russian schoolteacher Konstantin Tsiolkovsky around 1900, but it was not until the 1920s that American physicist Robert Goddard began to experiment with liquid propellants that had the potential power to reach space. These developments were followed with keen interest by a small German rocket society, the VfR, whose members included Wernher von Braun

« War rocket
The liquid-fuelled German V2 rocket was designed to be rapidly deployed and fired from mobile launch platforms in northern Europe. Targeted at London and southeastern England, it was an indiscriminate and frightening new weapon.

(1912–77). When the Nazis seized power in Germany, members of the VfR were recruited to work on military programmes that culminated in the first ballistic missile, the V2 rocket. Although the V2 had little effect on the course of the war, it clearly showed the potential for rockets, both as a weapon and as a means of peaceful exploration.

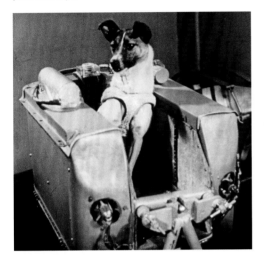

The Space Race

After Germany's defeat, the United States and the Soviet Union raced to capture as much German rocket technology as they could. Both sides saw rocket-powered ballistic missiles as the ideal method for delivering nuclear weapons. However, rocket scientists such as von Braun, working for the United States, and the Soviet Union's Sergei Korolev, both genuinely motivated by the desire to conquer space, were able to divert their respective countries' missile programmes towards other more ambitious goals.

Both countries aimed to launch a satellite in the International Geophysical Year of 1957. Political considerations led the United States to attempt launches with an underpowered naval research rocket rather than von Braun's more powerful military rockets. The Soviets had no such problems, and stole an early lead, successfully launching the first satellite, Sputnik 1, on 4 October 1957.

antenna

aluminium sphere 58cm (2ft) in diameter

⊼ **Red star**
Sputnik 1 transformed the world when it was launched in 1957. An 84-kg (185-lb) metal sphere, its main instrument was a radio beacon that transmitted a simple signal back to Earth to prove it had survived intact.

⊼ **The first animal in space**
Within a month of Sputnik 1, the Soviet team were ready to launch a far more ambitious satellite. Sputnik 2 weighed 508kg (1,120lb) and carried a passenger – a dog named Laika. She died after a few orbits when the spacecraft overheated.

≫ **KONSTANTIN TSIOLKOVSKY**

Konstantin Eduardovich Tsiolkovsky (1857–1935) is regarded as the founder of modern rocketry, although he never built a rocket himself (the picture shows him alongside a model). He proved the efficiency of liquid rocket fuels and multiple-stage rockets, and even worked out the principles of steering a rocket in flight. His work was not recognized until the foundation of the Soviet Union in 1917.

≪ **Playing catchup**
On 6 December 1957, US attempts to launch a satellite with a Vanguard rocket ended in an explosive fireball. Von Braun's military team were then told to prepare for launch, and the first US satellite, Explorer 1, reached space on 31 January 1958.

The race to the Moon

With the Soviet Union leading the Space Race, the next great challenge was to put humans in space. Here too, the Soviets had an advantage. However, a greater challenge lay in the race to the Moon, and it was this that the United States would ultimately win.

⚡ Man in space
Yuri Gagarin (1934–68) completed a single orbit of the Earth in 108 minutes aboard Vostok 1. He was killed in a plane crash while training for a return to space aboard Soyuz 3.

⚡ Mercury Seven
The Mercury Seven astronauts were feted as heroes even before the first launch. Early suggestions that women might be more suited to spaceflight were ignored for political reasons.

Humans in orbit

The Soviets had a flying start in the race to put humans in space – their rockets were powerful enough to launch comparatively massive satellites, while even the largest American rockets could only put a few kilos in orbit. Much of the challenge lay in how to bring an astronaut or cosmonaut home, and both countries carried out successful and unsuccessful missions with animals to test shielding and re-entry procedures.

Again, the Soviets worked under a veil of secrecy, selecting an elite group of potential cosmonauts from whom Yuri Gagarin was eventually picked. The Americans were caught by surprise when Moscow announced Gagarin's flight on 12 April 1961 (a considerable risk since Gagarin was still in orbit at the time, and was nearly killed during re-entry). A month later, Alan Shepard became the first American in space during a brief sub-orbital flight, but it was another nine months before John Glenn finally reached orbit.

Even before this, President Kennedy had announced the next leg of the race, vowing that America would put a man on the Moon by the end of the decade. This new challenge was to

》 SERGEI KOROLEV

Sergei Pavlovich Korolev (1907–66) was the mastermind of the early Soviet space programme. After working on liquid-fuelled rockets during the 1930s, he was imprisoned in 1938. Released after World War II, he was put in charge of the rocket programme. He was planning a Soviet lunar mission before his death during a routine operation.

push both sides to the limit. America launched the new two-man Gemini programme, which would rehearse many of the techniques needed for a successful lunar mission. The Soviets, meanwhile, suffered a series of setbacks that eventually pushed them out of the race altogether.

Apollo to the Moon

The Apollo programme began disastrously, when a fire killed all three crew members of Apollo 1 during launch rehearsals in 1967. After a series of unmanned tests and a mission to Earth orbit, however, Apollo 8 successfully completed a loop around the Moon in December 1968.

After two more rehearsal missions, Apollo 11 completed a flawless flight, and the lunar module Eagle touched down in the lunar Mare Tranquillitatis on 20 July 1969. Five more landers put astronauts on the Moon before the programme ended in 1972.

» Saturn V
Still the most powerful launch vehicle ever
built, the Saturn V stood 110m (360ft) tall
and used three stages to propel the Apollo
spacecraft towards the Moon.

« Last man
Astronaut Eugene
Cernan, the last
Apollo astronaut
to step on the
Moon, salutes the
US flag before his
departure on 14
December 1972.

» APOLLO MOON LANDINGS

APOLLO 11 Landed
20 July 1969 in the
Mare Tranquillitatis.
Neil Armstrong
became the first man
on the Moon, stepping
down from the lunar
module Eagle.

APOLLO 12 Landed 19 November 1969 in the
Oceanus Procellarum, allowing the astronauts
to inspect the nearby Surveyor 3 probe, which
had been on the Moon for two and a half years.

APOLLO 14 Landed 5 February 1971 in the Fra
Mauro region. The mission was commanded by
Alan Shepard, who had been America's first
man in space.

APOLLO 15 Landed
30 July 1971 near
Hadley Rille. A
modified lunar
module carried an
electric lunar rover
for the first time,
extending the range
of exploration.

APOLLO 16 Landed 20 April 1972 near the
crater Descartes. The only mission to explore
the lunar highland regions, it helped answer
many questions.

APOLLO 17 Landed 11 December 1972 in
Taurus Littrow region. The only lunar mission
to carry a qualified geologist, Harrison Schmitt.

Shuttles and stations

The decades since the Apollo Moon missions have seen exploration efforts refocus on near-Earth orbit. Space stations opened the way for long-duration spaceflight, while NASA's Space Shuttle, until its retirement, offered new ways of living and working in orbit.

⬆ Handshake in orbit
This mission patch celebrates US–Soviet collaboration in the joint Apollo–Soyuz mission of 1975.

At home in space

Once it became clear that America would win the race to the Moon, the Soviet Union redirected its space efforts closer to home. The duration of space missions had steadily extended throughout the 1960s, but if cosmonauts were to carry out long-term research in orbit, a semi-permanent space station would be needed.

The early years were troubled – Salyut 1's first crew were killed by a leak in their re-entry capsule on their return to Earth. However, the Soyuz capsules used to ferry people to and from orbit were soon made more reliable, and are still in use today. Salyuts 6 and 7 operated for four years each, and were replaced in 1986 by Mir, a much larger station that operated until 1999.

America's initial response was Skylab, a converted Saturn V rocket stage lifted into orbit in 1973. The first US space station had its share of problems, but hosted three crews for periods of up to 84 days. However, when Skylab was abandoned in 1974, the US had no immediate plans for a successor.

⬆ Skylab
This photo of the US Skylab shows a makeshift sunshade put in place by the first crew after the original shield was torn off during launch. The crew also had to pull open the main solar panel by hand.

⬇ Soviet success
The Mir space station used a "modular" design – new elements such as extra laboratories were added throughout its lifetime. The same idea has been applied to the International Space Station.

Routine spaceflight

The US space agency, NASA, was increasingly focused on developing the Space Shuttle – a "spaceplane" launched with the aid of a massive external fuel tank and two rocket boosters. The Shuttle made a successful maiden flight in April 1981, followed by five more space shuttles. But problems from the outset prevented it fulfilling its potential – problems with the fragile insulating tiles that protected the orbiter on re-entry and the spaceplane's position during launch alongside the fuel tank and rocket boosters left it vulnerable to disaster. The Challenger explosion of January 1986 led to a long suspension of flights and expensive redesigns.

« **Space laboratory**
Orbiting at a height of around 400km (250 miles) above Earth's surface, the International Space Station is as long as a football field, with a pressurized volume similar to a Boeing 747. Astronauts work in overlapping "expeditions" ranging from six months to a year or more.

All change

By the 1990s, priorities were changing – the end of the Soviet Union led to renewed cooperation between NASA and the newly formed Russian Space Agency, first with a series of shuttle missions to the Mir station, and later with plans for the ambitious International Space Station (ISS), a huge project that also involves contributions from the European Space Agency, Japan, and Brazil.

Construction on the ISS got underway in 1998, and the station welcomed its first permanent crew in 2001. However the schedule was derailed by the loss of a second Space Shuttle, Columbia, in 2003. The decision was soon made to bring the Shuttle programme to an end after completion of the ISS in 2011, with station crews delivered by Russian Soyuz capsules. While US policy encouraged space businesses such as Elon Musk's SpaceX to fill the gap in manned and unmanned launches to Earth's orbit left by the Shuttle's retirement, NASA's own development efforts have focused on a new launch system and spacecraft for manned exploration beyond Earth's orbit.

Re-usable rockets

The fact that most rocket hardware is used just once and then either lost or destroyed has always been a major barrier to lowering the cost of spaceflight. So when SpaceX successfully landed the lower stage of a Falcon rocket in 2016, upright and ready for re-fuelling and re-use, it marked a major leap forward. Some estimates suggest re-usable launch vehicles could soon cut the cost of access to space by 90 per cent.

« **Falcon 9**
On April 8 2016, the lower half of Falcon 9 landed intact on a landing pad in the Atlantic Ocean 200 miles offshore from the Kennedy Centre, Cape Canaveral, Florida, from which it had lifted off.

Alone in space
In 1984, the American Space Shuttle astronaut Bruce McCandless made this untethered spacewalk using a jet-propelled backpack. He flew to a distance of about 100m (330ft) from the Shuttle, further than any astronaut had ever ventured before from the safety of his ship.

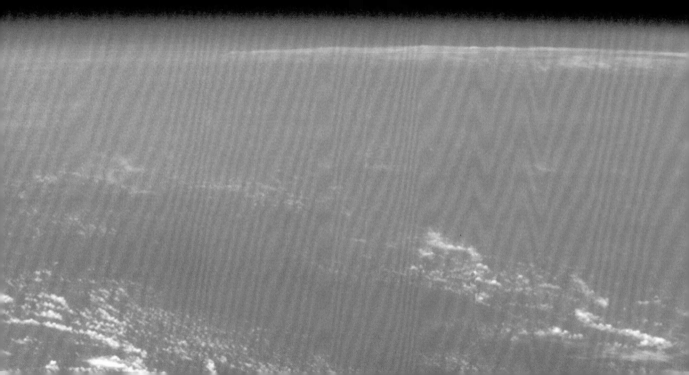

Exploring the Solar System

While humans have remained close to Earth, machines have ventured much further into space. Robotic probes have explored all the planets as well as several smaller bodies, returning spectacular images and transforming our view of the Solar System.

⌃ Pioneering
The early Pioneers, such as Pioneer 2, were attempts to reach the Moon. They were the first US probes into interplanetary space.

To other worlds

As early as 1959, the Soviet Union began to launch probes towards the Moon. The first Luna probe missed by thousands of kilometres, but the second made a direct hit, and the third successfully flew behind the Moon and returned pictures of the unseen far side.

The increasing power of rockets meant that the opposing nations of the Space Race were soon able to send probes out to the other worlds of the inner Solar System. The US Pioneers explored interplanetary space, while their series of Mariner probes, often built in pairs to insure against accidents, made the first successful flybys of Venus (1962), Mars (1965), and Mercury (1974). These first probes revealed just a glimpse of each planet – it was not until space scientists perfected the techniques for putting spacecraft in orbit and landing them on the surfaces of other planets (both pioneered by probes to the Moon) that our knowledge of the inner Solar System began to increase. The Soviets sent heavily shielded probes into the choking atmosphere of Venus, eventually receiving pictures from the surface in 1975, while

» Venus lander
The Soviet Venera 9 lander sent back data from the surface of Venus for 53 minutes after its landing on 22 October 1975.

NASA's Pioneer Venus and Magellan used radar to map the planet from orbit. NASA's Mariner 9 went into orbit around Mars in 1971, transforming our view of the planet. It was followed by the twin Viking probes of 1976, each comprising both an orbiter and a lander.

Grand tours and beyond

The first successful flybys of Jupiter and Saturn were made by Pioneers 10 and 11 in 1973 and 1979, but a unique alignment of planets brought the opportunity for a much more ambitious mission. The twin Voyager probes used a "gravitational slingshot" to tour the giant planets,

» Mariner 9
As the first spacecraft to orbit Mars, Mariner 9 put an end to ideas that it was just a cratered, Moon-like world.

» Cratered world
The photographs from Mariner 9 revealed canyons, volcanoes, and dried-up riverbeds.

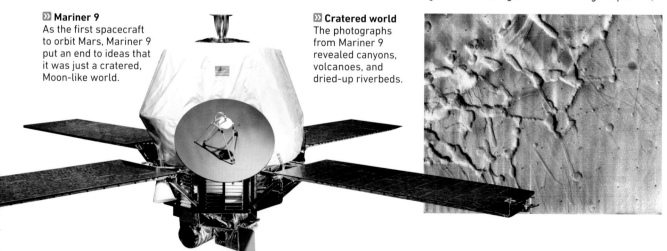

swinging past Jupiter in 1979 and Saturn in 1980–81. Voyager 2 went on to become the first and only probe to visit Uranus and Neptune.

1986 saw an international flotilla of space probes visit Halley's Comet, including the European Giotto Probe, which returned pictures of the nucleus. Subsequent missions to comets and asteroids have taken things further, orbiting these small bodies and even landing on them. Since the late 1990s, space probes have also returned to Mars with a series of orbiters, landers, and rovers revealing ever more Earth-like aspects to the red planet. NASA's Galileo, Juno, and Cassini probes have spent long periods orbiting Jupiter and Saturn, while in 2015 the New Horizons probe hurtled past Pluto at the edge of the solar system.

❯❯ CARL SAGAN

Famous for his TV series *Cosmos*, Carl Sagan (1934–96) was a NASA scientist involved in many planetary probes, and a pioneering researcher into possible life elsewhere in the Universe. He helped design plaques and laserdiscs for the Pioneer and Voyager spacecraft, in case they are found by an alien civilization in the distant future.

❯❯ Hyperion

This stunning false-colour photograph of Saturn's moon Hyperion was captured by the Cassini spacecraft in September 2005. Cassini's complex flightpath brings it within a few hundred kilometres of most of Saturn's major moons, revealing them in unprecedented detail.

Back to Mars

This view of the Martian landscape was taken by NASA's Opportunity in 2004. One of a pair of roving robots, it landed in the Meridiani Planum region and proved conclusively that it was once under water.

segment of exposed cliff face

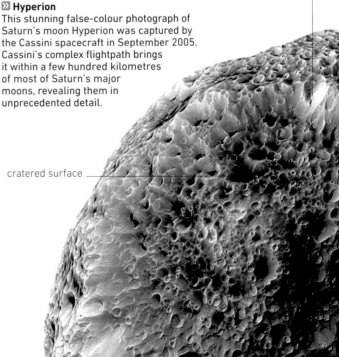

cratered surface

Saturn surveyor

The Cassini–Huygens Saturn probe, shown here in its protective fairing while awaiting launch, is the size of a bus and packed with state-of-the-art instruments to conduct a complete survey of the Saturnian system.

Unlocking the stars

In the early decades of the 20th century, astronomers had the techniques to study the characteristics of distant stars, and even their compositions, but the power source that lay within them and made them shine was still unknown.

Stellar furnaces

The discovery of radioactive elements in the 1890s opened the way for new dating techniques that suggested an age for the Earth of several billion years. Since it was generally accepted that the Sun and the planets had formed at the same time, this meant that the Sun too had been shining for billions of years, but there was no known energy source capable of sustaining it for that long (the previous favourite had been gravitational contraction and heating, which could have sustained the Sun for a few million years).

Fortunately, while nuclear physics had revealed the problem, it was also to produce the solution. As knowledge of reactions between atomic nuclei improved, astronomers such as Arthur Eddington began to realize that nuclear fusion (the joining of light atomic nuclei to make heavier ones) was a potential source of immense energy from the destruction of relatively little material. It still took until 1938 for German-born physicist Hans Bethe to work out the precise details of the hydrogen fusion chain that powers stars like the Sun. It is

⟫ ARTHUR EDDINGTON

British astronomer Arthur Eddington (1882–1944) led the 1919 expedition to the island of Principe to prove that Einstein's theory of general relativity was correct (see p.26). He went on to make the first direct measurements of stellar masses in binary stars, discovering the link between mass and luminosity for main-sequence stars. He also correctly suggested that fusion was the primary source of energy in stars.

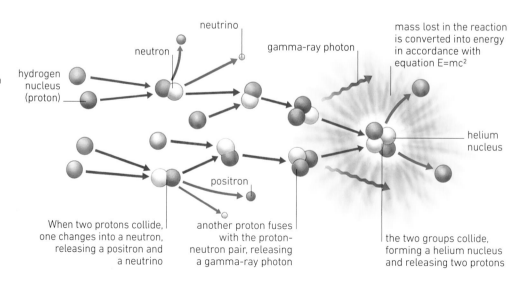

⟫ The hydrogen chain
The hydrogen or proton-proton chain that produces energy in the cores of Sun-like stars involves a sequence of reactions in which hydrogen nuclei (protons) fuse with each other and then undergo radioactive decay to eventually produce a helium nucleus (two protons and two neutrons).

neutrino

neutron

gamma-ray photon

mass lost in the reaction is converted into energy in accordance with equation $E=mc^2$

hydrogen nucleus (proton)

helium nucleus

positron

When two protons collide, one changes into a neutron, releasing a positron and a neutrino

another proton fuses with the proton-neutron pair, releasing a gamma-ray photon

the two groups collide, forming a helium nucleus and releasing two protons

(1910–95) suggested that they might be the collapsed cores of burnt-out stars, supported only by the pressure between their atoms. In 1932, Russian physicist Lev Landau realized that atomic physics put an upper limit on the mass of white dwarfs. Above a certain weight, known as the Chandrasekhar limit, the forces between the particles in the star would not be able to resist gravity, and the stellar remnant would collapse to an even denser state, a neutron star. Such stars were finally detected with the discovery of the first pulsar (see p.71) in 1967. And even neutron stars proved to have an upper limit, above which their particles would dissolve into even tinier quarks and they would collapse to form black holes.

>> STEPHEN HAWKING

Best known for his book *A Brief History of Time*, Stephen Hawking did much of his groundbreaking work in the 1960s and 1970s on the structure of black holes. Such objects, whose gravity prevents light escaping from them, had first been suggested in the 18th century, but were revived by the discoveries of particle physics in the 1960s. Hawking discovered many aspects of black hole behaviour, most famously the "Hawking radiation" that is generated around their boundaries.

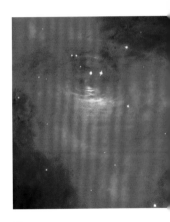

⌃ Over the limit
The Crab Nebula supernova remnant shows "ripples" caused by the rapidly rotating neutron star (pulsar) at its heart. Neutron stars are collapsed stars heavier than the 1.4-solar mass "Chandrasekhar limit" that marks the heaviest possible white dwarf. The limit was actually discovered by Lev Landau.

⌃ Particles from the Sun
Neutrino observatories such as this one in Canada use huge underground detector tanks to record the near-massless neutrino particles that are released by nuclear fusion in stars like the Sun.

now estimated that the Sun and stars like it have a sufficient mass of hydrogen to keep shining for about 10 billion years.

Extreme objects

The suggestion that stars were giant machines for releasing vast amounts of energy by turning hydrogen into helium triggered a burst of discoveries and wild new theories. Breakthroughs in the study of atomic physics on Earth proved to have implications for the nature and structure of some exotic stars. For example, astronomers had known for some time of the existence of superdense white dwarfs, but it was in 1927 that Indian astronomer Subrahmanyan Chandrasekhar

Pushing the limits

The late 20th century saw great advances in the tools at astronomers' disposal. Telescopes grew larger and larger and orbiting observatories studied the sky at wavelengths blocked from the surface of the Earth, while computers allowed data from them to be handled in new ways.

Looking deeper

From 1948 to 1991, the Hale Telescope on Mount Palomar in California, with its 5-m (200-in) mirror, was the largest functional telescope in the world. Since the 1990s, however, the development of thinner mirrors and computer-controlled systems that constantly correct distortions in a mirror's shape to ensure peak performance (so-called "adaptive optics") have led to a new generation of giants. Interferometry, a technique first used in the 1940s to combine the images from separate radio telescopes and compensate for their low resolutions (see p.42) is now used with optical telescopes, leading to huge "networked" telescopes such as the Very Large Telescope (VLT) in Chile.

✉ **Astronomers' peak**
The summit of Mauna Kea in Hawaii is studded with observatories. The largest are the twin Keck telescopes, one of which is seen here below. Both have 10-m (33-ft) mirrors, but their images can be combined by interferometry to simulate the resolution of a telescope with an 85-m (278-ft) mirror. The Keck I telescope was completed in 1992 and Keck II in 1996.

》 **LYMAN SPITZER**

American astronomer Lyman Spitzer (1914–97) made breakthroughs in the study of star formation, the interstellar medium, and the formation of planetary systems. One of the first to see the advantages of space-based observatories, he set out the benefits to both optical and invisible astronomy in a paper of 1946. It was largely due to his lobbying efforts that the Hubble Space Telescope was launched in 1990.

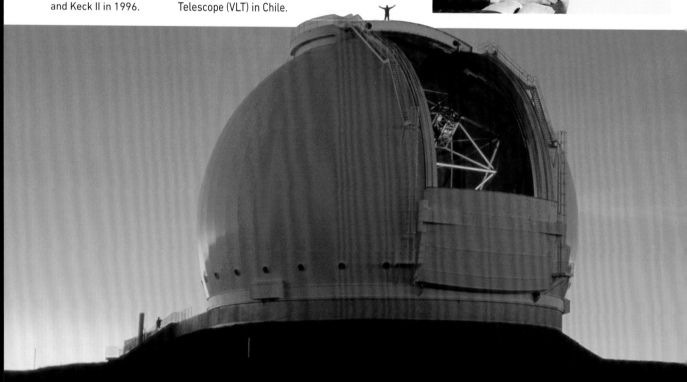

New horizons

The revolutionary new generation of telescopes brought entirely new types of object within the range of astronomers. While the Hubble Space Telescope's (HST's) mirror was not large, its position above the atmosphere meant it did not lose light through absorption, and its resolution was second to none. In contrast, giant ground-based telescopes such as the Keck and the VLT can see deeper and fainter, but not quite as clearly as the HST. Recording technology has changed too – electronic CCDs (see pp.162–63) respond to light in a different way from photographic plates, and because they record digitally, data from many short exposures can be combined into a long-exposure image, in a way that was impossible with conventional photography. Among the discoveries made in this new era have been the Kuiper Belt beyond Pluto, extrasolar planetary systems, and the deep structure of the Universe itself.

❯❯ The limits of the Universe

Astronomers can now measure the red shifts of thousands of distant galaxies. Plotting their distribution reveals features such as the filaments and voids of the large-scale Universe. The most detailed map so far was created by the Sloan Digital Sky Survey project.

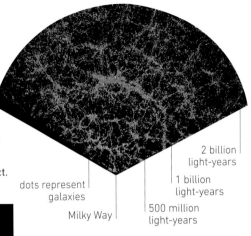

2 billion light-years

1 billion light-years

500 million light-years

dots represent galaxies

Milky Way

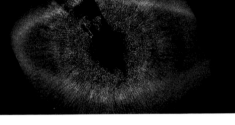

⌃ Worlds beyond
This HST photograph shows a ring of dust around the nearby Sun-like star Fomalhaut, within which planets may be forming. It was photographed using a black "occulting disc" to block light from the star itself – a technique that only works well in space.

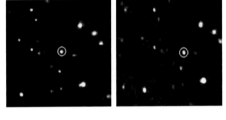

⌃ The edge of the Solar System
The latest telescopes have detected extremely faint objects in our cosmic backyard, such as the dwarf planet Eris, seen moving slowly against the background stars.

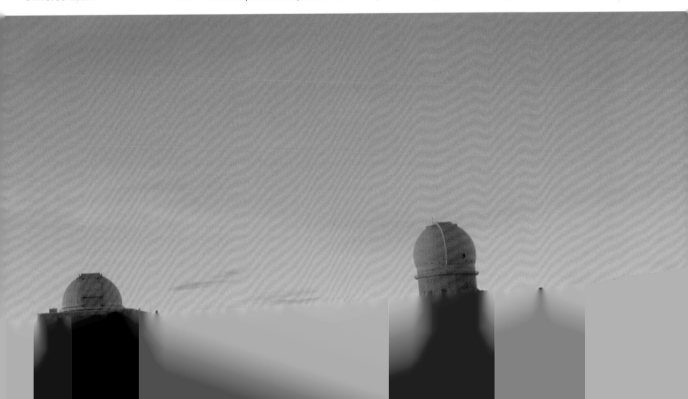

The invisible Universe

Until the middle of the 20th century, astronomers could only study celestial objects through the radiation they emitted at visible wavelengths. But visible light is only a tiny portion of the spectrum of electromagnetic waves, ranging from extremely long-wavelength radio waves to short-wavelength gamma rays. Short wavelengths in particular are very damaging to living tissue, but fortunately the Earth's atmosphere blocks out nearly all wavelengths except for visible light and some radio waves. Astronomers only discovered that objects like the Sun emitted energy at these other wavelengths in 1946, when they used rockets to launch the first detectors into the upper atmosphere.

Since then, the arrival of the space age has created new fields of invisible astronomy. Orbiting observatories have now mapped the Universe at wavelengths from gamma rays and X-rays, where some of the most violent events in the Universe can be detected, through ultraviolet light, where the hottest stars emit most of their radiation, and into the infrared (heat) waves primarily emitted by cool objects too dim to be seen in visible light.

Radio astronomy

The first of the new radiations to be explored, however, were the radio waves – for the simple reason that many of them reach Earth intact. Radio signals from the sky were first detected by US engineer Karl Jansky in 1932. He realized that the signals peaked when the Milky Way was in view, but took the research no further. The earliest dish antenna was built a few years later by another American, Grote Reber, but the first of the huge radio telescopes, pioneered in Britain by Bernard Lovell, were not built until the 1950s. Today, radio astronomers are able to use interferometry to combine images from telescopes scattered around the world. Among the major discoveries made by radio astronomy are the distribution of interstellar gas (mapped

》 SIR BERNARD LOVELL

British radio astronomer Bernard Lovell (1913–2012) was the driving force behind the world's first large radio telescope, the 76-m (250-ft) dish at Jodrell Bank in England. This paid off in 1957 when the beginning of the space age made the radio tracking of satellites a priority. Lovell's achievements include the first radio detection of meteors and studies of the radio activity of the Sun and of faint variable stars called "flare stars".

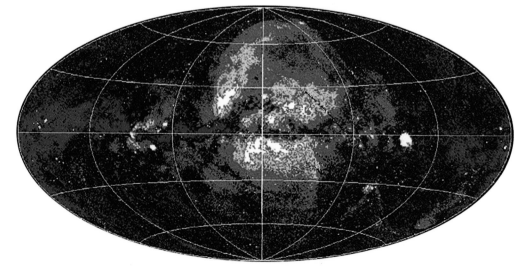

》 The hot Universe
Most maps of the sky at non-visible wavelengths can only be made from orbit. This X-ray chart of the whole sky, produced by the European ROSAT satellite, reveals hot gas bubbles in the solar neighbourhood in yellow and green, and intense radiation sources such as supernova remnants in blue.

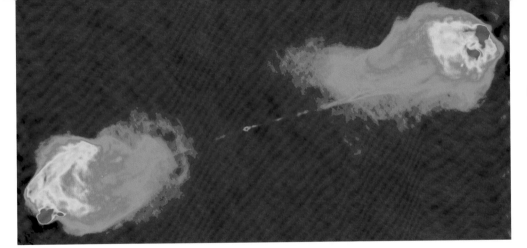

◀◀ **Active galaxies**
Radio galaxies are a type of active galaxy (see p.81) that forms when jets of high-energy particles from the galaxy's central region collide with material in intergalactic space. They are among the most intense radio sources in the sky.

through the radio emissions of hydrogen), radio emission from black holes, such as the one at the centre of our galaxy, and the active "radio galaxies".

Frontiers of astronomy

The science of astronomy continues to find new fields of inquiry even today. With the latest generation of telescopes, entire new classes of object are available for study. As thousands of new planets are discovered outside the Solar System, the search for life elsewhere in the Universe, and especially intelligent life, has intensified. A whole new science – astrobiology – studies the conditions under which life might survive among the stars. The Search for Extraterrestrial Intelligence (SETI) programme is constantly listening for tell-tale signals for alien civilizations. Powerful telescopes that allow us to look closer and closer to the Big Bang are also revealing the secrets of galaxy formation, and are on the verge of showing us the first generation of stars. In these and other ways, modern astronomy is bringing us ever closer to our origins.

▽ **Looking for life**
The 305-m (1,000-ft) radio telescope at Arecibo, Puerto Rico, beamed our first deliberate message to the stars in 1974. It has also discovered extrasolar planets and a binary pulsar.

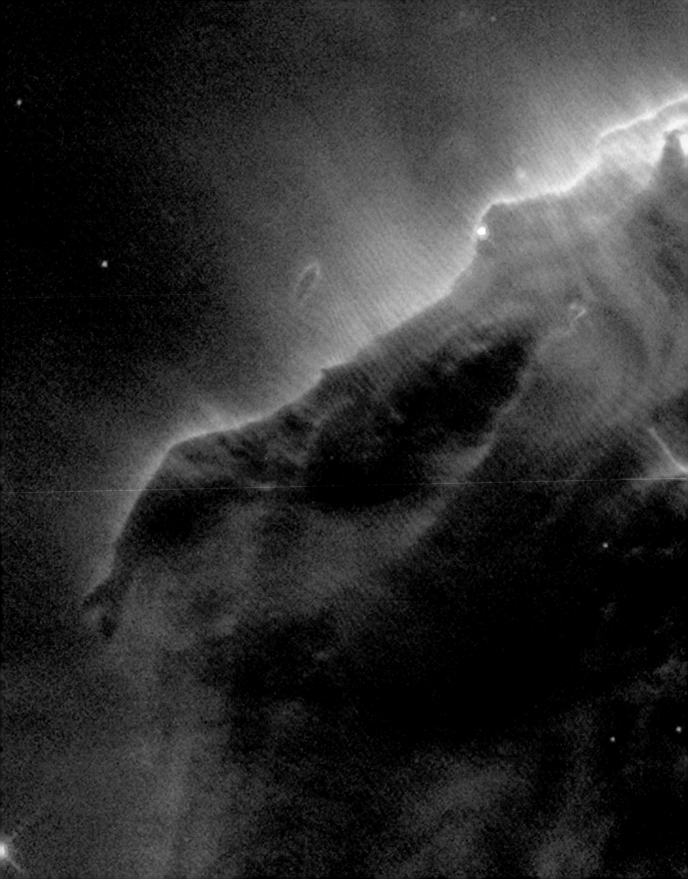

The Universe

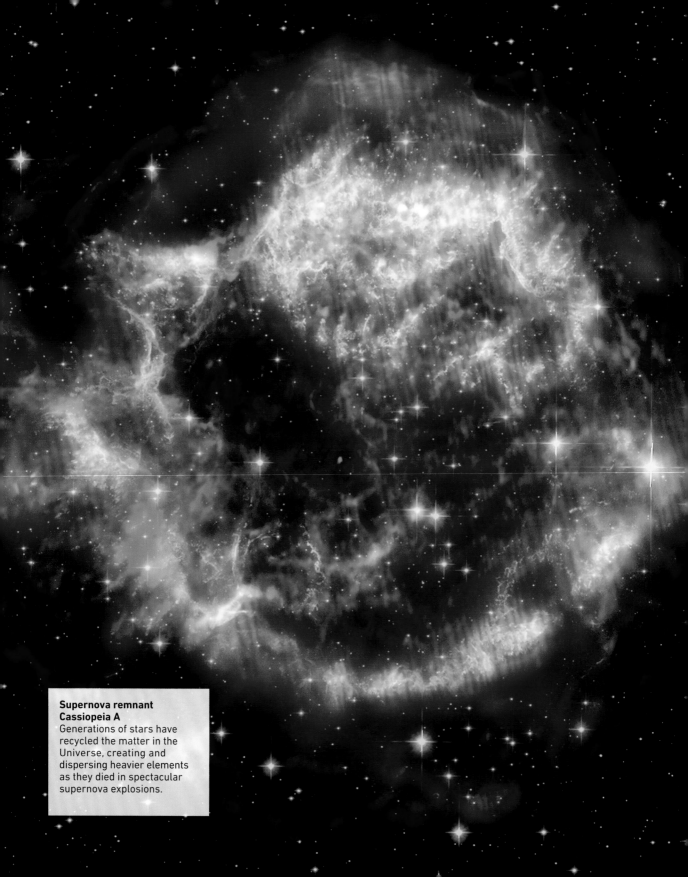

**Supernova remnant
Cassiopeia A**
Generations of stars have
recycled the matter in the
Universe, creating and
dispersing heavier elements
as they died in spectacular
supernova explosions.

Origins

The study of the Universe itself, its origins and evolution, is known as cosmology. It is a highly theoretical science, but one on which the foundations of astronomy all rest. Throughout history, astronomers and others have used various theories, some more scientific than others, to explain the Universe, but modern cosmology is based on the Big Bang theory. Developed in the mid-20th century, it is still the most successful explanation for the observed properties of the Universe and the laws of theoretical physics.

All the distant objects we can see in the sky are part of the same Universe, and we can only hope to understand them if they obey the same physical laws and are broadly similar to other objects close at hand. Cosmologists are limited by their inability to conduct experiments, and the fact that the only Universe they have to study is our own. Computer models have partly rectified this problem, but cosmology is still largely an effort to find theories that explain the features of the Universe as we see and measure them today. These include the fixed speed of light, the fact that the Universe is expanding, and the way in which matter is distributed through the cosmos.

The Big Bang

The Big Bang traces the expansion of the Universe and the creation of all matter back to a cataclysmic explosion 13.8 billion years ago. It accounts for many features of the present-day Universe, and the clinching evidence in its favour is that the entire sky is still glowing with faint radiation left over from the initial explosion.

However, the theory has had to be patched up a few times in order to explain new problems with the Universe itself. The largest of these patches is "inflation", developed in the 1970s when astronomers discovered that galaxies were distributed very unevenly through the Universe – the original Big Bang theory suggested that matter would be more evenly distributed. According to the theory, inflation was a brief but spectacular "growth spurt", in which the early Universe grew from the size of an atom to the size of a galaxy. In the process, microscopic variations in density were blown up to a scale where they can explain today's distribution of galaxies.

Looking into the future

Modern cosmology faces numerous challenges. One of the biggest is to predict the ultimate fate of the Universe. Recent measurements of the expansion of the Universe suggest that its growth is accelerating rather than slowing down, which means that it may continue to expand forever. This acceleration is due to a mysterious force called dark energy that is causing space itself to stretch apart. Just what dark energy is, and how it is best incorporated in current models of the Universe, are still open for debate.

◀ **The Hubble space telescope**
One of the telescope's primary aims was to measure the distance to remote galaxies, and help calculate the rate of expansion of the Universe.

The structure of the Universe

The Universe is everything – space, time, and all the matter and energy they contain, from the largest stars to the smallest subatomic particles. We can only study a small region in any detail, but from this, astronomers can form ideas about the scale and structure of the whole.

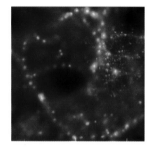

⌃ Filaments and voids
Plots of galaxy clusters and superclusters show patterns in the large-scale Universe. Clusters form string-like filaments and two-dimensional sheets around the edges of vast dark spaces or voids. This underlying structure must have originated in the earliest era of the Universe (see pp.54–55).

⊠ Cosmic distances
The vastness of the Universe can only be represented with a logarithmic scale. On the chart below, the first division represents 10,000km (6,200 miles). Each further division marks a 10x increase in scale on the previous one (not all divisions are labelled).

The bigger picture
The Universe stretches for unimaginable distances in every direction – perhaps to infinity itself. From our point of view on Earth, we are at the centre of an "observable Universe" whose most distant visible objects are now about 46.5 billion light-years away. Light from more distant parts has simply not had time to reach us since the Universe was created.

But the Universe stretches even further beyond our observable limits. Observers on a planet 13.8 billion light-years from Earth would have their own observable Universe

encompassing regions unknown to us, and so on. There is certainly no recognizable "edge" – there is nothing beyond the Universe for it to border – but because space and time can be warped by gravity, the Universe might "curve" in strange ways, perhaps even folding back on itself in places.

Planet Earth
Rocky planets such as the Earth typically have diameters of few thousand kilometres or more. Gas giant planets like Jupiter can grow much bigger – over 100,000km (62,000 miles) across.

⊠ The scale of the Universe
The Universe is so vast that it is almost impossible to grasp. One way to get a feel for its size is to use a series of astronomical stepping stones – from a planet, to a solar system, to a galaxy, and beyond.

The Solar System
The planets are lost in the space of the Solar System – orbiting the Sun at distances ranging from tens of millions to billions of kilometres. The Sun itself is a typical middle-aged star, with a diameter of 1.4 million km (869,000 miles).

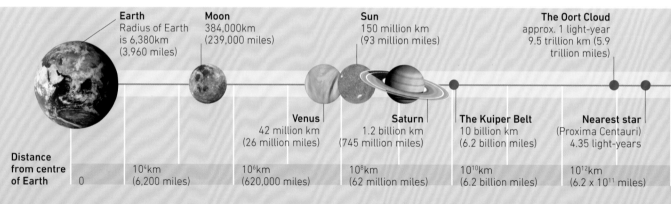

Earth
Radius of Earth is 6,380km (3,960 miles)

Moon
384,000km (239,000 miles)

Sun
150 million km (93 million miles)

The Oort Cloud
approx. 1 light-year 9.5 trillion km (5.9 trillion miles)

Venus
42 million km (26 million miles)

Saturn
1.2 billion km (745 million miles)

The Kuiper Belt
10 billion km (6.2 billion miles)

Nearest star
(Proxima Centauri) 4.35 light-years

Distance from centre of Earth

| 0 | 10^4km (6,200 miles) | 10^6km (620,000 miles) | 10^8km (62 million miles) | 10^{10}km (6.2 billion miles) | 10^{12}km (6.2×10^{11} miles) |

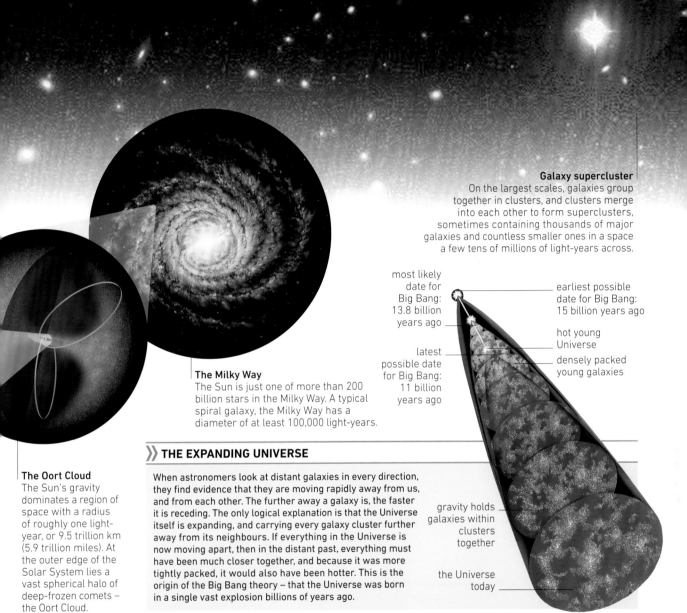

Galaxy supercluster

On the largest scales, galaxies group together in clusters, and clusters merge into each other to form superclusters, sometimes containing thousands of major galaxies and countless smaller ones in a space a few tens of millions of light-years across.

most likely date for Big Bang: 13.8 billion years ago

earliest possible date for Big Bang: 15 billion years ago

hot young Universe

densely packed young galaxies

latest possible date for Big Bang: 11 billion years ago

The Milky Way

The Sun is just one of more than 200 billion stars in the Milky Way. A typical spiral galaxy, the Milky Way has a diameter of at least 100,000 light-years.

gravity holds galaxies within clusters together

the Universe today

» THE EXPANDING UNIVERSE

When astronomers look at distant galaxies in every direction, they find evidence that they are moving rapidly away from us, and from each other. The further away a galaxy is, the faster it is receding. The only logical explanation is that the Universe itself is expanding, and carrying every galaxy cluster further away from its neighbours. If everything in the Universe is now moving apart, then in the distant past, everything must have been much closer together, and because it was more tightly packed, it would also have been hotter. This is the origin of the Big Bang theory – that the Universe was born in a single vast explosion billions of years ago.

The Oort Cloud

The Sun's gravity dominates a region of space with a radius of roughly one light-year, or 9.5 trillion km (5.9 trillion miles). At the outer edge of the Solar System lies a vast spherical halo of deep-frozen comets – the Oort Cloud.

The 1,000-light-year sphere
90 per cent of naked-eye stars are within 1,000 light-years of Earth

The Andromeda Galaxy
2.5 million light-years/
24 million trillion km
(15 million trillion miles)

Nearest quasar
1 billion light-years/
9.5 billion trillion km
(5.9 billion trillion miles)

 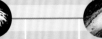

The Centre of the Milky Way
28,000 light-years

The Virgo Cluster
60 million light-years

The edge of the visible Universe
13.8 billion light-years
124 billion trillion km
(77 billion trillion miles)

10^{14}km
(62 trillion miles)

10^{16}km
(6.2×10^{15} miles)

10^{18}km
(6.2×10^{17} miles)

10^{20}km
(6.2×10^{19} miles)

10^{22}km
(6.2×10^{21} miles)

The Big Bang

The Big Bang is currently the best model of how the Universe began. About 13.8 billion years ago, the Universe came into being in a violent explosion. In a fraction of a second, all the energy and matter in the cosmos were created, and matter took on its present form.

In the beginning

The Big Bang was not an explosion in the conventional sense – it was an explosion of space itself, and the beginning of time. The theory does not and cannot attempt to explain what came "before", since time and space did not exist. All we can say is that the Universe was infinitely small, dense, and hot as it came into being. For the first 10^{-43} seconds, the so-called "Planck Time", the normal laws of physics did not apply.

From the Planck Time onwards, however, the theory is more successful. The density of energy was so high that particles of matter could form and decay spontaneously, in accordance with Einstein's famous $E=mc^2$ equation. As the Universe expanded, density and temperature dropped, and the mass of the particles that could form in this way grew smaller, until, after one microsecond (one millionth of a second), the temperature dropped below 1,000 trillion °C (1,800 trillion °F), and matter could no longer form.

》 GEORGE GAMOW

The Big Bang theory was developed by Russian-born physicist George Gamow (1904–68) and his colleagues in the 1950s. Gamow was an atomic physicist who helped explain how thermonuclear reactions power the stars. His knowledge of high-energy nuclear physics helped him explain how different particles formed in a "Hot Big Bang".

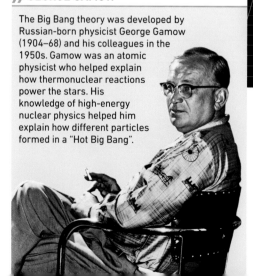

singularity at the start of time

Diameter	10^{-26}m / 3×10^{-26}ft	10m / 33ft
Temp.	10^{27}K (1,000 trillion trillion °C / 1,800 trillion trillion °F)	
Time	A 100-billionth of a yoctosecond / 10^{-35}secs	A 100-milli of a yoctose / 10^{-32} secc

quark

qua

The start of time

The first 10^{-35} seconds of time saw the sudden burst of inflation, accompanied by a dramatic drop in pressure and temperature, and a resurgence of temperature as inflation came to a halt.

›› INFLATION AND THE SEPARATION OF FORCES

Inflation – a brief period of sudden expansion in the first instant of creation, during which the Universe grew from smaller than an atom to bigger than a galaxy – is needed to explain the uniformity of the Universe as it appears today. The best suggestion as to what could have driven this growth spurt is that huge amounts of energy were released as four "fundamental forces", which have governed the Universe ever since, separated from a unified "superforce".

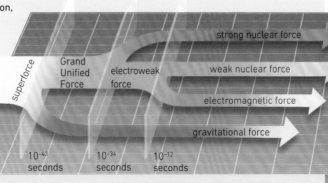

superforce

Grand Unified Force

electroweak force

strong nuclear force

weak nuclear force

electromagnetic force

gravitational force

10^{-43} seconds

10^{-34} seconds

10^{-12} seconds

10^5m (100km / 62 miles)		10^6m (1,000km / 620 miles)		10^9m (1 million km / 620,000 miles)		10^{12}m (1 billion km / 620 million miles)
10^{22}K (10 billion trillion °C / 18 billion trillion °F)		10^{21}K (1 billion trillion °C / 1.8 billion trillion °F)		10^{18}K (1 million trillion °C / 1.8 million trillion °F)		10^{15}K (1,000 trillion °C / 1,800 trillion °F)
1 yoctosecond / 10^{-24} seconds	1 zeptosecond / 10^{-21} seconds	1 attosecond / 10^{-18} seconds	1 femtosecond / 10^{-15} seconds	1 picosecond / 10^{-12} seconds	1 nanosecond / 10^{-9} seconds

quark

antiquark

quark-antiquark pair

X-boson

Higgs boson (theoretical)

photon

antineutrino

quark-antiquark forming and annihilating

graviton (theoretical)

decaying X-boson

particles and anti-particles

quark-antiquark pair

quark

antiquark

Particle soup

In the extreme temperatures, matter was created spontaneously from energy, forming a broiling mass of exotic particles. As well as familiar types of matter, these included heavy particles that scientists today can only make in high-energy particle accelerators.

More matter than antimatter

The Big Bang must have created equal amounts of matter and antimatter, yet today's Universe is dominated by normal matter.
The explanation may lie in a particle called the X-boson, which, as it decays, produces slightly more matter. In this way, a tiny imbalance (perhaps just 0.000001 per cent) was created.

particles and antiparticles (the same as normal particles, but with opposite electric charge) collide, "annihilating" each other in a burst of energy

Freeze out and annihilation

As particles and antiparticles collided, the material created in the Big Bang was nearly all converted back to energy, leaving a tiny excess of matter behind. The energy release sustained the Universe's temperature for a while, but eventually even light particles could not be spontaneously created, and the content of the Universe was fixed.

The early universe

After the first microsecond, falling temperatures meant that particles were no longer moving so rapidly and could bond together. The first to do so were the quarks – heavy particles that make up the protons and neutrons in the heart of all today's atoms. A small proportion of these then bonded to form the atomic nuclei of light elements. Photons

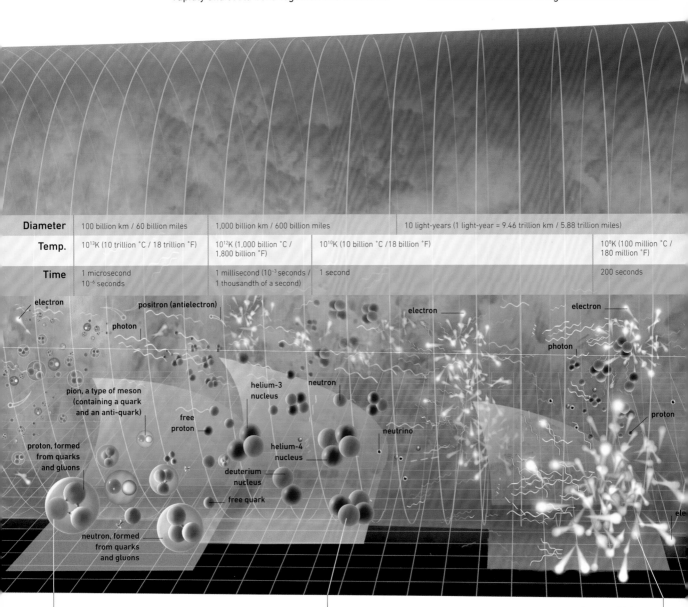

Diameter	100 billion km / 60 billion miles	1,000 billion km / 600 billion miles		10 light-years (1 light-year = 9.46 trillion km / 5.88 trillion miles)	
Temp.	10^{13}K (10 trillion °C / 18 trillion °F)	10^{12}K (1,000 billion °C / 1,800 billion °F)	10^{10}K (10 billion °C /18 billion °F)		10^{8}K (100 million °C / 180 million °F)
Time	1 microsecond 10^{-6} seconds	1 millisecond (10^{-3} seconds / 1 thousandth of a second)	1 second		200 seconds

electron

positron (antielectron)

photon

electron

electron

photon

pion, a type of meson (containing a quark and an anti-quark)

helium-3 nucleus

neutron

free proton

neutrino

proton, formed from quarks and gluons

helium-4 nucleus

proton

deuterium nucleus

free quark

neutron, formed from quarks and gluons

ele

The first protons and neutrons

The first of the familiar particles to form were protons and neutrons – the particles found in the nuclei of today's atoms. Two up quarks and a down quark coming together formed a positively charged proton, while two down quarks and an up quark formed an uncharged neutron.

Nucleosynthesis

As the temperature dropped, protons and neutrons formed stable bonds. Only the lightest atomic nuclei formed in this way, so the young Universe was dominated by hydrogen nuclei (unattached protons) and helium nuclei.

Opaque Universe

Nucleosynthesis reduced the numbers of free particles, but there were still countless leptons – light particles dominated by negatively charged electrons. Electromagnetic radiation, such as light photons, bounced back and forth between the leptons and nuclei, creating a fog.

bounced back and forth between particles of matter, preventing them from coalescing, and the cosmos was foggy rather than transparent. Dark matter, unaffected by the

photons, began to form structure. Finally, after 380,000 years, nuclei and electrons combined into atoms. The number of particles dropped rapidly, and the Universe fog cleared.

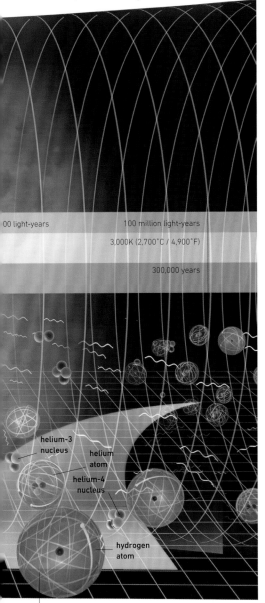

00 light-years   100 million light-years

3,000K (2,700°C / 4,900°F)

300,000 years

helium-3 nucleus

helium atom

helium-4 nucleus

hydrogen atom

The first atoms

380,000 years after the Big Bang, the temperature dropped to a point where atoms could remain stable. Electrons joined to atomic nuclei to form atoms of the light elements, and as the density of particles reduced, the fog cleared and photons were free to travel in straight lines.

INSIDE AN ATOM

An atom is made up of a nucleus of protons and neutrons (collectively termed nucleons), orbited by shells of electrons. Each proton or neutron is itself made up of three even tinier particles, called quarks. Every chemical element has a different number of protons and neutrons in the nucleus of its atoms.

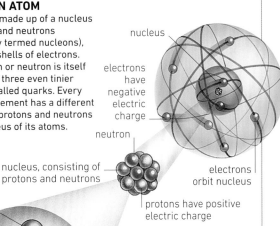

nucleus

electrons have negative electric charge

neutron

nucleus, consisting of protons and neutrons

blue quark

gluons

electrons orbit nucleus

protons have positive electric charge

proton, formed, like a neutron of quarks and gluons

》 WHAT IS DARK MATTER?

Ninety per cent of all matter is thought to be "dark". It makes its presence felt in gravitational effects within galaxy clusters and around individual galaxies, as revealed by the rotation of M81, shown here. Some may be formed from failed or dead stars, but most is thought to be undetectable Weakly Interacting Massive Particles (WIMPs). These began to coalesce in the aftermath of the Big Bang, forming a structure around which visible matter would later take shape.

Galaxy M81

The first stars and galaxies

The beginnings of structure in the Universe, and the origin of the first objects within it, lie hidden in a mysterious "dark age", just beyond the grasp of current telescopes, and are still largely theoretical.

Sowing the seeds

The Cosmic Microwave Background Radiation reveals the beginnings of structure in the Universe 380,000 years after the Big Bang. Two billion years later, galaxies were in an advanced state of formation. This implies that matter clumped together rapidly after the Big Bang. The presence of structure in the early Universe is the best evidence for inflation (see p.50). The idea is that, when the early Universe was blown up to form our observable cosmos, microscopic variations in temperature and density were blown up too, and became seeds of today's galaxy clusters and superclusters.

First light

Before galaxies could begin to form, the light gases from the Big Bang must have been through a certain amount of processing. Astronomers

» The long look back
The further away we look, the further back in time we are seeing. The most distant objects yet photographed are those captured in the Hubble Ultra-Deep Field – primitive galaxies less than a billion years after the Big Bang.

think this was done by a generation of giant stars – the "megasuns". Unlike today's stars, they would have been pure hydrogen and helium. This would have allowed them to grow much larger, reaching several hundred times the mass of the Sun. As they ran out of fuel and detonated in enormous explosions, they scattered heavier elements throughout the Universe and left black holes that might have been the starting points of galaxies.

The structure of the Universe

If all the matter in the Universe started out evenly spread and then fell together under the influence of gravity alone, galaxies would not have formed for billions of years. The large-scale structure of filaments and voids throughout the Universe would not have developed at all. There has simply not been enough time since the Big Bang for them to have been formed by gravity, so they must have been there from the very beginning.

Filaments and galaxies through time
Explaining the origin of the Universe's large-scale structure of filaments and voids is a key challenge for any theory of creation. This computer simulation shows how the filaments and voids might have developed and formed individual clusters and galaxies.

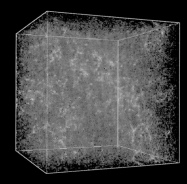

⌃ The Universe at 500 million years old
Just half a billion years after the Big Bang, the large-scale structure of filaments and voids is in place. The seeds of galaxy clusters are forming in the regions where material is most dense.

⌃ Typical early galaxy
A detail from the Hubble Ultra-Deep Field shows an indistinct blob with a bright core that appears to be sweeping up dust and gas from its surroundings.

⟫ Gravitational waves
While the impenetrable barrier of the Cosmic Microwave Background Radiation frustrates all attempts to track the Universe's early evolution with electromagnetic radiation, the 2016 discovery of gravitational waves may provide a new way of studying the infant cosmos. Ripples in space itself, caused by shifting masses and violent events of the Big Bang, have been spreading across the Universe ever since, and could be directly measured by future gravitational wave observatories such as at the Virgo Observatory near Pisa, Italy (see right).

waves passing through Earth change the relative length of the tunnels

wave detectors consist of two long tunnels at right angles

⟫ PLANET FORMATION

Planets like Earth are made mostly from heavy elements, which were created and scattered through the galaxies by countless supernovae. This makes it unlikely that any Earth-like planets formed with the early generations of stars. However, star formation would have provided left-over material that might have formed gas giant planets or faint "brown dwarf" objects (see p.83).

⟫ Potential planets
Five protoplanetary discs – objects made of gas and dust that surround recently born stars – were discovered by the Hubble Space Telescope in the Orion Nebula. They might evolve into planetary systems like ours.

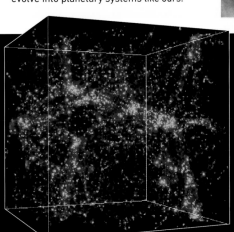

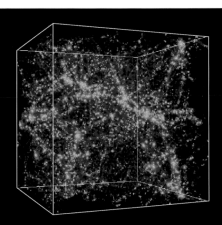

⌃ The Universe at 1.3 billion years old
Individual galaxies are now forming within the crowded filaments. Photographs of the early Universe show an excess of "blue" galaxies, rich in dust and young stars, at this time.

⌃ The Universe at 13.8 billion years old
Galaxies have evolved and much of the material that lay between the clusters has now fallen into them, but the overall pattern of today's Universe is similar to that of 10 billion years ago.

Hubble Ultra Deep Field
Captured by the Hubble Space Telescope, this image, known as the Ultra Deep Field, shows approximately 10,000 galaxies scattered across space and dating back almost to the beginning of time. The image shows a small patch of sky in the southern constellation Fornax, and is a composite of exposures that were taken at various wavelengths, from ultraviolet to near-infrared.

The expanding Universe

The discovery that distant galaxies are moving away from us at great speed has revolutionized our understanding of the cosmos. Galaxies are not moving away from each other because of a force that drives them – they are carried by the expansion of space itself.

Stretching space

Evidence that the Universe is expanding comes from the light of distant galaxies. When astronomers split this light up into a spectrum of colours and look for the tell-tale spectral lines that reveal the presence of gases in stellar atmospheres (see p.64), they find that the lines are in the wrong place – they are shifted towards longer wavelengths and appear redder than expected. The best explanation for this "red shift" is that it is a "Doppler effect" (see below), caused by the motion of the source galaxy away from our own position.

Different galaxies show different amounts of red shift. In the 1920s, Edwin Hubble (see p.27) compared the distances of galaxies with the red shifts in their light, and found that the two were related – the further away a galaxy is, the greater its red shift, and therefore the faster it is moving away from us. The only explanation for this is that the entire Universe is expanding – the further away an object lies, the larger effect the general

expansion has, and the faster it is carried away from us. The expansion of the Universe is, however, a large-scale phenomenon – in smaller regions of space, it can be modified or even overcome by the gravitational attraction of galaxies.

The precise measure of the rate at which the speed of a galaxy's retreat increases with increased distance is known as the Hubble constant. If the rate of expansion has remained constant throughout time, then this will also determine the age of the universe, since the expansion can be tracked back to the Big Bang. The best estimates of the Hubble constant so far, made by the Hubble Space Telescope and other experiments, put the age of the Universe at 13.8 billion years.

⟩⟩ LOCAL GRAVITY

If you think of the Universe as a stretching rubber sheet, the weight of galaxies creates localized dents that draw passing objects towards them. This analogy illustrates Einstein's theory of General Relativity. Our Local Group of galaxies is governed by the gravity of the Milky Way and Andromeda galaxies, which are drawing closer to each other and will eventually merge.

The Great Attractor
This huge concentration of mass lies more than 200 million light-years away in the direction of Centaurus. Its gravity affects our own galaxy and other galaxies for 400 million light-years around it.

☑ Red and blue shifts
The Doppler effect explains why light from distant sources is shifted to the red if the source is moving away from us, and to the blue if the source is moving towards us. The movement of the source affects the frequency at which the peaks and troughs of a light wave reach us, and so affects the wavelength we measure.

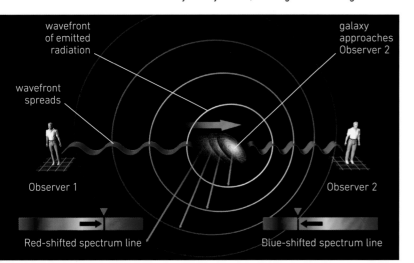

wavefront of emitted radiation

wavefront spreads

galaxy approaches Observer 2

Observer 1

Observer 2

Red-shifted spectrum line

Blue-shifted spectrum line

The expansion of space

This illustration gives an idea of how a region of space may have changed over a period of 9 billion years. Individual galaxies are like currants in a rising cake, carried apart by the expansion of the sponge between them. And our galaxy is not static at the centre of this motion – it is carried along with it. It is only because we have to measure motion relative to our own location that we appear to be static.

⟫ HENRIETTA LEAVITT

In 1912, US astronomer Henrietta Leavitt (1868–1921) identified a number of Cepheid variable stars (see p.76) at about the same distance in the Large Magellanic Cloud. She discovered that the periods of variation were related to their brightness. This "period–luminosity relationship" meant that the periods of Cepheids in other galaxies could be used to estimate their distances.

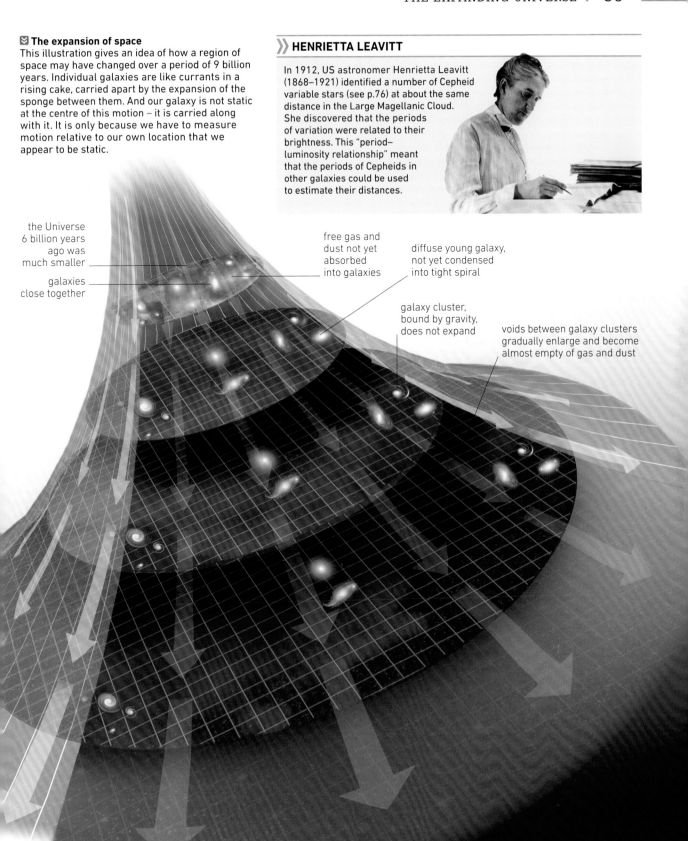

the Universe 6 billion years ago was much smaller

galaxies close together

free gas and dust not yet absorbed into galaxies

diffuse young galaxy, not yet condensed into tight spiral

galaxy cluster, bound by gravity, does not expand

voids between galaxy clusters gradually enlarge and become almost empty of gas and dust

The fate of the Universe

How the Universe will age and die is a subject of great interest for today's cosmologists. Although there are probably trillions of years until the Universe begins to change in any significant way, research into its ultimate destiny reveals hidden properties of today's cosmos.

Assessing the options

The fate of the Universe is governed by the balance between two forces – the outward push of expansion, and the inward pull of gravity. Today we know that the Universe is still expanding with energy it gained in the Big Bang, and the rate of this expansion has been measured. The strength of gravity, meanwhile, depends on the amount of matter in the Universe, and astronomers know that a substantial amount of this is undetected "dark matter". The best models predict that the amount of dark matter is enough to just balance expansion. If these were the only factors involved, gravity would slow expansion almost to nothing, but not reverse it.

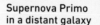

Cold death
In the "Big Chill" scenario, all galaxies will merge to form giant ellipticals like NGC 1316. Star formation will end, and eventually the last stars will go out. Over trillions of years, even the matter within them will disintegrate.

>> **SUPERNOVA COSMOLOGY**

The Supernova Cosmology Project searches for Type Ia supernovae detonating in remote galaxies. These supernovae, caused by white dwarfs collapsing to form neutron stars (see p.71), all have the same intrinsic brightness, so their distance can be calculated by their apparent brightness. Distances calculated in this way are consistently larger than the theoretical distances calculated from the redshifts of the supernovae host galaxies, suggesting that the expansion of the Universe is accelerating.

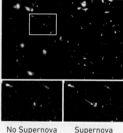

Supernova Primo in a distant galaxy

No Supernova   Supernova 10 Oct, 2010

Dark energy

However, in recent years a new factor has entered the equation. Cosmologists using a new method to cross-check the rate of universal expansion (see panel, above) discovered that the most distant galaxies are further away than they should be if expansion has been slowing down since the Big Bang – in other words, something is giving expansion a "boost". This "dark energy" is seen as a force that causes space itself to stretch apart at an increased rate. Improved measurements suggest that this force may be increasing over time, and has only overcome the general deceleration in the past 6 billion years. Its presence seems to indicate that the Universe will expand forever, but if its strength is increasing, it could have more drastic effects.

Possible destinies

Debates about the future of the Universe used to focus on two alternatives – "Big Crunch" and "Big Chill". The discovery of dark energy seems to have ruled out a Big Crunch, but has opened up two new possibilities. The four possible scenarios are shown here.

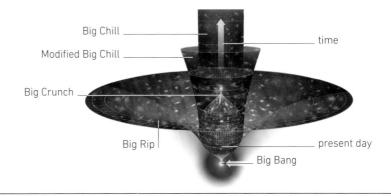

Big Chill

Modified Big Chill

Big Crunch

time

Big Rip

present day

Big Bang

Big Crunch

contracting Universe heats up

Universe contracts after reaching maximum diameter

expansion continues forever

dark energy overcomes gravity

The Big Crunch

In a Big Crunch scenario, the Universe's gravity is stronger than the force imparted by the Big Bang, so expansion slows to a halt and reverses. The Universe grows denser and hotter, and collapses back into a singularity. This might even give rise to a new Universe.

The Modified Big Chill

If the Universe's dark energy force remains steady, then the rate of expansion will steadily increase, overcoming the pull of gravity and pulling galaxies further and further away from each other. The Universe's ultimate fate, however, would still be a Big Chill.

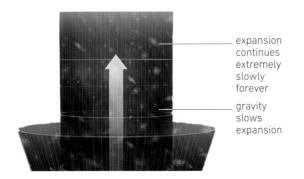

expansion continues extremely slowly forever

gravity slows expansion

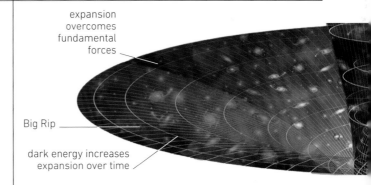

expansion overcomes fundamental forces

Big Rip

dark energy increases expansion over time

The Big Chill

In a Big Chill Universe, the amount of matter is too small for gravity to slow the expansion to a halt. The Universe continues to expand, but more and more slowly. Galaxies disintegrate, stars turn dark, and eventually atoms decay into their component particles.

The Big Rip

If the expansion effect of dark energy continues to increase, in billions of years it might grow to overcome not only gravity, but also the forces between and within atoms. Matter would be torn apart in a cataclysmic "Big Rip", and time itself would come to an end.

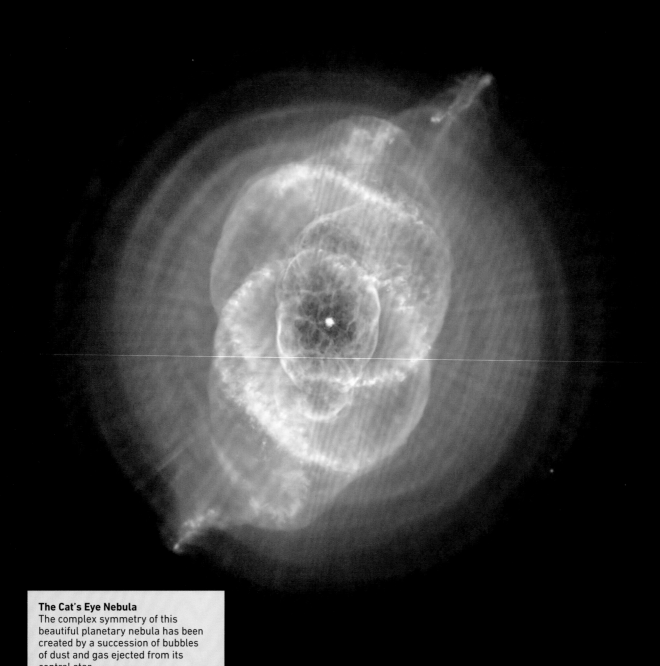

The Cat's Eye Nebula
The complex symmetry of this
beautiful planetary nebula has been
created by a succession of bubbles
of dust and gas ejected from its
central star.

Phenomena

The stars are the most obvious feature of the night sky, yet for millennia they were one of the most mysterious. To the untrained eye, the visible Universe still appears to be just a disorganized scattering of stars, but closer inspection reveals that it is host to an enormous variety of objects. Quite apart from the nearby worlds of the Solar System (covered in the next chapter), there are stars of many kinds, nebulae formed in a number of different ways, star clusters, and countless galaxies.

The stars are so far away that even the most powerful telescopes show them as nothing more than points of light. Until recently, there was no way to tell what they were – though some astronomers guessed correctly that they were stars like the Sun. It took a series of breakthroughs in the 18th and 19th centuries to create the science of astrophysics, the study of the physical properties of stars (see pp.24–25). Today astronomers are able to explain nearly all the phenomena they observe in stars as aspects of a single model of stellar evolution.

Stardust

Nebulae are simply clouds of gas and dust, but they can be created in various ways. Emission and reflection nebulae are the stuff of which stars are made – often enriched with the remains of previous stellar generations. They become visible when the fierce radiation from stars embedded within them makes them glow, or when they reflect the light of nearby stars. Dark nebulae are simply dense dust clouds that form a silhouette in front of more distant reflection nebulae or star clouds. Planetary nebulae and supernova remnants are the glowing material flung off by stars during events late in their lifetimes.

Galaxies and clusters

Our galaxy, the Milky Way, is made up of 200 billion or more stars, their associated gas and dust clouds, and the planetary systems that orbit many of them. The Milky Way is visible to the naked eye as a pale band across the sky and can be resolved into countless points of light by any optical instrument. Within the galaxy, many stars are found in clusters. Loose "open" ones – the aftermath of star formation – are found in the plane of the galaxy, while dense, globular clusters orbit above and below the galaxy, and have much more ancient origins.

Our own galaxy is just one of at least 200 billion. Many are spirals like the Milky Way, but there are other types, equally numerous. Ellipticals are ancient and sometimes huge balls of stars, while irregulars are chaotic havens of starbirth on a massive scale. Most galaxies are dwarfs, too faint to see except when they are on our cosmic doorstep. All these galaxies are typically found in clusters, but it is only recently that astronomers have begun to piece together the relationship between the different types.

The Tadpole Galaxy
This distant spiral galaxy, 420 million light-years from Earth, may have acquired its striking tail of stars, gas, and dust in a clash with another galaxy.

Star classification

Even the most casual glance into the night sky will reveal that the stars have different brightnesses and colours, and are distributed unevenly in the sky. These characteristics are in fact just the most obvious signs of fundamental differences between different types of star.

Gathering data

The stars are unimaginably distant from Earth – only the Sun is close enough to study in detail. Even through the largest telescopes, other stars appear as mere points of light. In order to discover the nature of these stars, astronomers use a variety of ingenious techniques, and scientific models pieced together from laboratory work and studies of the Sun.

≫ **Colourful starfields**
The rich starfields visible in the constellation Sagittarius, towards the centre of our own Milky Way Galaxy, reveal almost every variety of star, from dim dwarfs to bright giants, and from hot blue stars to cool red ones.

hot, bright blue star

intermediate white star

cool, bright red giant

Colour and the H–R diagram

A star's colour depends on its surface temperature. Just as an iron bar heated in a furnace glows first red, then yellow, then white hot, so the amount of energy heating each region of a star's surface controls its colour. However, stellar size can vary, and if a star grows larger at a certain phase of its life cycle, its surface area increases, and the amount of energy heating each region of the surface falls. So a star can grow larger and cooler at the same time. By plotting the relationships between luminosity and colour (or more accurately, spectral type) on a graph called the Hertzsprung–Russell diagram, astronomers have discovered rules that allow them to estimate the true brightness of a star.

≫ **THE USES OF SPECTROSCOPY**

Astronomers use an instrument called a spectroscope to split up the light from stars and study the brightness of different wavelengths. A spectrum can reveal the wavelengths at which the star is brightest, allowing its surface temperature to be calculated. Most spectra are crossed by a number of dark lines, where light of certain wavelengths is absorbed between us and the star. These "absorption lines" are caused by atoms of different elements in the star's own atmosphere absorbing radiation with certain energies. Each element absorbs or emits specific wavelengths, allowing astronomers to study the make-up of stars, planets, and nebulae.

dark absorption line

Spectrum

◨ **Annie Jump Cannon**
Working at Harvard College Observatory from the 1890s, Cannon catalogued thousands of stellar spectra and helped prove the link between star colours and elements in their atmospheres.

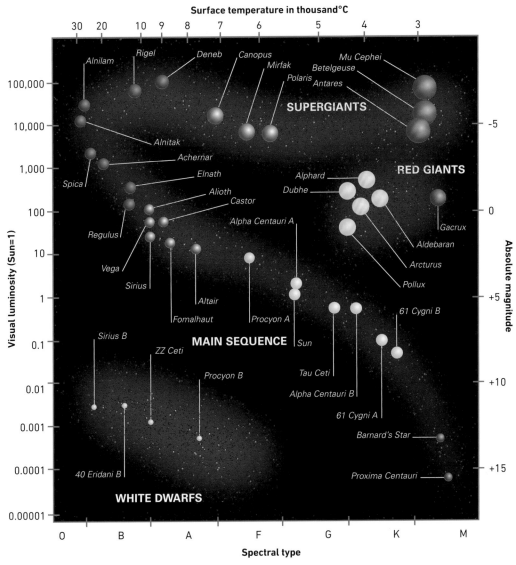

Surface temperature in thousand°C

SUPERGIANTS

RED GIANTS

MAIN SEQUENCE

WHITE DWARFS

Visual luminosity (Sun=1)

Absolute magnitude

Spectral type

Alnilam, Rigel, Deneb, Canopus, Mirfak, Mu Cephei, Betelgeuse, Polaris, Antares, Alnitak, Achernar, Spica, Elnath, Alioth, Castor, Regulus, Alpha Centauri A, Alphard, Dubhe, Gacrux, Aldebaran, Arcturus, Pollux, Vega, Sirius, Altair, Fomalhaut, Procyon A, Sun, 61 Cygni B, Sirius B, ZZ Ceti, Procyon B, Tau Ceti, Alpha Centauri B, 61 Cygni A, Barnard's Star, 40 Eridani B, Proxima Centauri

◄◄ Hertzsprung–Russell diagram
When stars are plotted on an H–R diagram according to their brightness and spectral type, patterns become obvious. The vast majority of stars lie along a diagonal from faint red to brilliant blue – the "main sequence". A smaller number of red stars (the red giants) are much brighter, while some stars (the white dwarfs) are faint but still white hot.

▼ The parallax effect
Direct measurement of the distance to stars relies on parallax – the movement of nearby objects against a more distant background as the observer's point of view shifts. The further away an object lies, the less it will appear to move against the "background" stars.

Distance and luminosity

The brightness of the stars in Earth's skies depends on their true luminosities, and also on their distance from Earth. A star's brightness in the sky is termed its apparent magnitude, while its true brightness or absolute magnitude is defined as its magnitude at a distance of 32.6 light-years. The only direct way of measuring a star's distance uses the parallax effect caused by the Earth's orbit around the Sun each year, but this only works for relatively nearby stars.

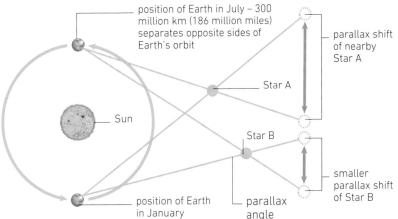

position of Earth in July – 300 million km (186 million miles) separates opposite sides of Earth's orbit

parallax shift of nearby Star A

Star A

Sun

Star B

smaller parallax shift of Star B

position of Earth in January

parallax angle

Stellar life cycles

Although stars are found in huge variety, nearly all can be seen as stages in a standard model of stellar evolution. In this model, a star's mass governs the nuclear reactions in its core, which in turn determine its other physical properties, its lifespan, and its eventual fate.

Life cycle
This sequence shows the early stages of stellar evolution. Most protostars with more than about 10 per cent of the Sun's mass will eventually reach the main sequence, but a significant number with less mass will peter out as dim brown dwarfs.

The birth of a star

New stars are created by the collapse of vast clouds of gas and dust within galaxies. Such collapses are frequent – they can be triggered by the gravity of nearby stars, the shockwaves from a supernova explosion, or simply the slow, regular rotation of a spiral galaxy. Depending on the size of the cloud and other factors, stars can be born in large clusters or smaller, looser groups. As the cloud collapses, it becomes denser and heats up. Any random motion in the initial cloud builds up until it becomes a spinning, flattened disc. More

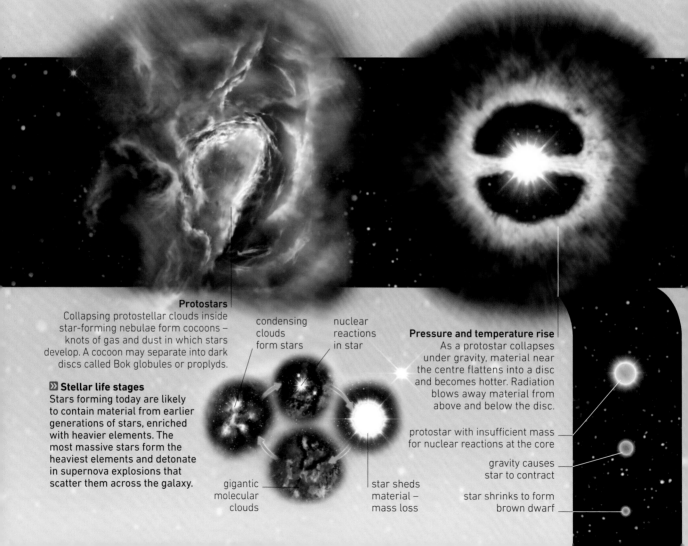

Protostars
Collapsing protostellar clouds inside star-forming nebulae form cocoons – knots of gas and dust in which stars develop. A cocoon may separate into dark discs called Bok globules or proplyds.

Stellar life stages
Stars forming today are likely to contain material from earlier generations of stars, enriched with heavier elements. The most massive stars form the heaviest elements and detonate in supernova explosions that scatter them across the galaxy.

condensing clouds form stars

nuclear reactions in star

gigantic molecular clouds

star sheds material – mass loss

Pressure and temperature rise
As a protostar collapses under gravity, material near the centre flattens into a disc and becomes hotter. Radiation blows away material from above and below the disc.

protostar with insufficient mass for nuclear reactions at the core

gravity causes star to contract

star shrinks to form brown dwarf

and more material is swept up by the gravity of the central part of the cloud, called a protostar, which heats up to the point where it begins to glow. Its core becomes denser and hotter, until it reaches the point where nuclear fusion reactions can begin, and it has truly become a star.

Adolescent stars

Young stars are still surrounded by a large cloud of gas and dust. Some of this will spiral into the star itself, but often just as much

will be ejected. Many young stars develop a magnetic field that traps material and spits it out in jets from the poles. The pressure of radiation may also be enough to blow away lighter elements such as hydrogen. Meanwhile, the star itself continues to collapse, and may go through a period of pulsation and instability known as a "T-Tauri" phase before settling down as a main-sequence star, where it will remain for most of its lifespan.

❯❯ STAR-FORMING NEBULAE

Some of the sky's outstanding emission nebulae are listed below. Since star formation is associated with our galaxy's spiral arms, it is unsurprising that the finest star-forming nebulae are scattered along the length of the Milky Way.

Name	Constellation
Eagle Nebula (M16)	Serpens
Lagoon Nebula (M8)	Sagittarius
Orion Nebula (M42)	Orion
Tarantula Nebula (NGC 2070)	Dorado
Trifid Nebula (M20)	Sagittarius
Omega Nebula (M17)	Sagittarius

strong stellar winds of charged particles

Main sequence
The star shines steadily for most of its life. Its brightness and lifespan are determined by the nuclear reactions inside the star, which are governed by its composition and mass.

Stellar adolescence
As the star approaches the main sequence, material from the disc continues to spiral inwards. The protostar's increasingly strong magnetic field can trap some of this material and eject it at the poles.

❯❯ Emission nebula
Newly formed stars often emit fierce ultraviolet radiation. This can be absorbed by surrounding gases, then re-emitted at visible wavelengths, creating spectacular emission nebulae such as the Trifid Nebula.

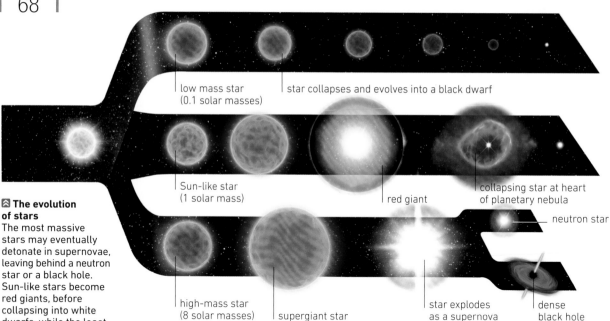

low mass star
(0.1 solar masses)

star collapses and evolves into a black dwarf

Sun-like star
(1 solar mass)

red giant

collapsing star at heart
of planetary nebula

neutron star

high-mass star
(8 solar masses)

supergiant star

star explodes
as a supernova

dense
black hole

**⌃ The evolution
of stars**
The most massive
stars may eventually
detonate in supernovae,
leaving behind a neutron
star or a black hole.
Sun-like stars become
red giants, before
collapsing into white
dwarfs, while the least
massive stars simply
dwindle away.

Star death

Main-sequence stars generate energy by nuclear
fusion of hydrogen in their cores. When supplies
are exhausted, they find a new source of power –
burning the helium they have spent their lives
generating, and perhaps other, heavier elements.
These changes in the star's energy supply make
it unstable, causing it to swell to giant size. The
star's ultimate demise is determined, like most
aspects of its evolution, by its mass.

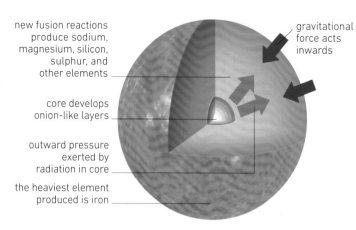

new fusion reactions
produce sodium,
magnesium, silicon,
sulphur, and
other elements

gravitational
force acts
inwards

core develops
onion-like layers

outward pressure
exerted by
radiation in core

the heaviest element
produced is iron

⌃ Red supergiant star
Gases in any "layer" are trapped between two
equal and opposing forces – the inward pull of
gravity and the outward pressure of radiation.
As the intensity of radiation varies, the star can
expand and contract.

⟫ GIANTS AND SUPERGIANTS

Many of the sky's brightest stars are giants – their
high luminosity makes them brilliant at far greater
distances than normal stars. This list includes red
giants as well as even more massive supergiants.

Name	Magnitude	Type	Constellation
Arcturus	0.0	orange giant	Boötes
Betelgeuse	0.5	red supergiant	Orion
Aldebaran	0.9	red giant	Taurus
Antares	1.0	red supergiant	Scorpius
Pollux	1.1	orange giant	Gemini
Eta Carinae	5.5	blue supergiant	Carina

Giant stars

When a star has exhausted the hydrogen at its
core, the hydrogen-burning process moves out
into a spherical shell. The star becomes much
brighter, but radiation from within makes its outer
layers balloon and cool. Stars like the Sun become
red giants, but more massive and luminous stars
can become supergiants of any colour. Inside the
giant, the core itself collapses, until it becomes hot
and dense enough to burn helium. The star then
stabilizes, shrinking back to a more normal size
for however long the helium lasts.

Planetary nebulae

Helium is the heaviest element a star like the Sun can burn, and its exhaustion marks the star's death throes. As the hydrogen- and helium-burning shells move further out, the star's instability increases, eventually causing the outer layers to blow away, creating the glowing shells of a planetary nebula – so named for their resemblance to ghostly, planet-shaped discs. Such nebulae are typically spherical, but are easily distorted by magnetic fields or by companion stars to form even more complex and beautiful shapes.

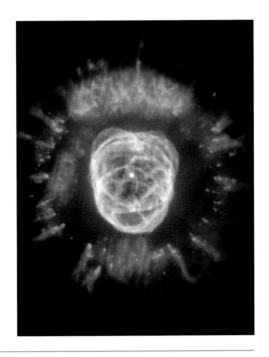

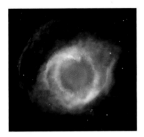

« The Helix Nebula
The ring-like appearance of this spherical planetary nebula is created where we look through the thickest parts of its gaseous shells.

« The Eskimo Nebula
One of the most spectacular planetaries, the Eskimo shows multiple shells of ejected matter, including comet-like outer trails. Its appearance is probably due to the interaction of the dying red giant with an unseen companion.

Supernovae

For a star with eight times or more the Sun's mass, helium burning is not the end. It can continue the fusion of successively heavier elements in its core. Eventually, the core develops an onion-like layered structure. However, the atomic structure of the different elements means that each new layer burns for a shorter time and produces less energy. Eventually, the sequence reaches iron, the first element whose fusion absorbs more energy than it releases. When the star attempts to fuse iron, its central energy source is cut off, and, with no outward pressure of radiation to hold it up, the layered core collapses under its own gravity. The rebounding shockwave tears the star apart and briefly allows fusion to form heavier elements.

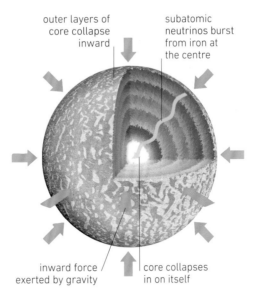

outer layers of core collapse inward

subatomic neutrinos burst from iron at the centre

inward force exerted by gravity

core collapses in on itself

« Supernova
When the outward force of radiation from the core cuts out as the star attempts to burn iron, the core collapses instantaneously, emitting a signature burst of neutrinos before the star is torn apart.

Before

During

« Brighter than a galaxy
The brightest supernova of recent times was SN1987A, in the Large Magellanic Cloud, one of the Milky Way's companion galaxies. For a few weeks, the supernova briefly outshone the whole of its host galaxy, as shown clearly in these two photographs of the same area. Astronomers have since identified its progenitor star – a blue supergiant of 18 solar masses.

⌃ Dog Star
The first white dwarf to be detected was Sirius B (circled here), companion of the brightest star in the sky. It could be detected from the wobble it causes in the motion of Sirius A.

⌄ The Crab Nebula
This spectacular photograph shows the Crab Nebula, remnant of a 1,000-year-old supernova, seen through the European Southern Observatory's Very Large Telescope in Chile.

White dwarfs

Once a Sun-like star has shed its outer layers as a planetary nebula, only the core region remains. Typically this is a ball of dense, hot matter, about the size of the Earth and glowing intensely. Such white dwarf stars have surface temperatures of 100,000°C (180,000°F), but their small size makes them very hard to spot. Some have atmospheres of carbon, oxygen, and other elements created in their helium-burning phase.

Most are doomed to fade slowly over many millions of years, eventually becoming black dwarfs. However, white dwarfs in close binary systems sometimes have sufficient gravity to pull material from their companion stars. This may result in the white dwarf becoming a cataclysmic variable (see p.77).

⟩⟩ WHITE DWARFS

White dwarfs are so faint that very few are within reach of amateur astronomers. In addition, two of the brightest (Sirius B and Procyon B) are companions of extremely bright stars. The easiest example to spot is Omicron^2 Eridani.

Name	Magnitude	Constellation
Sirius B	8.4	Canis Major
Procyon b	10.9	Canis Minor
Omicron^2 Eridani	9.5	Eridanus
Van Maanen's Star	12.4	Pisces
NGC 2440	11.0	Puppis
IP Pegasi	14.0	Pegasus

Supernova remnants

When a star goes supernova, most of its material is torn to shreds and blasted across surrounding space to form a glowing ghost of the former star. The most famous supernova remnant is the Crab Nebula, created by a stellar explosion recorded by Chinese and Native American astronomers in

1054 CE. Because they are composed of intensely hot gases, supernova remnants often emit most of their radiation as X-rays. As the hot gas speeds out across nearby space, it carries with it heavy elements from the supernova. These may collide with other interstellar gas clouds, perhaps even triggering another generation of starbirth.

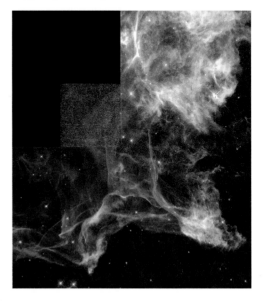

⌃ The Cygnus Loop
The Hubble Space Telescope captured this stunning image of the Cygnus Loop, remnant of a 5,000-year-old supernova.

Neutron stars

A supernova leaves a massive core, whose gravity is so strong that it shreds the star's individual atoms. Oppositely charged protons and electrons combine, leaving a fast spinning ball of neutrons, which typically stops collapsing at around the size of a city. The neutron star's magnetic field becomes highly powerful, channelling its radiation into two beams that emerge from the star's magnetic poles.

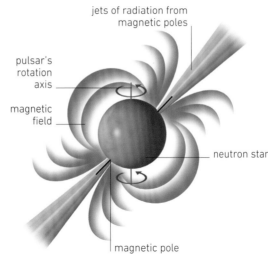

jets of radiation from magnetic poles

pulsar's rotation axis

magnetic field

neutron star

magnetic pole

◀◀ The origin of a pulsar
If a neutron star's magnetic and rotational poles are not aligned, its beams of radiation will sweep around like a lighthouse beam, so the star appears to wink on and off.

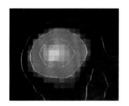

◀◀ Geminga pulsar
Geminga is one of the brightest gamma-ray sources in the sky – it emits nearly all its radiation at these high-energy wavelengths.

Black holes

When a star's core is truly massive, its collapse does not stop at a neutron star. The neutrons in turn are broken up into their component quarks, and the core becomes so dense that its gravity will not even allow light to escape. The result is a stellar-mass black hole. Sealed off from the Universe, black holes are among the strangest objects known to science. Their gravity affects the space around them but they are extremely difficult to detect.

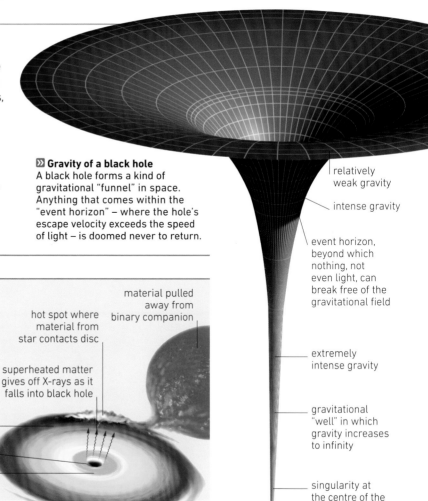

▶▶ Gravity of a black hole
A black hole forms a kind of gravitational "funnel" in space. Anything that comes within the "event horizon" – where the hole's escape velocity exceeds the speed of light – is doomed never to return.

relatively weak gravity

intense gravity

event horizon, beyond which nothing, not even light, can break free of the gravitational field

extremely intense gravity

gravitational "well" in which gravity increases to infinity

singularity at the centre of the black hole

▶▶ FINDING BLACK HOLES

A black hole is easiest to detect when it exists in a binary star system. In some cases, a hole can be detected by its gravitational effect on the visible star – if the mass of the companion exceeds the upper limit for a neutron star, it must be a black hole. In close systems, spectacular X-ray binaries can also form, as shown here.

material pulled away from binary companion

hot spot where material from star contacts disc

superheated matter gives off X-rays as it falls into black hole

accretion disc surrounding black hole

black hole at centre of disc

gas close to centre of disc is heated to 100 million °C (180 million °F)

Stellar flashbulb
This Hubble Space Telescope image shows a halo of dust around the star at the centre of the picture, V838 Monocerotis. The dust has been illuminated by a flash of light emitted during a brilliant outburst from the star, in which it brightened to more than 60,000 times the luminosity of the Sun.

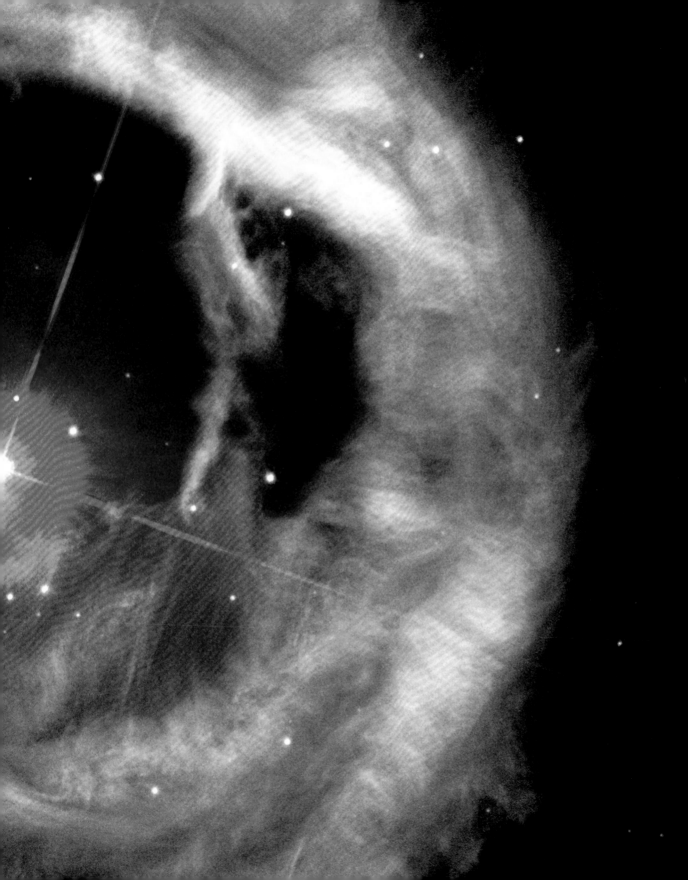

Multiple stars and clusters

Single stars such as the Sun are a minority within our galaxy – most are found in binary or multiple systems. Nearly all stars are born in substantial clusters – some of which hold together, while others slowly drift apart.

Multiple stars

Binary and multiple star systems form when a collapsing protostellar cloud separates into two or more individual clumps that are still bound to each other by gravity. Simple binaries are the most common type, but larger groups are also widespread. Because the protostellar cloud can separate unevenly, stars in a multiple system can develop with different masses, and therefore can follow different evolutionary pathways and age at different rates. This gives rise to spectacular binaries with strongly contrasting colours and brightnesses, and also to systems in which one star has become a stellar remnant while the other remains on the main sequence – a common recipe for some types of variable star.

Often, the stars in a multiple system are so close together that they cannot be separated by even the most powerful telescope. However, such systems can still give themselves away if their planes lie in the same direction as Earth and they form eclipsing variables (see p.76). In other cases, the only clue to a multiple system may be the presence of "double" absorption lines in the spectrum of an apparently "single" star.

≫ MULTIPLE STARS

The great majority of the stars in the sky are multiples – those listed below are some of the most spectacular that can be seen through a small telescope. Some of the multiple systems contain component stars that cannot be separated visually.

Name	Magnitude	Type	Constellation
Albireo	3.1	binary	Cygnus
Epsilon Lyrae	3.9	quadruple	Lyra
Mizar	2.0	quadruple	Ursa Major
Izar	2.4	binary	Boötes
Castor	1.6	sextuple	Gemini
Trapezium	4.7	quadruple	Orion

≫ **Stellar pairings**
Albireo, or Beta (β) Cygni, is one of the most colourful binary stars in the sky. It consists of an evolved yellow giant and a less massive blue-green star that is still on the main sequence.

≫ **Weighing multiple stars**
Multiple stars offer astronomers their only way to measure the relative masses of stars. All the stars in a system will orbit a common centre of mass, known as the barycentre. If the stars are of equal mass, the barycentre will be midway between them, but if one is more massive, then it will be closer to the barycentre.

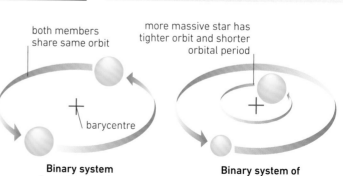

both members share same orbit

more massive star has tighter orbit and shorter orbital period

barycentre

Binary system of equal mass stars

Binary system of unequal mass stars

Stellar Jewel Box
The Jewel Box cluster (NGC 4755), in the constellation Crux, is one of the youngest known open clusters. Thought to be only about 10 million years old, it is dominated by hot blue stars, with one red supergiant.

OPEN AND GLOBULAR CLUSTERS

Open clusters are found close to the Milky Way, while globulars are mostly found above and below the plane of the Galaxy. Magnitudes given are estimated from the combined effect of all the member stars.

Name	Magnitude	Type	Constellation
Pleiades	1.6	open	Taurus
Hyades	0.5	open	Taurus
Jewel Box	4.2	open	Crux
Butterfly Cluster	4.2	open	Scorpius
Beehive Cluster	3.7	open	Cancer
Double Cluster	4.3, 4.4	open	Perseus
Omega Centauri	3.7	globular	Centaurus
M4	5.6	globular	Scorpius
47 Tucanae	4.0	globular	Tucana
M13	5.8	globular	Hercules

Open clusters

Groups of dozens or even hundreds of recently born stars are called open clusters, loosely clustered together, and often surrounded by traces of the nebulosity from which they were created. They are usually dominated by a few intensely hot, blue-white stars, the most spectacular of which are called "OB associations", from the spectral types of these stars. Such bright blue stars are rarely found elsewhere in the sky. They are so massive that they live and die in a few tens of millions of years and have no time to get caught up in the eventual dispersion of the cluster's more moderate stars.

CHARLES MESSIER

French astronomer Charles Messier (1730–1817) listed more than 100 of the brightest open and globular clusters in his famous Messier Catalogue of 1781. He drew up his catalogue to avoid confusion between newly discovered comets – he himself discovered 15 – and permanent features of the night sky. Nevertheless, it became the first standard reference for astronomers investigating so-called "deep-sky" objects, and the Messier or "M" numbers are still in use today.

Globular clusters

Much denser and more structured than open clusters, globular clusters are balls of thousands or even millions of yellow and red stars that form independent systems in orbit around our galaxy and others. Unlike most stars, they are not confined to the plane of our spiral galaxy – they mostly orbit above and below it in a spherical region called the halo. The spectra of the stars in these clusters indicates that they are unusual in having only tiny traces of heavy elements. Such clusters may form during the galactic collisions that give rise to spiral galaxies like our own.

Stellar relics
Small amounts of heavier elements mean that the stars in globulars age very slowly – those in the Milky Way, such as 47 Tucanae, are thought to be about 10 billion years old. Massive blue or white stars have long since died, leaving only lower-mass red and yellow stars.

Variable stars

Not all stars shine with a constant magnitude. Many are variable, their brightness changing in cycles that may last hours or years, or may be completely unpredictable. This variability can occur due to a number of different causes.

Pulsating variables

Many stars pass through a period of instability at some point in their lives. T-Tauri stars, which vary in size and brightness as they settle onto the main sequence, are one example. The most common type, however, is a star that leaves the main sequence and swells into a giant. Changes in its internal composition can make its outer layers transparent or opaque, altering how they are affected by the outward pressure of radiation, and causing the star to expand or contract. Frequently, a cycle of instability arises, with the star swelling and shrinking periodically. The change in size is accompanied by variations in brightness and in colour.

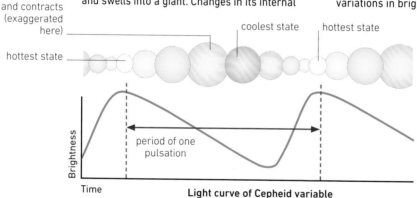

star expands and contracts (exaggerated here)

hottest state

coolest state   hottest state

Brightness

← period of one pulsation →

Time   **Light curve of Cepheid variable**

◀ **Pulsating variables**
Cepheid variables are yellow supergiants with more than three times the mass of the Sun. They pulsate in periods of a few days, and astronomers can plot their changing brightness over time on a "light curve" that shows a quite sharp rise to maximum brightness followed by a slower fall to minimum.

Eclipsing variables

Arising from a line-of-sight effect seen in some binary systems, eclipsing variables are one of the easiest types of variable to understand. They occur when the members of a binary star system pass in front of and behind each other as seen from Earth.

When the stars are seen alongside each other, their light output is combined, giving maximum brightness. When one star disappears entirely or partially behind its companion, the total output of light is reduced, and the star abruptly drops in brightness.

▶ **Eclipsing variables**
The distinctive light curve of an eclipsing variable is easy to identify. Typically, the light remains steady for most of the time, with two evenly spaced drops in brightness. Depending on the relative sizes and brightness of the eclipsing stars, one minimum may be shallower than the other, or they may be equal.

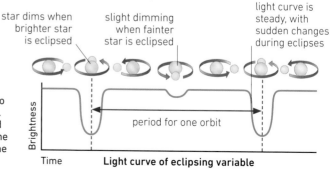

star dims when brighter star is eclipsed

slight dimming when fainter star is eclipsed

light curve is steady, with sudden changes during eclipses

Brightness

← period for one orbit →

Time   **Light curve of eclipsing variable**

Cataclysmic variables

Spectacular but generally unpredictable, cataclysmic variables are usually seen in close binary systems where the more massive star has evolved to a white dwarf and the less massive star is in its giant phase. In such cases, the white dwarf's gravity may be enough to pull material away from its companion onto itself. The dwarf star builds up a hot, dense atmosphere that can eventually detonate in a burst of nuclear fusion. The result is a nova, and the cycle can repeat many times, at predictable or unpredictable intervals. Even more spectacular are Type Ia supernovae. These occur when a particularly massive white dwarf in a nova system gains enough mass to tip it over the Chandrasekhar limit (see p.39), causing it to collapse into a neutron star.

⌃ Eta Carinae
This supergiant star is a highly unpredictable variable. Today, it shines at 5th magnitude, on the limit of naked-eye visibility, but in 1843, it reached magnitude -1, becoming the second brightest star in the sky.

Rotating variables

A recently recognized type of variable is a rotating star in which some areas of the surface are brighter than others – perhaps due to dark starspots (like our Sun's sunspots) or a bright "hotspot" like the one on Betelgeuse. As the star's rotation carries the darker or brighter region in and out of view, its brightness can vary. Another type occurs when stars are stretched into an ellipse by rapid rotation or the gravity of a binary companion. These vary in brightness as we see different amounts of the surface.

《 Computer map of Betelgeuse
The map shows a large bright spot, caused by hot material welling up from within. This may cause some variation in the star's brightness, although Betelgeuse is also a pulsating variable.

Galaxies

Galaxies are systems of billions of stars, held together by their own gravity, usually mingled with dust and gas. There are several types, and it seems that galaxies are transformed by collisions within galaxy clusters.

Our place in the Universe

The Milky Way is home to our Sun and all the other stars that we can see in the sky. It is a vast spiral system, and our celestial neighbourhood sits roughly two-thirds of the way from the centre, on the outer edge of a spiral arm. Because the galaxy is essentially disc-shaped, there are more stars in our line of sight as we look along the plane of the galaxy than there are when we look above or below the plane – the dense star clouds of the plane form the band of the Milky Way that wraps around the sky.

The Sun orbits the centre of the galaxy roughly once every 240 million years, but the Milky Way does not behave like a solid body, so the inner regions orbit the centre faster than the outer ones. The central region is dominated by a dense hub of old red and yellow stars, while the outer disc has a mix of stars. The brightest blue stars are concentrated in the spiral arms, usually in recently formed open star clusters. Stars in the disc and spiral arms of the Milky Way are known as Population I stars – they are typically relatively young. Older stars found in the hub and globular clusters form Population II.

Our celestial neighbours

The Milky Way is not alone in its region of space. For their size, galaxies are relatively crowded together, and our galaxy is a key member of a small cluster called the Local Group. There are two other major members – the Andromeda

Solar System

globular cluster
in spherical halo

dark halo | central bulge

galactic disc

⊠ **Our home galaxy**
Seen edge-on, the Milky Way Galaxy is a thin disc, with the hub forming a huge bulge of old red and yellow stars at the centre. The spiral arms and disc contain younger, bluer stars and gas clouds.

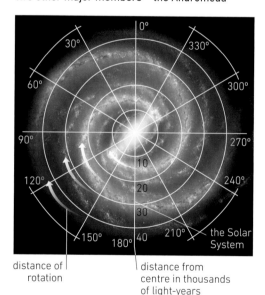

0°
30° 330°
60° 300°
90° 270°
10
120° 240°
20
30
150° 180° 40° 210° ×⟨ the Solar System

distance of rotation

distance from centre in thousands of light-years

⊠ **Galactic limits**
The Milky Way is a disc some 100,000 light-years across and a few thousand light-years deep. The hub is an elongated ball of stars roughly 25,000 light-years in diameter, longest along the axis that points towards the Sun – our galaxy may in fact be a barred spiral (see p.80).

The Milky Way
This long-exposure photograph of the Milky Way reveals the dense star clouds towards its centre. The dark areas are created by intervening dust clouds that obscure the brilliant stars behind.

and Triangulum spirals, which are about 2.5 and 2.7 million light-years away respectively, plus at least four dozen other small galaxies in the group. The Andromeda Galaxy is the biggest galaxy in the Local Group, twice the size of the Milky Way. The gravitational attraction between the two is so strong that the galaxies are moving together, doomed to collide and merge billions of years from now.

Around 160,000 light-years away orbit two shapeless "irregular" galaxies of moderate size, called the Magellanic Clouds. Even closer, a tiny, sparse "dwarf elliptical" galaxy is actually colliding with our own, on the far side of the galactic hub from us.

Globular cluster
Some of the oldest stars in the Milky Way are found in its globular clusters, such as NGC 5139, also known as Omega (ω) Centauri.

Companion galaxy
The Large Magellanic Cloud looks like a detached region of the Milky Way. It is rich in gas, dust, young stars, and pinkish star-forming nebulae.

THE MILKY WAY'S BLACK HOLE

While the galaxy is littered with stellar-mass black holes left over from supernova remnants, a much bigger secret lurks at its heart. The very centre of the Milky Way is home to a black hole with the mass of 3 million Suns. This supermassive hole has long since swept the region around it clear of stars and gas, but its gravity still affects the rotation of stars in the hub. Astronomers now think that most large galaxies have supermassive black holes at their centres.

Heart of the galaxy
Although the central black hole is dormant, the region around the Milky Way's centre contains many violent objects. This X-ray image reveals superhot stars, stellar remnants, and glowing material left by explosions around the black hole.

Types of galaxy

The four major types of galaxy in the Universe – spirals, barred spirals, ellipticals, and irregulars – are distinguished by more than just their shape. Each type possesses a unique balance of stars and other material within it, and they display a number of significant features that offer clues to how they may have evolved, and how they may be related.

The Andromeda Galaxy (M31)

Spiral galaxies

The hub, which is dominated by old red and yellow stars, is surrounded by gas- and dust-rich spiral arms. The space between the arms is not empty, but contains a scattered mix of stars – the arms are prominent only because they contain most of the brightest, short-lived stars. This is a clue that the arms are not permanent, but are regions of increased density that sweep around the disc, triggering starbirth as they go.

Galaxy M87

Elliptical galaxies

Ellipticals come in a huge range of sizes, encompassing the very smallest and the very largest galaxies. They are huge balls of mostly old red and yellow stars, each of which follows its own elliptical orbit around the centre. They contain very little gas and dust, so little or no star formation is occurring within them. The largest are found only in the centre of galaxy clusters – a clue to their origins (see p.82).

Galaxy NGC 1300

Barred spiral galaxies

In a barred spiral, the hub of a spiral galaxy is crossed by an elongated "bar" of stars, from which the spiral arms emerge. There is some evidence that the Large Magellanic Cloud, for example, is in fact a stunted spiral with a bar and a single arm. It now appears that our own Milky Way is also a barred spiral – the bar just happens to be aligned at an angle of about 45° to our line of sight.

The Large Magellanic Cloud

Irregular galaxies

Usually rich in gas and dust, irregular galaxies such as the Large Magellanic Cloud are more or less shapeless collections of stars. Some appear to have central black holes, bars, and the beginnings of spiral arms. Irregulars are frequently sites of intense starbirth activity, with large glowing emission nebulae. The most active of these "starburst" galaxies are forming stars at a much faster rate than normal spirals.

Active galaxies

Most galaxies are the sum of their stars – the radiation they emit comes from their individual stars. But for a substantial number, this is not the case. These "active galaxies" divide into four main types. Quasars and blazars are extremely distant galaxies in which most of the radiation comes from a small, rapidly varying region around the hub. Seyfert galaxies resemble normal spirals but have much brighter cores than expected, and radio galaxies are often insignificant galaxies surrounded by huge lobes of gas-emitting radio waves.

Astronomers think all these different types of activity are caused by the black holes at the centres of the galaxies. While in most galaxies the central black hole is dormant, starved of material to feed on, in active galaxies material is still falling inwards. In Seyfert and radio galaxies, the effects are relatively restrained, but quasars and blazars, at greater distances from us, are remnants from an earlier, more violent phase.

》 STAR-FORMING GALAXIES

This box lists some of the brightest galaxies of various types. The majority are in our Local Group of galaxies, although the Whirlpool and M87 are more distant but spectacular examples of their type.

Name	Magnitude	Type	Constellation
Andromeda	3.4	spiral	Andromeda
LMC	–	irregular	Dorado
SMC	–	irregular	Tucana
Triangulum	5.7	spiral	Triangulum
Whirlpool	8.4	spiral	Canes Venatici
M87	8.6	elliptical	Virgo

《 Multi-wave image of Centaurus A
One of the closest active galaxies, this elliptical has become active as it merges with a spiral. The image combines X-ray, radio, and optical pictures.

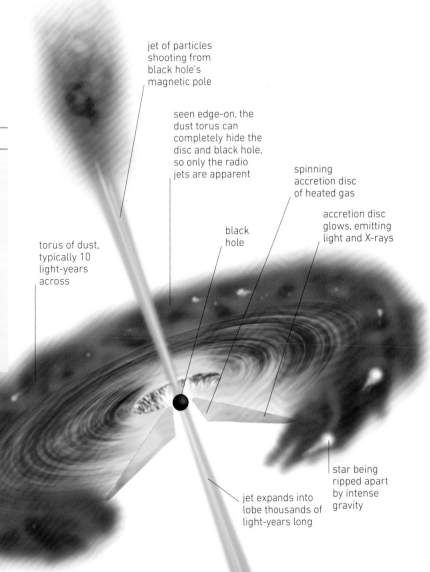

jet of particles shooting from black hole's magnetic pole

seen edge-on, the dust torus can completely hide the disc and black hole, so only the radio jets are apparent

spinning accretion disc of heated gas

accretion disc glows, emitting light and X-rays

black hole

torus of dust, typically 10 light-years across

star being ripped apart by intense gravity

jet expands into lobe thousands of light-years long

》 The core of an active galaxy
This illustration shows a black hole at the core pulling material into its accretion disc. Most active galaxies share the features shown here, but they may appear different when we see them from other angles.

Galaxy evolution and clusters

Galaxies are gregarious – they are usually found in clusters that merge together to form superclusters in the large-scale structure of the Universe. Inside the clusters, galaxies continually collide and merge in ways that are thought to explain how the different types of galaxy arise.

Galaxy evolution

Astronomers once thought that galaxies followed a simple evolutionary sequence from elliptical to spiral. Today, the sequence is thought to be much more complex. Studies of galaxy collisions have shown that gas is often stripped away from the galaxies, becoming hot gas that falls towards the centre of galaxy clusters. Robbed of their star-forming gas, merged galaxies are thought to become ellipticals, dominated by red and yellow stars as their blue and white ones burn out. However, cool gas pulled in by an elliptical can eventually rejuvenate a star-forming disc, allowing the cycle to repeat itself.

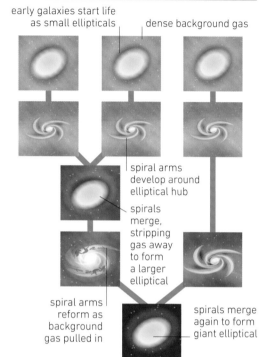

early galaxies start life as small ellipticals

dense background gas

spiral arms develop around elliptical hub

spirals merge, stripping gas away to form a larger elliptical

spiral arms reform as background gas pulled in

spirals merge again to form giant elliptical

»» How galaxies develop
The best modern theory for galaxy evolution suggests they go through a series of mergers and collisions against a steadily declining supply of cool "background" gas. Mergers begin to form elliptical galaxies that eventually come to dominate the central regions of galaxy clusters.

Galaxy clusters

Clusters of galaxies can contain anything from a few dozen spirals and irregulars (as in our own Local Group), to thousands of mostly elliptical galaxies, dominated by one or more giant ellipticals at the centre. Surprisingly, there is not much variation in size – most galaxy clusters occupy a space a few million light-years across. Although clusters frequently merge at their edges, each is distinctive as it is governed by its own local gravity.

»» Cluster formation
Clusters – and their galaxies – merge over time. A cluster's age can be estimated from how many large ellipticals it contains. Abell 1689 (left) is considered to be highly developed.

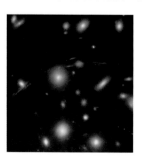

»» GEORGE ABELL

American astronomer George Abell (1927–83) carried out the first detailed survey of galaxy clusters, and developed the techniques required to distinguish clusters from randomly distributed background galaxies. A keen popularizer of science, Abell did most of his work at Mount Palomar observatory during the 1950s, using the most powerful telescope available at the time (right). Abell's catalogue is still the standard reference for galaxy clusters.

Planets of other stars

The Sun's Solar System is not unique, and since the 1990s, astronomers have finally been able to study planets and "brown dwarfs" orbiting nearby stars.

Planet hunting

While early infrared satellites discovered structures that looked like planet-forming discs around young stars, actual evidence for other fully formed planetary systems remained elusive, thanks in large part to their faintness. It was not until the mid-1990s that astronomers used new approaches to confirm "extrasolar" planets for the first time, and the decades since have seen a revolution in astronomy, with thousands of planets, and many more unconfirmed candidates, identified. While the original "radial velocity" planet-hunting technique (below) tends to find giant planets comparable to Jupiter, the more recent transit method can find much smaller worlds, and has confirmed several very Earth-like planets. Along the way, astronomers have also discovered large numbers of "brown dwarfs" – dense balls of gas somewhere between stars and planets.

⚝ **First photograph**
The first direct image of an extrasolar planet was captured in 2005 by the VLT telescope in Chile. The planet (left), with five times the mass of Jupiter, orbits a faint brown dwarf.

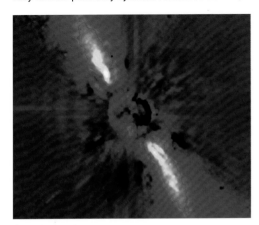

⚝ **Beta Pictoris**
Optical and infrared images of Beta (β) Pictoris, a Sun-like star 50 light-years away, show it ringed by a disc of softly glowing material, with a clearing at its centre about the size of our inner Solar System.

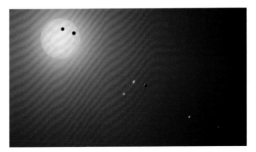

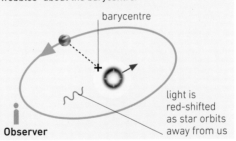

《 **Other Earths**
Small Earth-like planets like those of the TRAPPIST-1 system (illustrated left) are found by looking for dips in a parent star's brightness as they block its light. Future giant telescopes may be able to image such worlds directly.

》 DETECTING PLANETS

The original method used to detect extrasolar planets involved looking for their effect on stars. Just like the stars in a binary system, a star and its planets orbit their common centre of mass, or barycentre. In a planetary system this point is likely to be well within the parent star, but by studying its light spectrum at regular intervals, astronomers can detect minute red and blue shifts as the star "wobbles" about the barycentre.

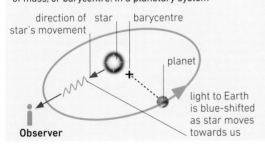

direction of star's movement | star | barycentre

planet

light to Earth is blue-shifted as star moves towards us

Observer

barycentre

light is red-shifted as star orbits away from us

Observer

Jupiter and Io
In this image captured by the
Hubble Space Telescope, the upper
atmosphere of Jupiter is the backdrop
to one of its moons, Io. The moon's
shadow is cast onto the Jovian
cloud tops.

The Solar System

Earth is a ball of rock that abounds with life, and it is a special place for humans because it is their home planet. But it is not the biggest, nor the most important object in its space neighbourhood. The Sun has that position. Earth is part of the Sun's family, the Solar System, which consists of eight planets, several smaller bodies called dwarf planets, over 180 moons, and billions of asteroids and comets. They have existed together for about 4.6 billion years within the Milky Way Galaxy, just one of the millions of galaxies that make up the Universe.

The Sun exerts its influence on a vast number of bodies and an immense volume of space. As the central, largest, and most massive member of the Solar System, all other members orbit around it. Mercury, Venus, Earth, and Mars orbit closest to the Sun. The largest planets: Jupiter, Saturn, Uranus, and Neptune, are beyond Mars. A number of large objects have also been discovered in the Kuiper Belt. The best known of these objects is the world of rock and ice named Pluto, which was regarded as a planet when it was discovered but has since been reclassified as a dwarf planet.

All the bodies within the system formed at the same time, about 4.6 billion years ago. Their origin was a nebula, a cloud of gas and dust many times larger than the present Solar System. By the time the system was formed, just two thousandths of the nebula's original mass remained. The rest had been blown, or pushed out into space. The Sun was made first, followed by the planets. Tiny particles of nebula material clumped together to form larger and larger pieces, chunks, boulders, and finally huge spheres, the rocky planets. The gas giants formed a solid core first, then these captured a gas atmosphere. Material between Mars and Jupiter failed to make a planet, and became the Main Belt of asteroids. Remaining material beyond Neptune became objects in the Kuiper Belt, and the comets.

View from Earth

The closest objects seen in Earth's sky belong to the Solar System. The Sun brightens Earth's day, and illuminates the Moon and planets to shine at night. All the planets, except Neptune and the dwarf planet Pluto, can be seen with the naked eye if conditions are good. The Sun, Moon, and six of the planets – Mercury, Venus, Earth, Mars, Jupiter, and Saturn – have been known since ancient times; as have comets, although their nature was not then understood. Other objects, such as moons, were discovered once the newly invented telescope was turned skywards in the early 17th century. Uranus was discovered in 1781, the first asteroid in 1801, Neptune in 1846, Pluto in 1930, and the first Kuiper Belt object in 1992. Space probes have been revealing close-up details of these remote, fascinating worlds to us for almost 60 years. They are set to continue, along with ever improving Earth-based observing techniques, to reveal more, as well as new members of the system.

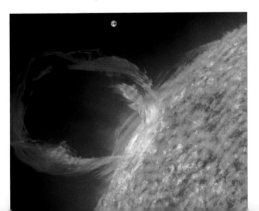

《 Local star, the Sun
The Sun is the only star we can see in detail. Here, an enormous prominence is seen shooting into space from its surface; the Earth is shown to scale above it.

Our Solar System

The Solar System consists of the Sun and its large family of space objects. These include the planets and moons, and countless smaller bodies such as the asteroids and comets. The number of known members rises monthly as smaller and fainter objects are found.

Structure of the Solar System

The Sun, which is the largest and most massive member, dominates the system. Its central position and strong gravity keeps the whole together. Each other member moves along a path around the Sun. One complete circuit of this path is one orbit; each object spins as it travels. The planetary part of the system is disc-shaped; it is almost flat and nearly circular. Beyond this is the domain of the comets. They orbit the Sun in any plane, from close to the planetary plane, to above and below the Sun, or anywhere in between. They make a vast sphere around the planetary part, stretching to about 1.6 light-years from the Sun. Beyond is interstellar space.

Earth
The only place in the Solar System known to harbour life

Mercury
The innermost planet, and the one with the shortest and fastest orbit

Jupiter
The largest and most massive planet, it has the fastest spin

Mars
This is half Earth's size but more distant and so colder

Venus
The second planet from the Sun but the hottest due to its thick atmosphere

Main Belt
This ring of asteroids separates the inner and outer planets

Formation of the Solar System

The Solar System formed about 4.6 billion years ago from a vast, spinning cloud of gas and dust termed the Solar Nebula. Material collapsed into the centre and made the Sun. A spinning disc of unused material surrounding it produced the planets. Rocky and metallic material near the Sun formed the rocky planets. In the cooler, outer regions rock, metal, gas, and ice formed the outer gas giants.

⊻ The orbits of the planets

The planets orbit the Sun in roughly the same plane, and travel in the same direction, anticlockwise when seen from above the North Pole. The length of an orbit and the time to complete one increases with distance. The planets and their orbits are not shown to scale.

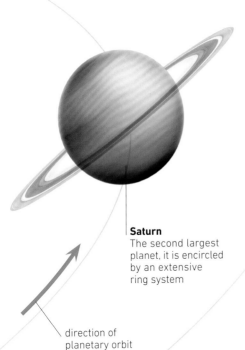

Neptune
The smallest and most distant of the four gas giants

Uranus
Twice as far from the Sun than inner Saturn, it is tilted on its side

Saturn
The second largest planet, it is encircled by an extensive ring system

direction of planetary orbit

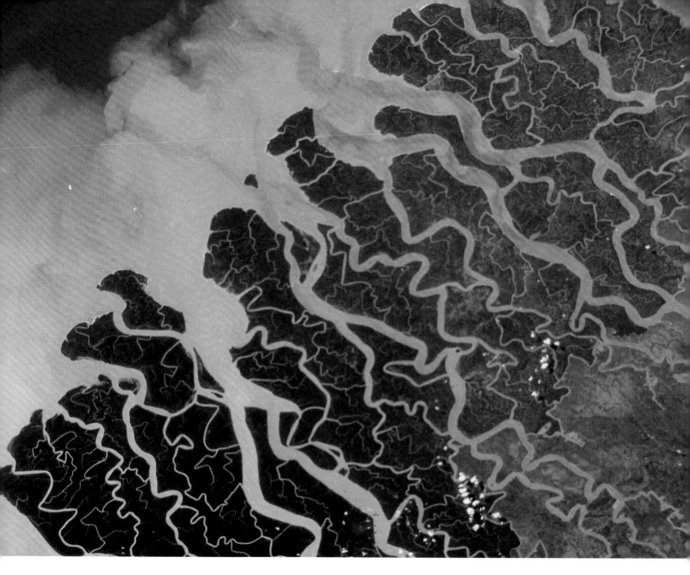

The Ganges Delta, Earth
Only Earth has liquid water on its surface. Nearer to the Sun, Earth's water would boil, if further away, it would freeze.

Inner rocky planets

The four inner planets – Mercury, Venus, Earth, and Mars – are collectively known as the rocky planets. In fact, this is something of a misnomer as they are all rocky-metallic. A slice through any one of them would reveal a metal core surrounded by a rocky mantle and crust. On the surface they are worlds of contrast. Craters scar Mercury's dry, grey landscape, and the thick poisonous atmosphere enveloping Venus hides its volcanic terrain. Oceans of water cover more than 70 per cent of Earth, and dry, cold Mars is home to the Solar System's largest volcanoes.

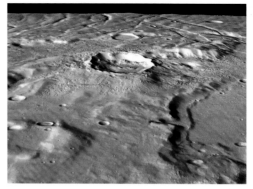

Surface of Mars
The Martian surface has been shaped by tectonic activity and by flowing water. It is also scarred by tens of thousands of impact craters.

Outer gas giants

The four largest planets, which lie beyond the Main Belt, are known as the gas giants, although they are only made in part of gas. Excluding the Sun, these planets (Jupiter, Saturn, Uranus, and Neptune) are the biggest objects in the Solar System. The visible surface of each is the top of their cloudy upper atmospheres. All four of these distant, cold worlds have large families of moons and are encircled by ring systems.

⚈ **Saturn's rings**
Saturn has an extensive but thin system of rings. This image from the Cassini probe shows individual ringlets, made up of thousands of pieces of dirty ice.

⚈ **Jupiter's stormy surface**
The top layer of Jupiter's dense atmosphere forms different-coloured bands and contains raging storms.

⟫ MOONS, ASTEROIDS, AND COMETS

Billions of small objects exist in the Solar System. Of these, over 180 are moons orbiting seven of the planets. The two largest are both bigger than Mercury; the smallest are potato-shaped lumps just a few kilometres across. Billions of space rocks, termed asteroids, make up the Main Belt between Mars and Jupiter. The Kuiper Belt, which lies outside Neptune's orbit, consists of many thousands of icy rocky objects, and over a trillion comets exist in the Oort Cloud.

Comet Ikeya-Zhang

Asteroid Gaspra

The Sun

The Sun is a huge ball of hot, luminous gas. In its core, hydrogen is converted to helium, releasing energy that is felt on Earth as heat and light. The Sun has existed in this form for some 4.6 billion years and should stay the same for another 5 billion or so years.

Inside the Sun

The Sun is immense; its diameter is 109 times that of Earth, and it contains 99 per cent of all the material in the Solar System. It is made mainly of hydrogen (about 71 per cent by mass) and helium (about 27 per cent) with tiny amounts of around 90 other elements. Gravity keeps the Sun's gas together, pulling it towards the centre. The temperature increases and the pressure builds up with depth. Pressure tries to push the gas out and prevent it from becoming ever more tightly packed. As long as the balance between gravity and pressure is maintained, the Sun will keep its present size and globe shape.

In the core, which contains about 60 per cent of the Sun's mass, it is about 15 million °C (27 million °F). Here, nuclear reactions convert the hydrogen to helium at the rate of about 600 million tons each second. In the process, nearly five million tons is released as energy. The Sun is not solid, so – unlike Earth, which spins as a whole – some parts of the Sun spin at different speeds from others. The equator spins in 25 days, while the polar regions take about 10 days longer.

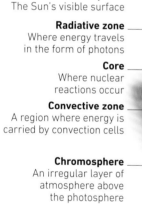

Photosphere
The Sun's visible surface

Radiative zone
Where energy travels in the form of photons

Core
Where nuclear reactions occur

Convective zone
A region where energy is carried by convection cells

Chromosphere
An irregular layer of atmosphere above the photosphere

《 The Sun's interior
Hydrogen is converted to helium in the Sun's core. Energy produced moves towards the surface by radiation and then by convection, before leaving the Sun through the photosphere.

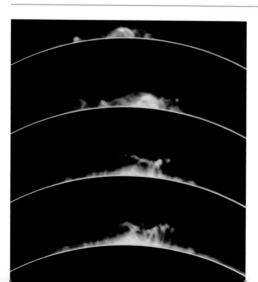

The Sun's surface

The visible surface of the Sun is a layer, 500km (310 miles) deep, called the photosphere. It is made of cells of rising gas called granules, which give it a mottled appearance like that of orange peel. These convection cells, which are about 1,000km (620 miles) across, are short-lived and constantly renew themselves. Spicules are

《 Solar flares
Gas continually erupts and shoots up from the Sun's surface in jets and sheets. In this series of photographs taken over an 8-hour period, solar flares explode out from the photosphere.

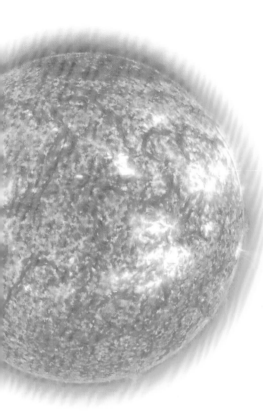

<< **Our local star**
In this SOHO spacecraft image, the Sun's mottled surface is clearly visible. The white regions are faculae, which are bright active regions associated with sunspots. The image was taken in ultraviolet light, which is why the Sun appears dark orange. The Sun's usual distinctive yellow colour is due to the temperature of the photosphere.

>> SUN DATA

Diameter 1.4 million km (869,400 miles)

Average Distance From Earth
149.6 million km (92.9 million miles)

Rotation Period (Equatorial)
25 Earth days

Surface Temperature 5,500°C (9,932°F)

Core Temperature 15 million °C (27 million °F)

Size comparison

The Sun

Earth

short-lived jets of gas that stand out from the surface. They look tiny compared with the Sun's disc but can be thousands of kilometres long. Much larger and more substantial solar flares and loops of hot gas are often associated with sunspots.

Dark spots measuring hundreds or thousands of kilometres across appear periodically on the surface and typically last for a few weeks at a time. These sunspots are cooler areas where hot gas cannot reach the surface. They lie within a region that stretches from about 40° either side of the equator. The spots materialize according to an 11-year cycle. They appear and disappear in positions ever closer to the equator, with the number of spots changing during the cycle.

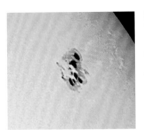

<< **Sunspots**
These are about 1,500°C (2,730°F) cooler than the rest of the photosphere. The darkest, coolest central zone is the umbra; around it is the lighter, less cool penumbra.

a huge solar prominence arches out from the Sun —

☑ **Prominences**
These are giant clouds and sheets of gas that reach far out from the visible surface. They extend hundreds of thousands of kilometres into the Sun's outer atmosphere, the corona, and can last for days or weeks.

Atmosphere

An extensive atmosphere consisting of two main layers stretches beyond the photosphere. The chromosphere, just above the photosphere, extends out about 5,000km (3,100 miles), its temperature increasing with height to about 20,000°C (36,000°F). The outer layer, the corona, which extends into space for millions of kilometres, is hotter still, at 2 million °C (3.6 million °F).

Such high temperatures are unexpected, and at present there is no generally accepted explanation for them.

The atmosphere is not usually visible but lost in the dazzling light of the Sun. The pinkish chromosphere and the pearly white corona can be seen when the disc is obscured by, for example, the Moon during a total eclipse. Invisible from view is the solar wind of atomic particles streaming away from the Sun.

》 Chromosphere
The irregular, thin dark pink arc is the chromosphere; below it is the brighter white glow from the photosphere. The disc of the Sun has been blocked out. A red flame-like protuberance shoots out from the chromosphere into the corona (not seen here).

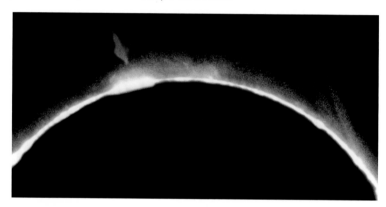

》 SDO
NASA's Solar Dynamics Observatory (SDO), launched in 2010, studies the Sun's photosphere, atmosphere, and solar-wind production.

Studying the Sun

Over 20 space probes have studied the Sun and its effect on the space around it. Some have investigated our local star from orbit around the Earth, while others work from solar orbit. Ulysses followed a path that took it over the Sun's poles. One craft, Genesis, brought solar-wind particles back to Earth. The probes have studied parts of the Sun, including solar flares and the corona, in a range of wavelengths.

》 TEIDE OBSERVATORY

Earth-based astronomers use specialist telescopes to study the Sun. They are housed in tall towers at high-altitude sites where interference from the Earth's atmosphere is at a minimum. A moving mirror (heliostat) at the top of the tower reflects the sunlight into a fixed telescope. The light is directed to measuring instruments in underground rooms, kept cool because of the intense heat from solar radiation. The Vacuum Tower Telescope at the Teide Observatory (the tallest tower seen to the left) has the air removed from the tower to limit image distortion by the Sun's heat.

Teide Solar Observatory, Tenerife

⟩⟩ SOLAR ECLIPSES

When the Sun, Moon, and Earth are in alignment, the Moon covers the disc of the Sun and blocks it from view. Such a solar eclipse occurs at new Moon, but not every new Moon, only typically twice a year when the three worlds are directly aligned. The Moon's shadow is cast on to Earth. Anyone within the inner part of this shadow, the umbra, will see a total solar eclipse; those in the outer shadow will see a partial eclipse.

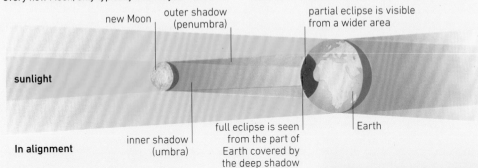

new Moon

outer shadow (penumbra)

partial eclipse is visible from a wider area

sunlight

In alignment

inner shadow (umbra)

full eclipse is seen from the part of Earth covered by the deep shadow

Earth

The stages of a solar eclipse

It takes about an hour and a half for the dark disc of the Moon to cover the bright disc of the Sun (below). Totality, or total eclipsing of the Sun (right), usually lasts for two to five minutes but it can last more than 7 minutes. The Sun's outer halo, its corona, can then be viewed.

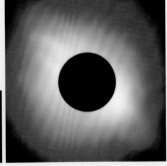

Observing the Sun

The Sun should never be observed directly by naked eye or with any instrument, as its intense light can damage eyes. Astronomers adopt two methods to observe the Sun with instruments. The protection method uses, for instance, filters to block harmful rays. The projection method (the image is projected onto a screen) is safer, because it does not mean looking at the Sun.

⟩⟩ Binocular projection

Cover one lens so sunlight can only pass through the other. Direct the Sun's image through the binoculars. Adjust the viewing card until the Sun's image is in focus.

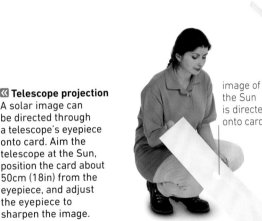

image of the Sun is directed onto card

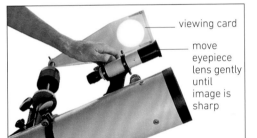

viewing card

move eyepiece lens gently until image is sharp

⟨⟨ Telescope projection

A solar image can be directed through a telescope's eyepiece onto card. Aim the telescope at the Sun, position the card about 50cm (18in) from the eyepiece, and adjust the eyepiece to sharpen the image.

Solar flare
To the unaided eye, the Sun appears in our sky as a featureless disc. However, instruments on Earth or in space reveal more detail. This image taken by the Earth-orbiting TRACE satellite shows a violent release of energy known as a solar flare in the Sun's corona.

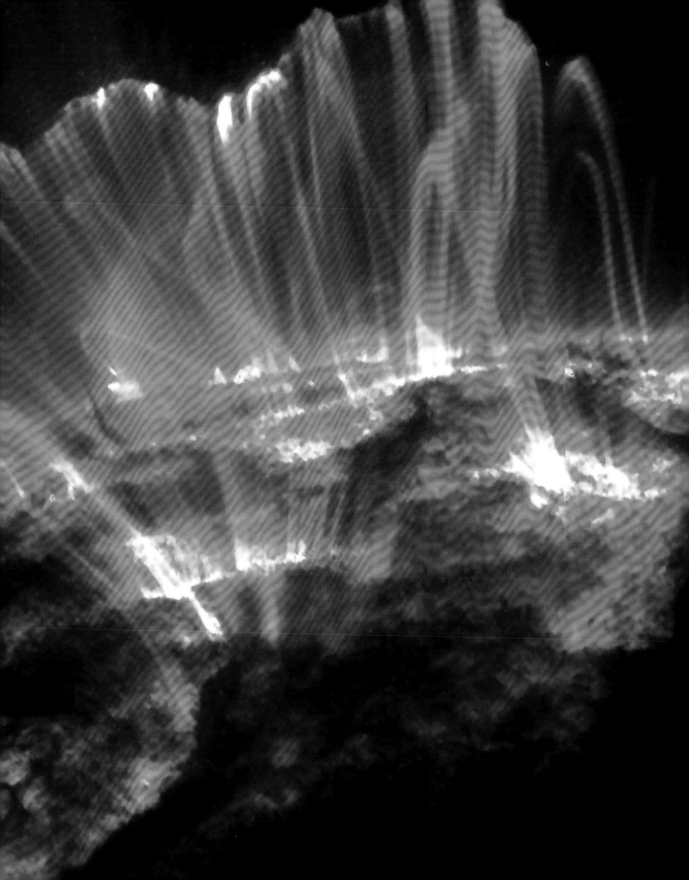

Mercury

Moving along an eccentric orbital path, Mercury is the closest planet to the Sun. It is a dry, rocky, cratered world that feels the full force of the Sun's heat during the day, but, with only the barest of atmospheres, experiences freezing cold nights.

Structure and atmosphere

Below Mercury's silicate rock surface is a solid rocky mantle about 550km (340 miles) thick. This layer would have been liquid when Mercury was young and the source of volcanic eruptions. The mantle has now cooled and solidified, and during the past billion years volcanic eruptions have ceased. Below the mantle is a large iron core, formed when heavy iron sank within the young planet. The core is partially molten and is the source of the planet's weak magnetic field.

Elements from Mercury's surface, such as sodium, along with helium from the solar wind, form a very thin atmosphere. This is temporary and needs to be continuously replenished, because Mercury's gravitational pull cannot hold on to the gases.

crust of silicate rock
iron core
rocky, silicate mantle

⏩ The planet's interior
Mercury is extremely dense compared with the other rocky planets, which signifies it is rich in iron. Its huge iron core is about 3,600km (2,235 miles) in diameter. This is surrounded by a mantle and crust, both made of silicate rock rich in iron and titanium.

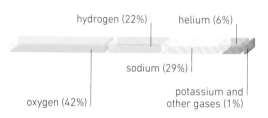

hydrogen (22%)   helium (6%)

sodium (29%)

oxygen (42%)

potassium and other gases (1%)

⏶ Atmosphere
The composition of Mercury's thin, temporary atmosphere varies with time as gases are lost and replenished. Oxygen, sodium, and helium are, however, the most abundant elements.

Surface features

Mercury is covered in thousands of impact craters, formed when meteorites hit the surface. The oldest date from about 4 billion years ago when meteorites bombarded the young planet. They range in size from small bowl-shaped ones to the Caloris Basin, a quarter of Mercury's diameter. The basin formed when an impacting body, probably about 100km (60 miles) across, hit the planet. Shockwaves buckled the surface, forming rings of ridges around the site and crumpled terrain on the opposite side of the planet. The surface also has plains formed by volcanic lava and cliff-like ridges shaped when the young, hot planet cooled and shrank.

⏶ Pockmarked surface
The Caloris Basin is a huge impact feature over 1,500km (930 miles) across, as seen by the Messenger probe. It has been flooded with volcanic lava and subsequently pockmarked by smaller impact craters.

《 **Moon-like planet**
Mercury is the smallest major planet and the fastest to orbit the Sun. Its volcanic plains are dotted with meteorite impact craters.

》 MERCURY DATA

Diameter 4,879km (3,032 miles)

Average Distance from Sun
57.9 million km (36 million miles)

Orbital Period 88 Earth days

Rotation Period 59 Earth days

Surface Temperature −180°C to 430°C
(−292°F to 806°F)

Number of Moons None

Size Comparison

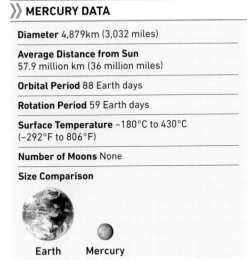

Earth   Mercury

Missions to Mercury

Mercury is only ever seen near Earth's horizon, where our planet's atmosphere is turbulent. This makes it extremely hard to study the surface from Earth. The first spacecraft to visit Mercury, Mariner 10, flew by three times in 1974–75. Its images revealed a crater-covered world similar to our Moon. A second probe, Messenger, went into orbit around Mercury in March 2011. Among other things, it has provided more detailed maps of the surface.

sunshade protects body and scientific instruments

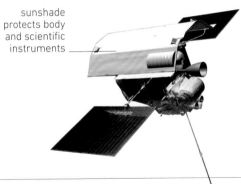

《 **Messenger**
After three close flybys between 2008 and 2009, Messenger went into orbit around Mercury in 2011 in order to map the planet's surface and study its geochemistry, atmosphere, and magnetic field.

Observing Mercury

Mercury has phases like the Moon but is difficult to see because it is never far from the Sun. It is found low in the twilight sky, before sunrise or after sunset. It is best seen at greatest elongation, when it is furthest from the Sun, six or seven times a year. Several times a century, Mercury appears to cross over the Sun's face as it passes between the Sun and Earth.

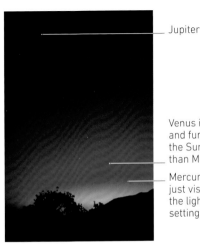

Jupiter

Venus is brighter and further from the Sun's light than Mercury

Mercury is just visible in the light of the setting Sun

》 **Mercury and Venus in the night sky**
Mercury is seen in the evening twilight. The Sun has set and Mercury will also soon be below the horizon. The planet appears star-like and is within naked-eye visibility.

Venus

Similar to Earth in size and structure, Venus is our inner neighbour, the planet that gets closest to us and appears brighter than any other in our sky. Yet, it reveals little of itself – only a top layer of thick, unbroken blanket of cloud is visible.

Structure and atmosphere

In diameter, Venus is only about 650km (400 miles) smaller than Earth and its internal layers are of similar size and composition. Below its silicate crust is a rocky mantle, and below this a core, which is solid in the centre. It spins on its axis more slowly than any other planet, one spin taking longer than one Venusian orbit. It also spins from east to west, the opposite direction from most other planets.

The 80-km- (50-mile-) deep atmosphere is predominantly carbon dioxide. A thick cloud deck of sulphuric acid droplets reflects away over 65 per cent of the sunlight hitting Venus. The clouds also trap heat from the Sun. As a result, Venus is an overcast place with a surface temperature higher than that of any other planet.

silicate crust

molten iron and nickel outer core

rocky mantle

solid iron and nickel inner core

◀◀ **The planet's interior**
Venus is a dense, rocky world whose material has settled to form layers. Its iron and nickel core has cooled and partly solidified. Molten subsurface mantle material is released onto the planet's surface as volcanic lava.

Surface features

Volcanic features dominate the surface of Venus. About 85 per cent of it is low-lying plain covered by volcanic lava. The remainder consists of three highland regions, the largest of which is Aphrodite Terra. The volcanic surface is relatively young in geological terms. Venus's hundreds of volcanoes and their extensive lava fields may be no more than 500 million years old.

Some volcanoes may still be active. Venus has unique flat-topped mounds of lava – pancake domes – and spider-like volcanic features, known as arachnoids. Its surface is pitted with hundreds of impact craters. Along with other features on Venus, these are named after women.

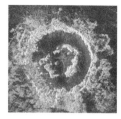

▲ **Barton Crater**
This crater is 52km (32 miles) across, and is named Barton Crater after Clara Barton, founder of the Red Cross.

▼ **The volcanic landscape**
Venus's volcanoes range widely in size. Maat Mons is a shallow-sloped shield volcano that grew with successive eruptions; it rises 8km (5 miles) above the surface.

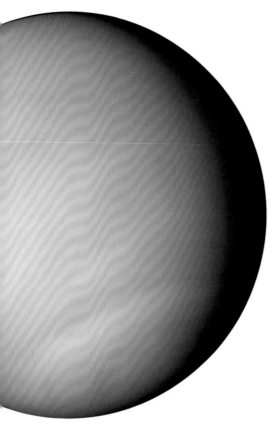

carbon dioxide (96.5%)

nitrogen and trace gases (3.5%)

Cloudy Venus

The thick layers of cloud would prevent anyone on the planet seeing out. Fierce winds blow the clouds westwards, in the same direction as – but much faster than – the planet's spin. They circle the planet every four days. Space probes have revealed that below the clouds lies a gloomy, stifling world.

Atmospheric composition

The atmosphere around Venus is rich in carbon dioxide, with a small amount of nitrogen and trace elements of other gases, such as water vapour, sulphur dioxide, and argon.

⟩⟩ VENUS DATA

Diameter 12,104 km (7,517 miles)

Average Distance from Sun
108.2 million km (67.2 million miles)

Orbital Period 224.7 Earth days

Rotation Period 243 Earth days

Surface Temperature 464°C (867°F)

Number of Moons None

Size Comparison

Earth   Venus

⟩⟩ MISSIONS TO VENUS

Since 1962, over 20 space probes have flown past, orbited, or landed on Venus. The landers have had to survive the corrosive clouds, infernal surface heat, and strong surface pressure (92 times that of Earth). The spacecraft Magellan (1990–94) and Venus Express (2006–14) have orbited the planet and mapped the entire surface using radar and infrared. Venus Express also studied the planet's thick atmosphere.

Soviet Venera lander probe

⟩⟩ Venus from Earth

Venus, seen here with the crescent Moon, is easy to spot. Its bright presence at the start or end of the day has earned it the nicknames Morning Star and Evening Star. Its brightness and apparent size vary with time.

Observing Venus

Venus shines brightly in Earth's sky because of its reflective cloud-top surface, and because of the planet's closeness to us. At its brightest, Venus is magnitude –4.7 and is then only outshone by the Sun and Moon. It goes through a cycle of phases like the Moon. When close to Earth, only part of the planet facing us is lit. Venus is best seen at greatest elongation; either in the evening sky, after sunset, when it is shrinking from half phase to crescent, or in the morning, before sunrise, when growing from crescent to half phase.

Earth

The largest of the four rocky planets, Earth is the only place known to support life. A dynamic world, unlike any other, it has an abundance of liquid water, and its surface undergoes constant change. Its one companion in space is the lifeless Moon.

Structure

Earth was formed about 4.6 billion years ago. Since then its material has differentiated into layers. The central core, which is hot and dense, has solidified and consists of iron, with some nickel. Above this is a solid rock mantle, and then a thin crust made up of many different types of rocks and minerals, but predominantly silicate rock. This crust is broken into seven large solid plates and some smaller ones. They float on partly molten under-lying mantle. Earth's continents, oceans, and air all support life.

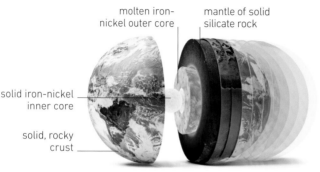

molten iron-nickel outer core

mantle of solid silicate rock

solid iron-nickel inner core

solid, rocky crust

◄◄ The planet's interior
Movement within the outer core, partly driven by Earth's rotation, produces a magnetic field that deflects solar wind.

Surface features

Earth's crust varies in thickness; the thickest parts form the seven major continental landmasses. The rest of the crust, which amounts to over 70 per cent of Earth's surface and is generally thinner, is covered in water. Almost all of this is in liquid form and makes Earth's five vast oceans. Just 2 per cent is in the form of ice, in the caps around the north and south poles. The plates in Earth's crust move against and away from each other, and at plate boundaries give rise to such features as mountain ranges, deep-sea trenches, and volcanoes.

⌃ Shaped by water
The Tigre, a tributary of the Amazon, Earth's greatest river, cuts through the Peruvian rainforest. The Amazon carries more water each year than any other river. Water, wind, and life forms have all shaped Earth's surface to varying degrees.

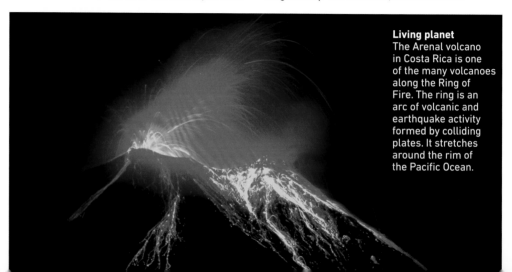

Living planet
The Arenal volcano in Costa Rica is one of the many volcanoes along the Ring of Fire. The ring is an arc of volcanic and earthquake activity formed by colliding plates. It stretches around the rim of the Pacific Ocean.

>> **EARTH DATA**

Diameter 12,756km (7,921 miles)

Average Distance from Sun
149.6 million km (93 million miles)

Orbital Period 365.26 Earth days

Rotation Period 23.93 hours

Surface Temperature 15°C (59°F)

Number of Moons 1

《 **Blue planet**
When Earth is seen from space, it is easy to
appreciate just how much of it is covered by water.
Water also moves constantly between the surface
and the atmosphere, forming white clouds of water
vapour. Forest covers about a third of the land.

>> **Envisat**
Scientists use orbiting spacecraft to
monitor Earth just as they use them
to study other planets. The European
Envisat (shown here) was the largest
observation satellite ever built.

Atmosphere and weather
A layer of gas, rich in nitrogen and oxygen,
surrounds Earth. The oxygen sustains life,
and high in the atmosphere forms ozone,
which acts as a shield against solar radiation.
The atmosphere reaches about 500km (310 miles)
above the surface, but most of it is within about
16km (10 miles) of the planet and this is where
the weather occurs. The Sun heats Earth unevenly,
so producing a variation in air pressure. Winds
develop as a result and these drive the air and
moisture over the planet.

nitrogen (78.1%)   oxygen (20.9%)

argon and trace gases (1%)

⌃ **Atmospheric composition**
The atmosphere is dominated by nitrogen.
Together, with the second abundant element,
oxygen, it makes up 99 per cent of dry air.
The water vapour content varies; it can be
up to about 4 per cent.

⌃ **Aurora Borealis**
Solar wind particles entering Earth's upper atmosphere can produce
spectacular displays in the night sky, known as aurorae. The colourful
lights are a result of atmospheric gas interacting with the solar particles.
The Aurora Borealis, or Northern Lights, can be seen north of about latitude
50°N. Similar lights in the southern sky are known as the Aurora Australis.

Earth at night
In this view of the Earth at night, which has been created from satellite images, artificial lights reveal the locations of the main areas of human habitation on our planet. Individual cities and major roads can be picked out, along with features such as the river Nile in Egypt. Artificial lights cause trouble for astronomers by spreading light pollution over the night sky.

The Moon

The Moon is a cold, dry, lifeless ball of rock with virtually no atmosphere. Grey and crater-covered, it is the fifth-largest moon in the Solar System. It looms large in our sky, and was an early target for space exploration. It is the only other world that humans have visited.

⌃ Full Moon
The Moon is fully lit but the strong, direct sunlight has swamped some of its surface features.

Structure and atmosphere

This rock ball has an outer layer of calcium-rich, granite-like rock, about 45km (28 miles) thick on the near side and 60km (37 miles) thick on the far side. Below is a solid rocky mantle rich in silicate minerals, which becomes partially molten with increased depth. A small iron core may be in the centre.

The atmosphere is very thin. In total, it is equivalent to the amount of gas released by a landing Apollo craft. The Moon's gravity, one-sixth of Earth's, cannot hold on to the atmosphere, but it is replenished by solar-wind material.

≫ The terminator
In this view, only the left side of the Moon is sunlit. The boundary between this and the unlit part is called the terminator. Here, shadows are longest and surface features, such as craters, are seen to best effect.

≫ Interior
The Moon is about a quarter the size of Earth. Its upper crust is cracked. Below is solid crust, and below this is a rocky mantle. Its average density indicates it may have a small iron core.

crust of granite-like rock

possible small metallic core

thick rocky mantle

≫ MOON DATA

Diameter 3,475km (2,158 miles)

Average Distance from Sun
384,400km (238,700 miles)

Rotation Period 27.32 Earth days

Surface Temperature −150°C to 120°C (−240°F to 248°F)

Size Comparison

Earth   Moon

THE MOON | 105

Formation of the Moon

Most astronomers think that the Moon was formed when a Mars-sized asteroid collided with Earth about 4.5 billion years ago. Molten rock from the two bodies splashed into space and formed a ring of material round the young Earth. In time, the material clumped together to form one new, large body. This was the young Moon, which then cooled, solidified, and formed a surface crust.

⊗ 1. Collision with an asteroid
An asteroid collides with the young Earth, giving it a glancing blow. Material is pushed out of the mantle.

⊗ 2. Massive cloud forms
Debris creates a massive cloud of gas, dust, and rock. It quickly begins to cool.

⊗ 3. Ring of debris
Most of the debris moves into orbit round young Earth. It forms a dense, doughnut-shaped ring.

⊗ 4. Moon is born
Pieces collide and form a single large body, the young Moon, which sweeps up remaining material.

Surface of the Moon

About 4 billion years ago the young Moon was bombarded by asteroids, and craters formed all over its surface. The impact of the rocks also pushed up mountains. In time, lava seeped from inside the Moon through surface cracks and flooded the large craters. Geological activity has long ceased and, with no erosion from wind or water, the landscape has barely changed in 2 billion years. Today the Moon is a dead world covered by a porous blanket of rubble. Micro-meteorites still bombard it; astronaut footprints will eventually be erased.

⊗ Open plains
Lava filled the floors of the largest craters and produced dark, flat plains. When observed from Earth, these look like seas and were called maria (plural of the Latin for sea, *mare*).

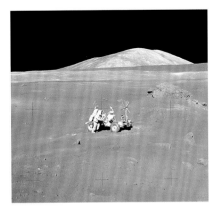

⊗ Impact craters
Impact craters range from bowl-shaped ones, less than 10km (6 miles) across, to those larger than 150km (90 miles), filled by solidified lava. Middle-sized ones have outer walls and peaks.

⊗ Fine dust
The lunar soil, termed regolith, is fine-grained, fragmented bedrock. It is dust-like on the surface but the grains get larger with depth. Astronauts' boots left crisp footprints.

Earth and Moon partnership

The Moon takes 27.3 days to complete one elliptical orbit around Earth. As it moves, the Earth orbits around the Sun. It takes 29.5 days for the Moon to return to the same position relative to the Sun in Earth's sky, and complete its cycle of phases. The Moon's gravity pulls on Earth, making the oceans on either side of the planet bulge, in turn producing the tides. Tidal forces slow down Earth's spin, with the result that the Moon is moving away from Earth by about 3cm (1in) each year.

The Moon spins round once every 27.3 days. This is the same length of time that it takes to make one orbit around Earth. As a result of this synchronization of the Moon's rotation and orbit, the Moon keeps the same side, termed the near side (marked here by a red dot), facing Earth.

imaginary point always faces the Earth

Earth's spin causes tidal bulges to sweep over surface

tidal bulges

Moon's orbit

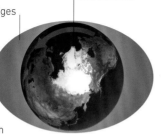

gravitational pull of the Moon

inertial force

⌃ Tidal bulges
Two bulges in Earth's oceans (exaggerated here) are created by the gravitational interaction of the Earth and Moon. As Earth spins, the bulges of water sweep over the planet's surface, creating changes in sea level – high and low tides.

⌃ Low tide in Oregon, USA
The time of high tide depends on the position of the Moon in the sky. Tide height changes during the lunar cycle.

When the Sun, Earth, and Moon are aligned, a lunar eclipse can occur. The Moon can move into the shadow cast by Earth. If it is completely within the shadow, the Moon is totally eclipsed. If only partly covered by shadow, the Moon is in partial eclipse. An eclipse is visible from anywhere on Earth, as long as the Moon is above the horizon.

>> **How lunar eclipses occur**
Earth blocks the Sun's light and casts a shadow into space. The Moon moves into the shadow. In the darkest part, it is totally eclipsed, and can appear reddish when sunlight is bent into the shadow by Earth's atmosphere. Lunar eclipses happen up to three times a year.

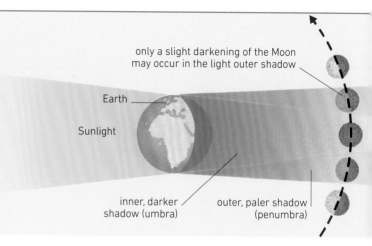

only a slight darkening of the Moon may occur in the light outer shadow

Earth

Sunlight

inner, darker shadow (umbra)

outer, paler shadow (penumbra)

Phases of the Moon

Sunlight always lights up one half of the Moon, just as it lights up one half of Earth. The half that is lit changes, as the Moon spins and moves along its orbit. From Earth, just one side of the Moon is visible. This is at times fully lit, partially lit, or unlit, giving the impression of a different shape. The changing views we have of the Moon are known as its phases; a complete cycle of phases takes 29.5 days.

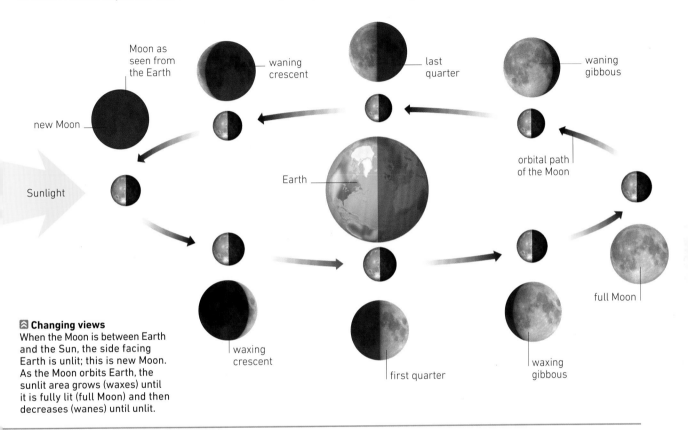

Moon as seen from the Earth

waning crescent

last quarter

waning gibbous

new Moon

orbital path of the Moon

Sunlight

Earth

full Moon

waxing crescent

first quarter

waxing gibbous

⌃ Changing views
When the Moon is between Earth and the Sun, the side facing Earth is unlit; this is new Moon. As the Moon orbits Earth, the sunlit area grows (waxes) until it is fully lit (full Moon) and then decreases (wanes) until unlit.

⌃ Onset of eclipse
The Moon has moved into Earth's shadow; part of its disc is obscured.

⌃ Red moon
As the Moon moves further into Earth's shadow, its face turns a pinkish red.

⌃ Over halfway
Earth is blocking the sunlight from most of the Moon. The crescent left shines brightly.

⌃ Close to totality
The eclipse is about to be total. Except for a sunlit slither, the Moon is within Earth's shadow.

Man on the Moon
Twelve men from six different Apollo space missions walked on the Moon between July 1969 and December 1972. They collected over 380kg (838lb) of lunar rock and soil from six different sites.

Missions to the Moon

Over 60 craft have journeyed to the Moon. About half of these flew in the decade from 1959 to 1969, when both the United States and the Soviet Union were preparing to land a man on the Moon. This aim was fulfilled by the United States in 1969 (see pp.30–31). About the same time, Soviet probes returned samples of the Moon's soil and two robotic craft, Lunokhods 1 and 2, roved across its surface. No further craft were sent until the 1990s. Now, more probes to study the Moon from orbit and land on its surface are planned.

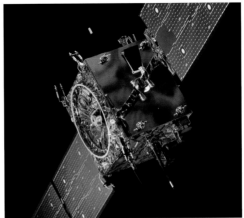

《 SMART-1
The European Space Agency's SMART-1 is the first European craft to be sent to the Moon. Launched in September 2003, one of its tasks is to make an inventory of chemical elements in the lunar surface.

Near side

Far side

⌃ Lunar topography
Accurate height measurements were made of the surface by the Clementine probe in 1994. In these maps, made from Clementine data, blue is low, green medium, and red high. The Moon's far side is right, the near left.

Mapping the Moon

Humans have observed the Moon ever since they looked skywards. Its full bright disc lit up the dark night hours for early people, and its changing face and movement marked the passage of time. The newly invented telescope was turned moonwards in the 17th century, when the first lunar maps were made. The first photographic atlas came in 1897. Decades later, the space age offered the chance to see the far side and observe more detail. During 1966–67, five Lunar Orbiter craft mapped 99 per cent of its surface.

Observing the Moon

The Moon, which shines by reflecting sunlight, is easily spotted in Earth's sky. It only goes unnoticed near new Moon, when its near side is unlit. But even then, the unlit part can receive some light reflected from Earth, known as Earthshine. The Moon is close enough and bright enough for us to see detail. Two types of terrain are clearly visible: the large, dark plains (*maria*); and the brighter, heavily cratered highland regions.

◀◀ Visible by day
For part of every month, the Moon can be seen in the daytime sky. It does not stand out so well against the brighter background, but it is possible to make out the dark regions and lighter patches on its surface.

⌃ Viewed by naked eye
The Moon is easy to see, and moves quite rapidly. Dark and light features are apparent.

⌃ Viewed by binoculars
The Moon is still a whole disc but the image is larger and more detailed with more features.

⌃ Viewed by telescope
A portion of the disc is now seen. Innumerable craters and individual mountain ranges are in view.

Montes Jur
Sinus Iridum
Aristarchus Crater
Mare Imbri
Oceanus Procellarum
Montes Carpatus
Copernicus Crater
Kepler Crater
Grimaldi Crater
Mare Cognitum
Mare Orientale
Gassendi Crater
Mare Nubiun
Mare Humorum
Darwin Crater
Palus Epidemiarum
Tycho Crater
Longomontanus Crater

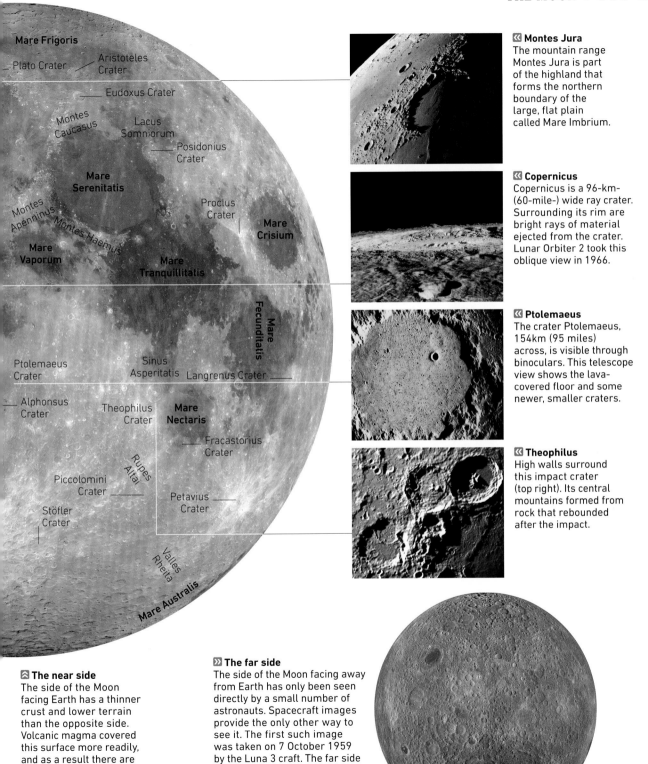

Mare Frigoris

Plato Crater — Aristoteles Crater

Eudoxus Crater

Montes Caucasus — Lacus Somniorum

Posidonius Crater

Mare Serenitatis

Proclus Crater

Montes Apenninus

Montes Haemus

Mare Crisium

Mare Vaporum

Mare Tranquillitatis

Mare Fecunditatis

Ptolemaeus Crater — Sinus Asperitatis — Langrenus Crater

Alphonsus Crater

Theophilus Crater

Mare Nectaris

Fracastorius Crater

Rupes Altai

Piccolomini Crater

Petavius Crater

Stöfler Crater

Valles Rheita

Mare Australis

◀ Montes Jura
The mountain range Montes Jura is part of the highland that forms the northern boundary of the large, flat plain called Mare Imbrium.

◀ Copernicus
Copernicus is a 96-km- (60-mile-) wide ray crater. Surrounding its rim are bright rays of material ejected from the crater. Lunar Orbiter 2 took this oblique view in 1966.

◀ Ptolemaeus
The crater Ptolemaeus, 154km (95 miles) across, is visible through binoculars. This telescope view shows the lava-covered floor and some newer, smaller craters.

◀ Theophilus
High walls surround this impact crater (top right). Its central mountains formed from rock that rebounded after the impact.

⌃ The near side
The side of the Moon facing Earth has a thinner crust and lower terrain than the opposite side. Volcanic magma covered this surface more readily, and as a result there are more dark plains and fewer craters.

▶ The far side
The side of the Moon facing away from Earth has only been seen directly by a small number of astronauts. Spacecraft images provide the only other way to see it. The first such image was taken on 7 October 1959 by the Luna 3 craft. The far side is completely pockmarked by craters and has no large maria.

Hadley Rille
During the Apollo 15 mission in 1971, astronauts David Scott and James Irwin drove the first electrically powered lunar roving vehicle to the edge of a valley called Hadley Rille to collect rock samples. This image shows the partially shadowed rille, with Scott working at the lunar rover in the foreground.

Mars

Rust-red Mars is the fourth planet from the Sun. This dry, cold world is about half the size of Earth. Giant volcanoes, deep faults, rock-strewn plains, and dried-up riverbeds mark its surface. Like Earth, it has polar ice caps and seasons.

⌃ Martian terrain
The distinctive red of the terrain comes from the iron oxide (rust) in the rocks and soil. The rover Spirit took this view of a rock outcrop near Gusev Crater in 2004.

Structure and atmosphere

This ball of rock and metal is the outermost rocky planet. When it was young and molten, its material differentiated to form a core and layers. The heavy iron sank to the centre, and the lighter silicate rocks formed a mantle around the metallic core. The least dense material formed the crust. Mars then started to cool and solidify from the outside in. The core is probably solid now, since the relatively small size of Mars and its distance from the Sun would suggest it has cooled more than Earth's core.

Mars takes almost two Earth years to orbit the Sun. Its axis of rotation is tilted by 25.2° to the plane of its orbit and, like Earth, Mars experiences seasons and a day that lasts about 24 hours. Its orbit is more eccentric than Earth's; there is a difference of about 42 million kms (26 million miles) between its closest and farthest distance from the Sun. When closest, it receives 45 per cent more solar radiation, which leads to higher surface temperatures.

» The planet's interior
A thick mantle of solid silicate rock surrounds a metallic core, which probably consists of iron and nickel. The mantle was a source of volcanic activity in the past. Above the mantle is a rocky crust just tens of kilometres deep.

core, probably of iron and nickel

rock crust

mantle of silicate rock

upper atmosphere

carbon dioxide (95.3%)

argon (1.6%)

nitrogen (2.7%)

oxygen, carbon monoxide, and trace gases (0.4%)

⌃ Atmospheric composition
A thin carbon-dioxide rich atmosphere surrounds the planet. Iron oxide dust particles suspended in the atmosphere colour it pink. Frozen carbon dioxide and water ice form thin clouds.

The red planet
The Valles Marineris canyon system slashes across the centre of Mars's disk. The three dark circular patches on its left limb are three giant shield volcanoes on the Tharsis Bulge.

» MARTIAN DATA

Diameter 6,792km (4,218 miles)

Average Distance from Sun
227.9 million km (141.5 million miles)

Orbital Period 687 Earth days

Rotation Period 24.62 hours

Surface Temperature −125°C to 25°C
(−193°F to 77°F)

Number of Moons 2

Size Comparison

Earth   Mars

Surface features

Much of Mars's northern hemisphere is covered by relatively smooth, low-lying volcanic plains. The southern-hemisphere terrain is older and typically crater-covered highland. Mars's major surface features are found in a roughly 60° band that is centred on the equator. The most striking feature is the Valles Marineris, a complex canyon system over 4,000km (2,500 miles) long. The canyons were formed about 3.5 billion years ago when internal forces in young Mars split its surface apart. They have since widened and deepened through water and wind erosion, or by collapse. Internal forces also formed raised areas such as the Tharsis Bulge. Olympus Mons and other large shield volcanoes, formed by successive lava flows, dominate this region.

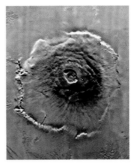

» Olympus Mons
The caldera on the summit of Olympus Mons is surrounded by lava flows. This giant volcano is the largest and tallest in the Solar System, at 21km (13 miles) high.

Vera Rubin Ridge
This amalgamation of 70 photos taken by the Mars rover Curiosity on 13 August 2017 shows clear and distinctive layering in the sedimentary rock in Vera Rubin Ridge.

⟫ EXTRATERRESTRIAL LIFE

Mars has long been thought of as a possible haven for life. The search started in mid-1976, when Viking 1 and 2 looked for signs of life in the soil of Mars. In the first decade of the 21st century, the rovers Spirit and Opportunity located sites where liquid water, generally regarded as an essential element for life, was once present. Since mid-2012, the rover Curiosity has worked in Gale Crater, which was home to a lake more than 3 billion years ago. Although the lake had the ingredients for microbial life to form, no proof of past or present life has been found.

◁◁ Hematite pebbles
Opportunity has found hematite-rich pebbles on Mars. Hematite is an iron-rich mineral, which on Earth almost always forms in the presence of liquid water.

Phobos and Deimos

PHOBOS

DEIMOS

Two small moons orbit Mars. The bigger of the two, Phobos, is 26km (16 miles) long and orbits 9,380km (5,825 miles) from Mars. Deimos is just 16km (10 miles) long and two and a half times farther away. These dark, rocky bodies are believed to be asteroids captured by young Mars's gravity.

⟫ The orbit of the moons
Both moons follow near-circular orbits around Mars's equator. Phobos, the innermost moon, is so close to Mars and orbits so quickly that it rises and sets three times every Martian day.

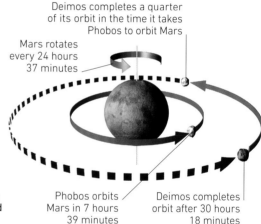

Deimos completes a quarter of its orbit in the time it takes Phobos to orbit Mars

Mars rotates every 24 hours 37 minutes

Phobos orbits Mars in 7 hours 39 minutes

Deimos completes orbit after 30 hours 18 minutes

Water on Mars

Mars is a cold planet and most of its water exists as ice locked in polar ice caps or vapour. However, in July 2018, a European Space Agency radar detected what is likely to be a 20km- (12 mile-) wide body of liquid water underneath Mars' south polar ice cap. Dry river valleys and ancient flood plains are evidence that water once flowed across the planet. That was about 3–4 billion years ago, when Mars was a warmer place.

⟫ Polar ice
In 2005, Mars Express imaged this circle of water ice on the floor of an unnamed crater not far from Mars's northern polar cap. The ice is about 12km (7.5 miles) wide.

Observing Mars

With an average magnitude of about −2.0, Mars is one of the easiest planets to see with the naked eye. It is in Earth's sky for much of the year but is best observed at opposition, when it is close to Earth, and at its largest and brightest. Opposition occurs approximately once every two years and two months. Mars's elliptical orbit brings it particularly close at some oppositions; such favourable conditions occur every 15 or 17 years.

« Naked-eye view
To the naked eye, Mars appears as a bright reddish spot, which moves from night to night against the background stars. Features such as its dark surface markings, white polar caps, dust storms and white clouds over Olympus Mons require fairly large amateur telescopes to be seen (below).

⌃ Mars through binoculars
The disk shape is clearly seen through binoculars, but surface features are not yet visible.

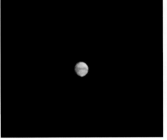

⌃ With a small telescope
Mars's orange-red colour and some surface features, including the white polar caps, emerge.

⌃ With a large telescope
Light and dark markings are now clearly in view and Mars's tilt is apparent from the visible pole.

Missions to Mars

Space probes were first sent to Mars in the early 1960s. Over 30 craft have now made successful missions; they have flown by it, orbited it, landed on it, and driven over it. The earliest probes gave us our first close-up views of Mars. Later craft made more detailed studies. Once only the target of probes from the USA or the Soviet Union, Mars has now been visited by missions from six countries.

The entire planet has been surveyed by orbiters, notably Mars Express, working since 2003, and Mars Reconnaissance Orbiter, which was launched in 2005. Opportunity and Curiosity are the latest of four rovers to study Mars. They roam across its surface, making on-the-spot investigations.

« Exploring the surface
This "selfie", shot by Curiosity (a rover that is exploring Mars's surface), was taken in January 2018. At the time, the rover was investigating the Vera Rubin Ridge.

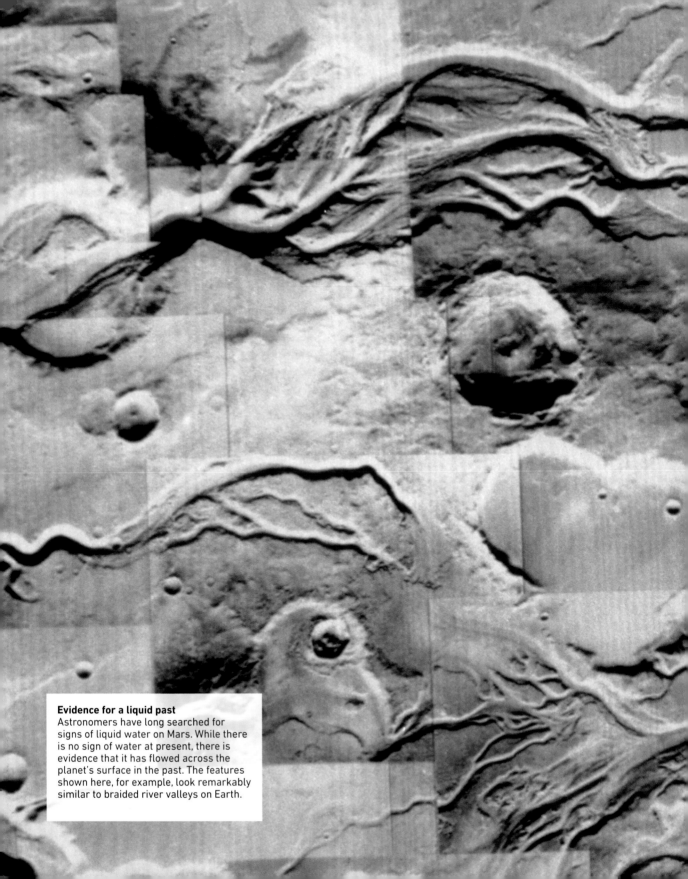

Evidence for a liquid past
Astronomers have long searched for signs of liquid water on Mars. While there is no sign of water at present, there is evidence that it has flowed across the planet's surface in the past. The features shown here, for example, look remarkably similar to braided river valleys on Earth.

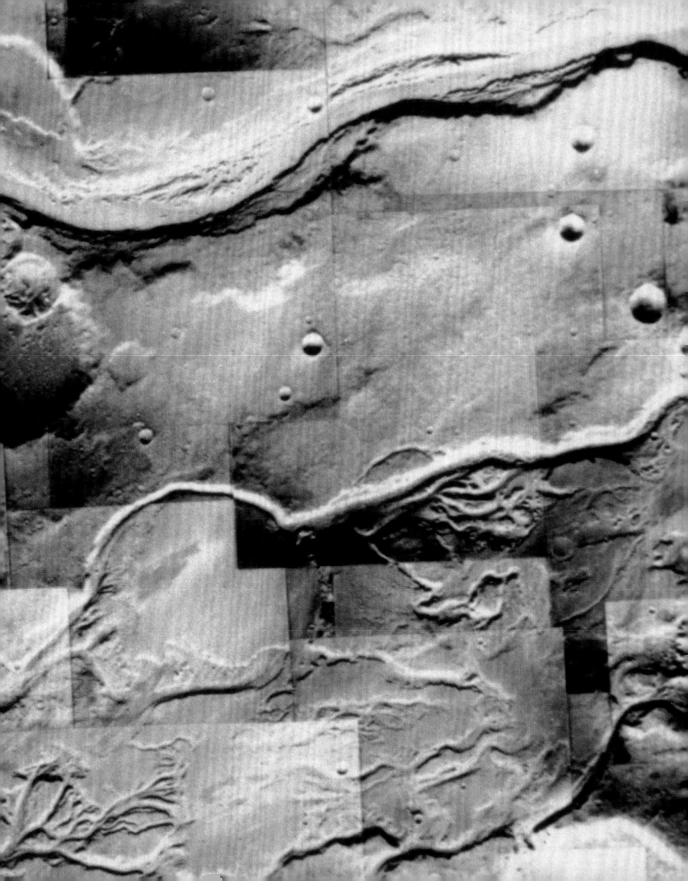

Jupiter

Jupiter is huge – it has a mass 2.5 times the total mass of all the other planets combined. It does not have a solid surface and, when we look at it, we see the top layer of its thick atmosphere. A thin, faint ring system encircles it, and it has the largest family of moons.

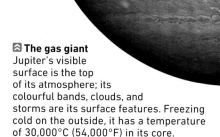

Structure

Jupiter is made predominantly of hydrogen, with a much smaller amount of helium. Its abundance of hydrogen means its composition is more like the Sun's than that of the other Solar System planets. Had it been made of about 50 times more hydrogen, it would have turned into a star. The hydrogen is gaseous in the planet's outer layer, its atmosphere. The state of the hydrogen changes with depth, as the density, pressure, and temperature increase. A solid core about 10 times Earth's mass is in the centre of the planet.

⬆ The gas giant
Jupiter's visible surface is the top of its atmosphere; its colourful bands, clouds, and storms are its surface features. Freezing cold on the outside, it has a temperature of 30,000°C (54,000°F) in its core.

》 JUPITER DATA

Diameter 142,984km (88,793 miles)

Average Distance from Sun
778.4 million km (483.4 million miles)

Orbital Period 11.86 Earth years

Rotation Period 9.93 hours

Cloud-top Temperature −110°C (−166°F)

Number of Moons 79

Size Comparison

Earth   Jupiter

⬅ The planet's interior
Jupiter has a layered structure, but with no rigid boundaries between the layers. Its outer layer is its 1,000-km- (600-mile-) deep hydrogen-rich atmosphere. Below, the hydrogen acts as a liquid; deeper still, it is compacted and acts like molten metal.

core of rock, metal, and hydrogen compounds

inner layer of metallic hydrogen and helium

outer layer of liquid hydrogen and helium

gaseous hydrogen and helium

The planet's orbit and spin

Jupiter lies at an average distance of about 778 million km (483 million miles) from the Sun. Its axis is almost perpendicular to its orbit, tilted by just 3.1°. Jupiter is the fastest-spinning planet; material is thrown outwards, causing it to bulge at the equator.

⊟ Jupiter's orbit
Jupiter's orbit is elliptical. The difference between its closest (perihelion) and furthest (aphelion) distance from the Sun is 76.1 million km (47.3 million miles).

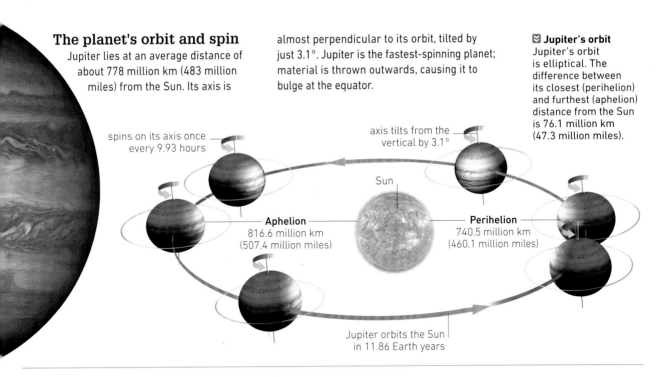

spins on its axis once every 9.93 hours

axis tilts from the vertical by 3.1°

Sun

Aphelion
816.6 million km
(507.4 million miles)

Perihelion
740.5 million km
(460.1 million miles)

Jupiter orbits the Sun in 11.86 Earth years

Atmosphere and weather

Jupiter's fast spin and rising heat from inside the planet disturb the atmosphere, producing hurricanes and raging storms, which can last for years at a time, and winds in its equatorial region can reach speeds of 250mph (400 km/h).

Hydrogen compounds condense to form different coloured clouds at different altitudes in the upper atmosphere. These are channelled into forming the bands round the planet. The white, bright bands are zones of rising gas, and the red-brown bands are belts of falling gas.

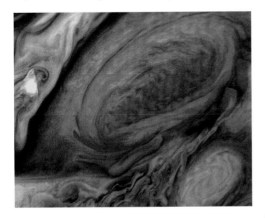

⌃ The Great Red Spot
The largest and most obvious feature on Jupiter's surface is the Great Red Spot. This vast storm has been observed for over 150 years. It is constantly changing its size, shape, and colour. It is bigger than Earth and has been up to three times wider than Earth.

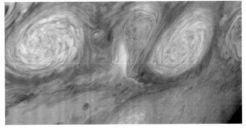

⊠ White oval storms
Two white storms have formed in a region of turbulent atmosphere. Such storms are fixed within a particular belt or zone but travel around the planet, to the east or west.

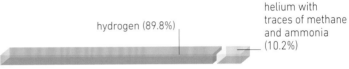

hydrogen (89.8%)

helium with traces of methane and ammonia (10.2%)

⌃ Atmospheric composition
The atmosphere is mainly hydrogen. Trace compounds in the atmosphere include simple ones such as methane, ammonia, and water, and more complex ones such as ethane and acetylene.

›› THE GALILEAN MOONS

Jupiter has over 60 moons. Most are small and irregular in shape with distant orbits – they could be fragments of an asteroid. Eight moons are much closer; four of these are large and round, and formed at the same time as the planet. These are Jupiter's so-called Galilean moons, observed by the Italian astronomer Galileo Galilei in 1610.

⌄ Ganymede
The largest moon in the Solar System, Ganymede is bigger than both Pluto and Mercury. This 5,262km- (3,268 mile-) wide moon has a rocky interior and an upper mantle of ice. Its icy crust has contrasting dark and bright areas.

Contrasting surfaces of Ganymede

⌄ Io
Io circles Jupiter every 42.5 or so hours. It is the innermost of the Galileans, and contrasts sharply with the other three. They are worlds of rock and ice, whereas Io is the most volcanic body in the Solar System. The surface is constantly renewed by eruptions of molten material through the moon's hundreds of volcanoes and vents.

Colourful lava flows on Io

⌄ Callisto
Second in size to Ganymede, Callisto is the outermost Galilean and the eighth moon out from Jupiter. It is a ball of ice and rock with a smooth surface, which is scarred by impact craters. Ice on the crater floors and rims make them shine brightly against the otherwise dark moon.

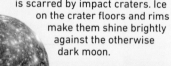

dark areas of Callisto lack ice

Pitted surface of Callisto

›› Europa
A little smaller than Earth's Moon, Europa is the smallest of the Galileans. Its thin water-ice crust may be covering a liquid layer that could harbour life. Its surface consists of smooth ice on the bright polar plains and dark disrupted regions where the crust has broken up and floated about.

Dark grooves and ridges running across Europa's water-ice surface

Magnetic field

Electric currents within Jupiter's metallic hydrogen layer generate a magnetic field about 20,000 times stronger than Earth's and stronger than any other planet's. It is as if a large bar magnet was embedded inside Jupiter, tilted by about 11° to the spin axis. As a result, radiation belts surround Jupiter and solar-wind particles are channelled into the planet's upper atmosphere around its magnetic poles. These interact with the gases to produce glowing aurorae.

⌃ **Shimmering aurorae**
This image, taken by NASA and combined with another image taken in ultraviolet light, shows the brilliant extent of aurorae encircling Jupiter's north pole for thousands of kilometres.

》 MISSIONS TO JUPITER

Jupiter was the first of the outer planets to be visited by spacecraft. Pioneer 10 flew by in 1973, then Pioneer 11 a year later. Two more probes, Voyagers 1 and 2, flew by in 1979. Galileo, the first craft to orbit Jupiter, arrived in 1995. Over the next eight years, it studied the planet and flew by its Galilean moons in turn. Juno started its polar orbit of Jupiter in July 2016, using its instruments to probe beneath its clouds and help determine how the planet formed.

nuclear-powered generators provide electricity

》 Galileo
Galileo has made the longest and most in-depth study of Jupiter. It also revealed details of its major moons, such as Ganymede (see p.122). On arrival at Jupiter, Galileo released a smaller probe, which plunged into the planet's atmosphere.

Observing Jupiter and its moons

Even though Jupiter is hundreds of millions of kilometres from Earth, sunlight reflects brightly off its atmosphere and it is easily spotted in the night sky. It is brightest (magnitude up to -2.9) at opposition, which occurs every 13 months. It then appears all night long, rising at sunset, being highest in the middle of the night, and setting at sunrise. Jupiter is visible for about 10 months each year. It spends about 12 months in one zodiac constellation before moving into the next.

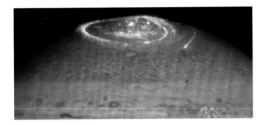

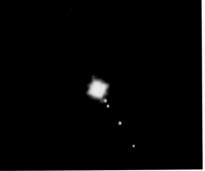

⌃ **To the naked eye**
Jupiter appears to be a particularly bright, creamy-white star. At most times, it is the second brightest planet, after Venus.

⌃ **Through binoculars**
The planet's Galilean moons come into view, ranged along an imaginary line stretching either side of Jupiter's equator. They change position as they orbit.

⌃ **With a large telescope**
Jupiter's banded appearance is now visible. The view changes as the planet spins, and large surface features, such as the Great Red Spot, can be seen.

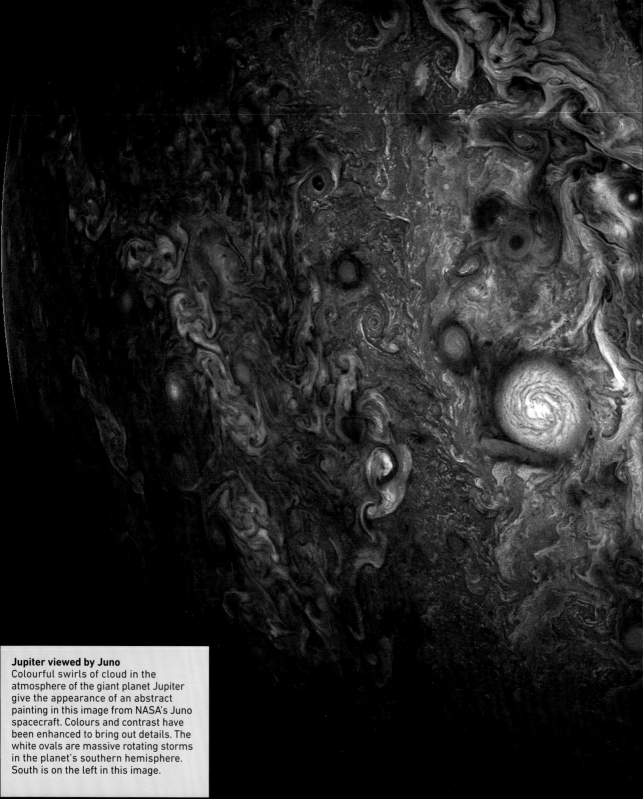

Jupiter viewed by Juno
Colourful swirls of cloud in the atmosphere of the giant planet Jupiter give the appearance of an abstract painting in this image from NASA's Juno spacecraft. Colours and contrast have been enhanced to bring out details. The white ovals are massive rotating storms in the planet's southern hemisphere. South is on the left in this image.

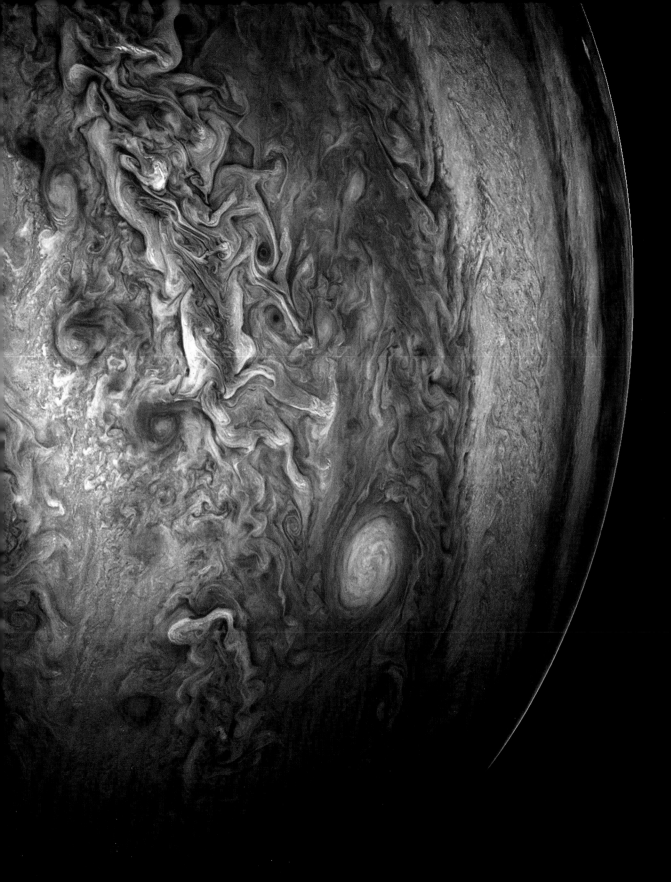

Saturn

The sixth planet from the Sun, Saturn is twice as far from Earth as Jupiter. Its most distinctive feature is the complex system of rings that surrounds it. The planet, which has a muted, banded surface, has a large family of moons.

The structure of the planet

Saturn is the second largest and least dense of all the planets. It is made of 95 times the amount of material in Earth, but it occupies a much greater volume. About 764 planets the size of Earth could fit inside Saturn. The planet is composed of hydrogen and helium. In the planet's outer layer, these elements are gaseous. Inside the planet, where temperature and pressure increase with depth, the hydrogen and helium act like a fluid, and then deeper still, as a liquid metal. Saturn's central rock-and-ice core is about 10–20 times Earth's mass.

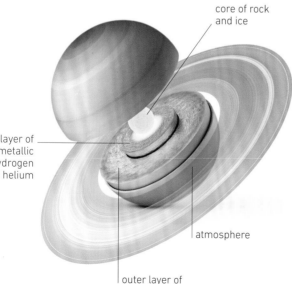

core of rock and ice

inner layer of liquid metallic hydrogen and helium

atmosphere

outer layer of liquid hydrogen and helium

>> **SATURN DATA**

Diameter 120,536km (74,853 miles)

Average Distance from Sun
1.43 billion km (888.56 million miles)

Orbital Period 29.46 Earth years

Rotation Period 10.67 hours

Cloud-top Temperature −180°C (−292°F)

Number of Moons 62

Size Comparison

Earth   Saturn

≪ **The planet's interior**
Saturn is made of hydrogen and helium, layered according to their state. The changes between the layers are gradual. As the planet makes its rapid spin, material is flung outwards to form a bulging equator, about 10 per cent wider than at the poles.

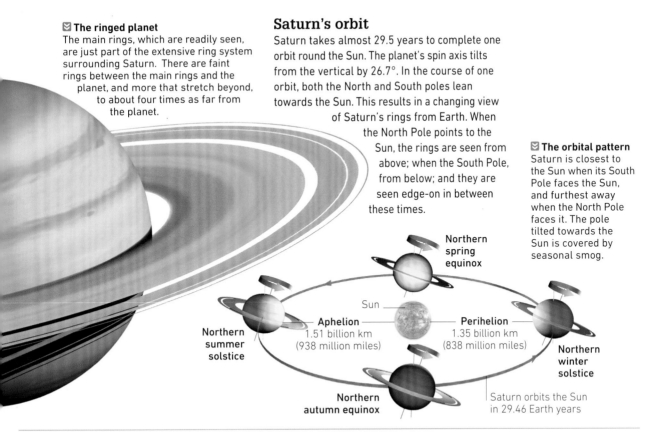

The ringed planet
The main rings, which are readily seen, are just part of the extensive ring system surrounding Saturn. There are faint rings between the main rings and the planet, and more that stretch beyond, to about four times as far from the planet.

Saturn's orbit
Saturn takes almost 29.5 years to complete one orbit round the Sun. The planet's spin axis tilts from the vertical by 26.7°. In the course of one orbit, both the North and South poles lean towards the Sun. This results in a changing view of Saturn's rings from Earth. When the North Pole points to the Sun, the rings are seen from above; when the South Pole, from below; and they are seen edge-on in between these times.

The orbital pattern
Saturn is closest to the Sun when its South Pole faces the Sun, and furthest away when the North Pole faces it. The pole tilted towards the Sun is covered by seasonal smog.

Northern spring equinox

Sun

Aphelion
1.51 billion km
(938 million miles)

Perihelion
1.35 billion km
(838 million miles)

Northern summer solstice

Northern winter solstice

Northern autumn equinox

Saturn orbits the Sun in 29.46 Earth years

Atmosphere and weather
Saturn's pale yellow surface is the top of its thick atmosphere. It is covered by a thin, smoggy haze. Helium raindrops within the metallic layer generate heat as they fall. This heat is transported to the lower atmosphere, where, coupled with the planet's rotation, it generates fierce winds. Near the equator, these can reach 1,800kph (1,200mph). Giant storms are a feature of the upper atmosphere.

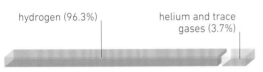

hydrogen (96.3%)

helium and trace gases (3.7%)

Atmospheric composition
The atmosphere is predominantly hydrogen; trace gases include methane, ammonia, and ethane. The visible layer is made of ammonia ice crystals, with ammonium hydrosulphide below.

storm

grey bands are layers of high cloud

Dragon storm
The pinkish feature seen here is a giant storm that was given the name "Dragon Storm". This part of Saturn's southern hemisphere is dominated by storm activity and has been nicknamed "storm alley".

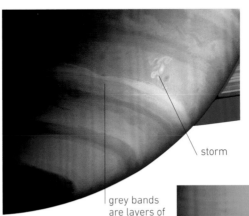

Aurora at South Pole
An oval aurora, invisible to the human eye, was captured by the Cassini probe in June 2005. Solar-wind particles cause hydrogen to glow blue in ultraviolet light.

》 SATURN'S RINGS AND MOONS

Saturn has more than 60 moons. Most have been discovered in the last 25 years, and it is expected that more will be found. Titan is the largest, followed by a few other large, spherical moons such as Dione. Most are small irregularly shaped ones, some of which, like Phoebe, travel backwards compared to the others. The moons are mixes of rock and water ice, in varying proportions. About a third of its moons, and more than 150 "moonlets", lie within the ring system.

《 Surrounding Saturn
The ring system stretches for hundreds of thousands of kilometres into space but is only kilometres deep. The rings are made of pieces of dirty water ice that follow their own orbits round Saturn. Ranging in size from boulders several metres across down to dust grains, they reflect sunlight well, making the rings easy to see. The conspicuous gap in the rings is known as the Cassini Division.

⬇ Titan
About the size of Mercury, Titan is the only moon in the Solar System to have a substantial atmosphere. Nitrogen-rich smog covers the moon, but the surface was revealed in Cassini images.

Euphemus Crater

》 Phoebe
Phoebe is one of the outer moons, orbiting 12.95 million km (8.04 million miles) from Saturn. Its surface is deeply scarred with impact craters.

Jason Crater

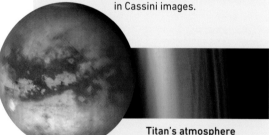

Titan's atmosphere

⬆ Daphnis
One of Saturn's moons, Daphnis, is one of many moons that disrupt the rings with their gravity. As a result, Daphnis has earned the nickname the "wavemaker moon".

⬆ Dione
Whizzing round within the ring system, Dione orbits in 2.74 days. Its surface is marked by ice cliffs and impact craters. Saturn's fourth-largest moon, it shares its orbit with two small, irregularly shaped moons – Helene moving ahead of it, and Polydeuces behind.

Missions to Saturn

Four craft have journeyed to Saturn. The first three, Pioneer 11, which arrived in 1979, and Voyagers 1 and 2 (1980 and 1981) were all flyby missions. The fourth, Cassini–Huygens, was an orbiter which made an in-depth study of Saturn, and its rings and moons. An American-European venture, it consisted of the main probe, Cassini, and Huygens, which was attached to Cassini for the journey.

The joint probe arrived in mid-2004 after a 7-year journey. Within months, Huygens was released on its 21-day journey to Titan, where it descended to the moon's surface. Cassini, which was designed to work on Saturn for four years, had its mission extended twice. In total, it made 294 orbits of Saturn and 162 flybys of its moons, finally ending its exploration in September 2017 with a deliberate plunge into Saturn's atmosphere.

◀◀ Cassini–Huygens
The Cassini spacecraft, with the Huygens probe mounted on the right, is assembled before its launch from Cape Canaveral.

》 CHRISTIAAN HUYGENS

The Huygens space probe is named after the versatile Dutch scientist Christiaan Huygens (1629–95), who invented the pendulum clock, proposed the wave theory of light, and made excellent telescopes. In 1655, Huygens not only discovered Titan, he also explained the mystery of Saturn's rings – that they are a band of material whose appearance changes according to the planet's position with respect to Earth.

◀◀ The moon Mimas from Cassini
Mimas, which orbits Saturn in the outer part of the ring system, is 185,520km (115,208 miles) from the planet. The lines beyond are shadows from its rings.

Observing Saturn

Saturn is visible to the naked eye for about 10 months of each year, and it looks like a bright yellowish star. At its brightest, it reaches magnitude –0.3. This happens when the rings are face-on to us and more light is reflected.

A telescope will show the rings and is needed to reveal any surface detail. The best time to see Saturn is at opposition, which happens annually, about two weeks later each year. In the course of its orbit, Saturn spends about 2.5 years in each zodiacal constellation in turn.

⌃ Through binoculars
The disc is visible through any binoculars. Powerful ones also show the rings, when full-on to Earth, as a bump at either side.

⌃ With a small telescope
The ring system, seen here tilted towards Earth, looks like two ear-like lobes, or handles, one on each side of Saturn's disc.

⌃ With a large telescope
The main rings and the Cassini Division are now seen clearly. A number of Saturn's largest moons can be seen as dots of light.

Studying Saturn
The Cassini spacecraft orbited
Saturn 22 times, weaving its way
between the planet's many rings and
atmosphere. Cassini's mission ended
with a final, deliberate plunge into
Saturn on 15 September 2017.

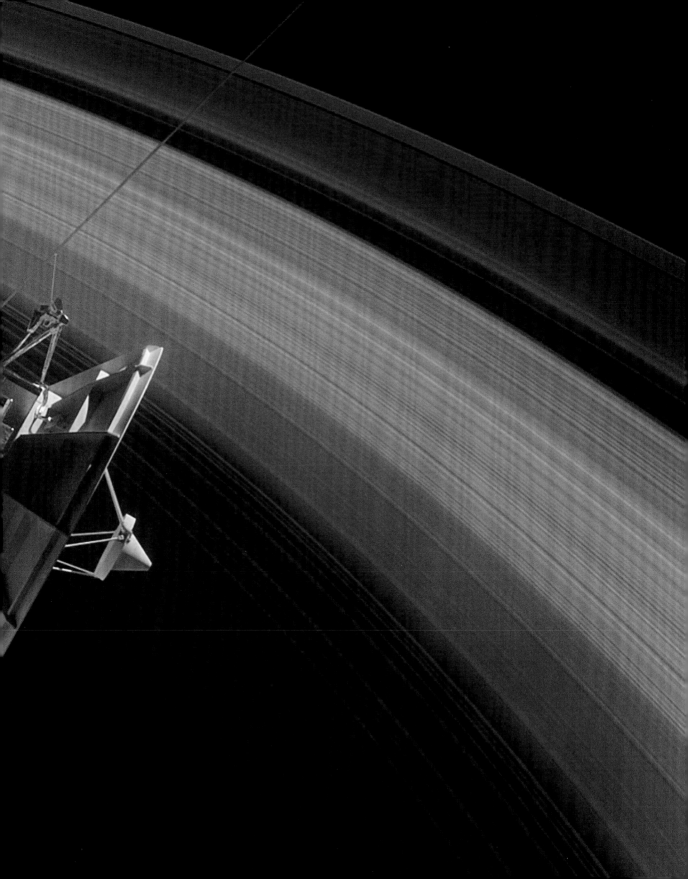

Uranus

Pale blue Uranus is the seventh planet out from the Sun. Most of what we know of it has come from the one probe to visit it, Voyager 2. Uranus is tipped on its side so its ring system and family of moons seem to encircle it from top to bottom.

Structure

Uranus is the third largest planet. It is about four times the size of Earth and its volume is 63 times that of Earth. Uranus is, however, made of only 14.5 times our planet's mass, and so its material is not as dense as Earth's. Below its gas atmosphere, Uranus consists of a deep layer of water, methane, and ammonia ices. It is thought that electric currents within this layer generate the planet's magnetic field. The boundaries between the layers are not rigid, with one layer merging into the next.

» Seen from space
The featureless southern hemisphere of Uranus is lit by the Sun in this Voyager 2 image dating from 1986.

» The planet's interior
Like Jupiter, Saturn, and Neptune, Uranus is a gas giant. The top of its hydrogen-rich atmosphere is its visible surface. Below is a layer of ices; below this is its core.

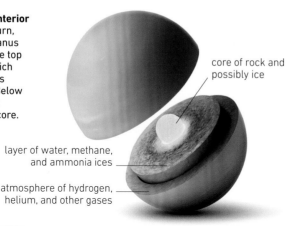

core of rock and possibly ice

layer of water, methane, and ammonia ices

atmosphere of hydrogen, helium, and other gases

» URANUS DATA

Diameter 51,118km (31,744 miles)

Average Distance from Sun
2.87 billion km (1.78 billion miles)

Orbital Period 84 Earth years

Rotation Period 17.24 hours

Cloud-top Temperature −214°C (−353°F)

Number of Moons 27

Size Comparison

Earth   Uranus

Atmosphere and weather

Uranus appears featureless but this is in part due to haze in the upper atmosphere obscuring our view. The haze is produced by ultraviolet sunlight interacting with atmospheric methane. Uranus has no complex weather system but ammonia and water clouds are carried around the planet by its winds and rotation. The cloud-top temperature is around −214°C (−353°F).

« Clouds
This infrared image of Uranus reveals the cloud structure. The highest clouds appear white; the mid-level ones bright blue; and the lowest darker blue. Image processing has turned the rings a colourful red.

hydrogen (82.5%)   methane (2.3%)

helium (15.2%)

⬆ Atmospheric composition
Hydrogen is the main element in the atmosphere. The methane gives Uranus its blue colour; it absorbs incoming red wavelengths of light and reflects the blue.

Spin and orbit

Uranus's spin axis is tilted so far (98°) from the vertical that it lies almost in the plane of Uranus's orbit. This means that from Earth the planet is seen pole-on, side-on, or in between, as its poles and equator face the Sun during the course of its 84-year-long orbit. A collision with a planet-sized body in the distant past probably knocked young Uranus into its sideways position.

▣ Orbit and spin
Uranus appears to orbit the Sun on its side. Its long orbit and high tilt mean each hemisphere faces the Sun for about 42 years at a time.

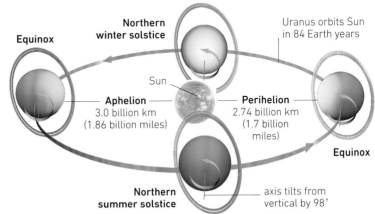

Northern
winter solstice

Equinox

Sun

Aphelion
3.0 billion km
(1.86 billion miles)

Perihelion
2.74 billion km
(1.7 billion miles)

Uranus orbits Sun
in 84 Earth years

Equinox

Northern
summer solstice

axis tilts from
vertical by 98°

The rings and moons

Thirteen rings, separated by wide gaps, surround Uranus. They are made of dark carbon-rich pieces of material ranging from dust to possibly a few metres across. Twenty-seven moons orbit Uranus; more probably exist. Five are major moons, dark, rocky bodies with icy surfaces. The rest are much smaller, mainly tens of kilometres across.

⏭ False-colour view of rings
The bright, colourless ring (far right) is Uranus's most prominent ring. Five more rings, coloured blue-green, and three, off-white, are to its left.

⏭ Surface of Ariel
Ariel is about a third the size of Earth's Moon and orbits Uranus every two and a half days. It was discovered from Earth but is only seen in detail thanks to Voyager 2. Long, broad faults cut across the moon's icy surface.

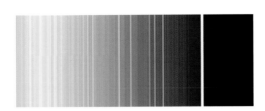

Observing Uranus

Uranus is twice as far from the Sun than its inner neighbour, Saturn. Its great distance makes it difficult to see but at magnitude 5.5, it is just visible with the naked eye. It looks like a star to the eye or through basic binoculars. More powerful ones or a small telescope will show it as a disc. Following the planet's slow progress against the background stars will confirm its identity.

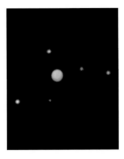

◖◖ Uranus in the night sky
The disc of Uranus and the planet's five major moons – Ariel, Umbriel, Oberon, Titania, and Miranda – are visible here through an Earth-based telescope. The blue colouring of the featureless globe is unmistakable.

Neptune

Faraway Neptune was only discovered in 1846 and remained poorly known until Voyager 2 flew by in 1989. The probe revealed a cold, blue world encircled by rings and moons, with a surprisingly dynamic atmosphere.

》 The planet's interior
Neptune has no solid surface; its visible surface is the top of its atmosphere. Inside, the planet is layered, with each layer merging into the next.

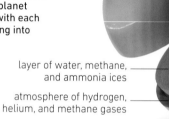

core of rock and possibly ice

layer of water, methane, and ammonia ices

atmosphere of hydrogen, helium, and methane gases

Structure

Neptune is the smallest and most distant of the four gas giants. It is almost four times the size of Earth with a structure most like that of its inner neighbour, Uranus. Neptune's outer layer is its atmosphere, which is made mostly of hydrogen. Below is a deep layer of water and ices, and below this a core of rock and possibly ice. Due to the planet's fast spin, material is pushed outwards to form a bulging equator.

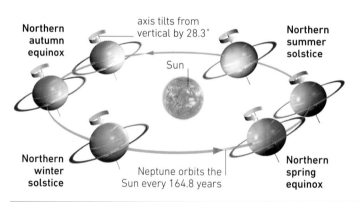

Northern autumn equinox

axis tilts from vertical by 28.3°

Northern summer solstice

Sun

Northern winter solstice

Neptune orbits the Sun every 164.8 years

Northern spring equinox

《 Spin and orbit
Neptune's path around the Sun is less elliptical than that of any other planet, except Venus. This means there is no huge difference between its closest (perihelion) and furthest (aphelion) distance from the Sun. Neptune's spin axis is not perpendicular to its orbit, but is tilted from the vertical by 28.3°.

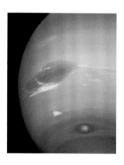

⌃ Great dark spot
Voyager 2 images revealed a dark storm-like cloud, since named the Great Dark Spot, in Neptune's atmosphere in 1989. Bright clouds of methane ice surround the spot, which was almost as big as Earth. It had disappeared by 1994. A similar but new dark spot was observed in 2016.

Atmosphere and weather

Even at 30 times further from the Sun than Earth, Neptune is affected by the Sun's heat and light. The north and south poles lean sunwards in turn as Neptune orbits the Sun, and the result is seasonal change. Each season on Neptune lasts roughly 40 years. Yet, the Sun's heat alone is not enough to drive the dark storm-like features and the ferocious equatorial winds. Perhaps the planet is warmed by internal heat.

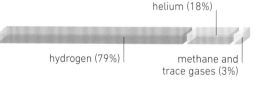

helium (18%)

hydrogen (79%)

methane and trace gases (3%)

⌃ Atmospheric composition
The atmosphere is predominantly hydrogen. But it is the relatively small amount of methane in the upper atmosphere that creates the distinctive blue colour. The methane absorbs red light and reflects blue.

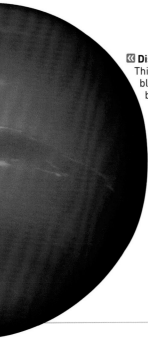

《 Distinctively blue
This gas giant's vividly blue colour is caused by its atmosphere. The white streaks are clouds of frozen methane.

》 NEPTUNE DATA

Diameter	49,528km (30,757 miles)
Average Distance from Sun	4.5 million km (2.8 million miles)
Orbital Period	164.8 Earth years
Rotation Period	16.11 hours
Cloud-top Temperature	−200°C (−328°F)
Number of Moons	14

Size Comparison

Earth   Neptune

Rings and moons

Fourteen moons are known to orbit Neptune, and more small ones probably exist. Just one of the moons, Triton, is large and round. The innermost four lie within the ring system. This consists of five complete rings, the outer of which contains three dense regions of material, as well as a sixth partial ring. The rings are sparse and made of tiny pieces of unknown composition. Voyager 2 revealed the rings but earlier observations made from Earth had predicted their existence.

《 The rings of Neptune
Four rings are visible in this Voyager 2 view. A diffuse band of material, named Lassell, lies between the two bright rings, Adams and Le Verrier. Closest to Neptune is the Galle ring.

》 Triton
The icy moon Triton was discovered just 17 days after Neptune. It is bigger than Pluto. Dark surface patches are the sites of geyser-like eruptions.

Observing Neptune

Neptune is far too distant to be seen by the naked eye alone. At its brightest, it reaches magnitude 7.8 and can then be seen through binoculars or a small telescope. Even then the planet will resemble a faint star. A telescope larger than about 150mm (6in) aperture will resolve the planet into a disc shape. However, only the most powerful instruments will reveal any detail, such as changes in surface brightness.

《 View through a telescope
Neptune's disc is seen twice here, next to Triton. Images taken on separate nights are combined to show their movement against the fixed stars.

Pluto

A dark, desolate world of rock and ice, Pluto is difficult to observe from Earth. It remained largely unknown until the space probe New Horizons flew by in 2015, revealing one of the most colour-contrasted bodies in the Solar System. Pluto is the largest dwarf planet – a body that is much like a planet, but that has not cleared the region around its orbit of other materials.

Structure and atmosphere

About 60 per cent of Pluto's mass is made up of the dwarf planet's huge rocky core. Surrounding this is a comparatively thin mantle of water ice. Evidence from New Horizons indicates that the mantle may include a water-ice ocean. On top of this is an icy crust of frozen nitrogen, water, carbon monoxide, and methane. Pluto's insubstantial nitrogen atmosphere has a pressure just one-hundred-thousandth of Earth's atmosphere. Seen from a distance, it appears as a high-altitude, blue, smog-like haze surrounding the planet. The haze is thought to be the result of sunlight acting on methane and other chemicals.

⊡ **Atmospheric content**
Pluto's atmosphere is almost entirely nitrogen but includes trace elements of methane and carbon monoxide.

trace gases (0.03%)   nitrogen (99.97%)

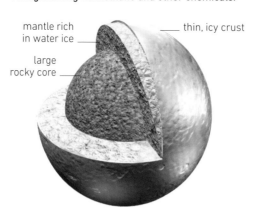

mantle rich in water ice — thin, icy crust

large rocky core

⌃ **Pluto's interior**
Pluto is thought to be 70 per cent rock and 30 per cent water ice. The rock has sunk to its centre, and is surrounded by an ice mantle, topped by ice.

≫ **MISSION TO PLUTO**

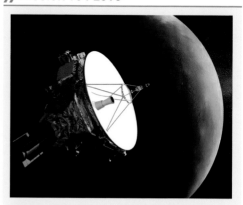

After its nine-year journey from Earth, the New Horizons spacecraft flew by Pluto in 2015. Its suite of instruments imaged both Pluto and Charon, and tested Pluto's atmosphere. The spacecraft was then directed onto a course for a flyby of Kuiper Belt object 2014 MU69 in early 2019. It will be the most distant object visited by a spacecraft.

Surface features

The New Horizons spacecraft revealed that Pluto has a frozen landscape of ice plains, impact craters, and rugged, snow-capped mountains as high as Earth's rocky mountains. It also showed that the surface is young, at only about 180,000 years old, and that it is refreshed through continuous glaciation. The huge heart-shaped region dominating its surface is a giant glacier – the largest in the Solar System. Named Sputnik Planitia, this 1,000-km (620-mile) wide nitrogen ice plain is renewed by fresh material moving up from underneath.

Charon, Pluto's largest moon

Pluto has five moons. Charon, which is about half Pluto's size is, by far the largest. Like Pluto, it is an ice-and-rock ball and has mountains and plains. Its most distinctive features are a red polar cap and a huge canyon, which in places is 9-km (5.5-miles) deep – five times the depth of Earth's Grand Canyon. It is speculated that the pair formed from one body of material that broke up early in their history. Charon is locked in an orbit around Pluto that keeps the same side of the dwarf planet always facing the same side of Charon.

⌃ Colourful world
Pluto had usually been depicted as colourless until NASA's New Horizons spacecraft captured its true hues – swathes of deep red, pale beige, and ice-blues – in 2015.

⟫ PLUTO DATA

Diameter 2,376km (1,476 miles)

Average Distance from Sun 5.9 billion km (3.7 billion miles)

Orbital Period 247.9 Earth years

Rotation Period 6.39 Earth days

Surface temperature -230°C (-382°F)

Number of Moons 5

Size comparison

Pluto

Earth

red polar cap

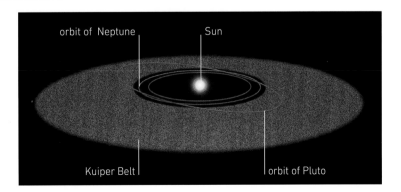

◀ Tiny moon
In 2015, the New Horizons spacecraft also captured startling new images of Charon, revealing a red north pole.

The Kuiper Belt and the Oort Cloud

A flat belt of rock-and-ice objects measuring 3 billion km (2 billion miles) from edge-to-edge stretches out beyond the orbit of Neptune. The Kuiper Belt, as it is known, is thought to consist of hundreds of thousands of these objects. Most are classed as Kuiper Belt objects, some are comet nuclei, and the largest are dwarf planets such as Pluto and Eris. The outer edge of the belt merges with the Oort Cloud. This is the vast sphere of comet nuclei that surrounds the Solar System. It contains over a trillion comets, and its outer edge is 1.6 light-years away, not quite halfway to the closest stars.

orbit of Neptune   Sun

Kuiper Belt   orbit of Pluto

⌃ The Kuiper Belt
Encircling the planetary region of the Solar System, most of the Kuiper Belt's comet-like objects take more than 250 years to orbit the Sun. They are the main source of short-period comets. The small objects are difficult to see, but over 900 have been found so far.

Comets and meteors

Comets are dirty snowballs that originate in the Oort Cloud. They are of special interest as they are made of pristine material from the birth of the Solar System some 4.6 billion years ago. If a comet travels in towards the Sun, it sheds dust that can produce meteors in Earth's sky.

How comets change

A comet is a solid, irregularly shaped mix of two-thirds ice and snow, and one-third rocky dust. When one of these mountain-sized snowballs, termed a nucleus, travels closer to the Sun than Mars, it is affected by the Sun's heat. Some of the ice is turned to gas and, along with dust that is released, forms a coma – a vast, expanding cloud, which can be many times the diameter of Earth. Gas and dust are pushed from the coma and form tails, typically 100 million km (62 million miles) long.

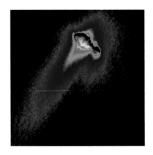

⌃ Comet Borrelly
The comet's 8-km- (5-mile-) long nucleus is shaped like a bowling pin. Gas and dust jets are emitted from its surface. Borrelly is a short-period comet; it returns every 6.86 years.

» The orbit of a comet
Comets only develop a coma and tails near the Sun. Over 500 comets return in less than 20 years and are described as short-period. Long-period comets take hundreds or thousands of years to return.

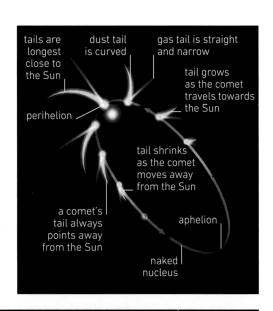

tails are longest close to the Sun

dust tail is curved

gas tail is straight and narrow

tail grows as the comet travels towards the Sun

perihelion

tail shrinks as the comet moves away from the Sun

a comet's tail always points away from the Sun

aphelion

naked nucleus

Comet Hale–Bopp
Comets with a large coma and tails are big enough and bright enough to be seen in Earth's sky. Comet Hale–Bopp was seen in 1997 as its orbit brought it close to the Sun. Its gas tail is blue, and its curved dust tail, white. It will return in about 2,400 years.

⟫ MISSIONS TO COMETS

The first five missions to comets all flew to Comet Halley. Giotto was the most successful. In March 1986, it gave us our first view of a cometary nucleus. Twenty years later, Stardust returned to Earth with particles of Comet Wild-2. The most ambitious mission to date, Rosetta arrived at Comet Churyumov-Gerasimenko in 2014. It orbited the comet, monitoring the evolution of its coma and tails, and released Philae to make the first landing on a comet's nucleus.

⟨⟨ Comet 67P
As it passed Comet 67P, Rosetta took photos of a dust plume emanating from its surface. It is currently unknown what is causing this particular plume to erupt.

⟫ Rosetta spacecraft
Two solar panels, measuring 32m (105ft) from tip to tip, keep Rosetta powered on its long journey out into space.

From comet to meteor

The dust particles shed by comets are termed meteoroids. If one enters Earth's atmosphere, it heats up and produces a short-lived trail of light – a meteor, also sometimes called a shooting star. A really bright meteor is termed a fireball. When Earth passes through a stream of dust particles shed by a comet along its orbital path, a meteor shower will result.

A bright meteor streaks past Orion

Observing comets and meteors

About 3,500 comets have been detected, many by the SOHO spacecraft. From Earth, they are, in the main, faint and only observable with a telescope. A few can be seen each year through binoculars, and one or two every decade with the naked eye.

These appear like a fuzzy, possibly elongated, patch of light. Meteors, which typically last for less than a second, can be viewed by the naked eye. They occur on any night of the year but are best spotted during one of the 20 or so annual meteor showers.

⟨ Leonid meteor shower
The Earth ploughs into the dust stream of Comet Tempel–Tuttle every November. The dust particles burn up and produce meteors while speeding through Earth's atmosphere.

⟫ PROMINENT ANNUAL METEOR SHOWERS

Name	Date	Constellation
Quadrantids	1–6 Jan	Boötes
Lyrids	19–24 Apr	Lyra
Eta Aquariids	1–8 May	Aquarius
Delta Aquariids	15 Jul–15 Aug	Aquarius
Perseids	25 Jul–18 Aug	Perseus
Orionids	16–27 Oct	Orion
Taurids	20 Oct–30 Nov	Taurus
Leonids	15–20 Nov	Leo
Geminids	7–15 Dec	Gemini

Asteroids and meteorites

Asteroids are dry, dusty lumps, made of rock, metal, or a mix of both, that orbit the Sun. If their paths cross, asteroids can collide, break up, and be set on a path towards Earth. A chunk of asteroid landing on Earth is termed a meteorite.

Asteroid profile

There are over a billion asteroids and more than 750,000 have been discovered so far. Asteroids are material that failed to form a rocky planet some 4.6 billion years ago when the Solar System's planets formed. Asteroids are, mostly, irregular in shape and range in size from several hundred kilometres across, down to boulder, pebble, and dust size. About 100 are larger than 200km (125 miles).

Over 90 per cent of asteroids are in the Main Belt, also known as the Asteroid Belt. Ceres, the largest object in the Main Belt, was reclassified as a dwarf planet in 2006. Although the belt has more than a billion measuring over 2km (1.25 miles) long, it is not crowded; thousands of kilometres separate asteroid from asteroid. The Near Earth Asteroids have orbits that take them outside the belt. The Trojans are two groups that follow Jupiter's orbit, one ahead of the planet, the other behind.

》 Asteroid orbits
The Main Belt, the Trojans, and some individual asteroids are shown here. All asteroids orbit close to the planetary plane, in the same direction as the planets. Asteroids in the Main Belt take about 3–6 years to orbit. They spin as they orbit, in just hours.

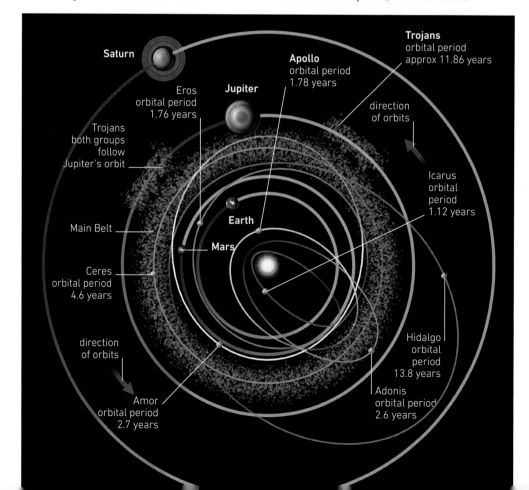

Saturn

Eros
orbital period
1.76 years

Trojans
both groups
follow
Jupiter's orbit

Jupiter

Apollo
orbital period
1.78 years

Trojans
orbital period
approx 11.86 years

direction
of orbits

Icarus
orbital
period
1.12 years

Main Belt

Earth

Mars

Ceres
orbital period
4.6 years

direction
of orbits

Hidalgo
orbital
period
13.8 years

Adonis
orbital period
2.6 years

Amor
orbital period
2.7 years

Ida and Dactyl
The asteroid Ida, which is 60km (37 miles) long, has a tiny moon, Dactyl (lower right). Ida spins once every 4.6 hours, so the Galileo probe could take pictures of most of it as it flew by.

⚈ **Fractured surface**
The Dawn spacecraft captured images of vein-like fractures on Ceres's surface. It is unknown how they formed, although they could be a result of an as-yet unknown material upwelling beneath its rocky surface.

Missions to asteroids

The Galileo probe took the first close-up of an asteroid in 1991, while en route to Jupiter. NEAR was the first to study an asteroid in depth, first orbiting and then landing on Eros in 2001. The first sample of an asteroid was collected from Itokawa by the spacecraft Hayabusa. These samples were returned to Earth in 2010. The spacecraft Dawn has orbited the two most massive asteroids. First, Vesta in 2011–12, and Ceres, from 2015.

Meteorite impacts

Pieces of asteroid too large to burn up in Earth's atmosphere land on its surface. Over 3,000 such meteorites, each weighing over 1kg (2lb), reach Earth every year. Most fall in the sea, the rest hit land. About 190 impact craters have been identified, measuring from just a few metres to about 140km (90 miles) across. Most were formed more than 100 million years ago.

⚈ **Barringer crater**
This 1.2-km- (0.75-mile-) wide crater in Arizona, USA was formed when an iron meteorite, probably some 30m (100ft) wide, hit Earth about 50,000 years ago.

》 METEORITE COMPOSITION

Meteorites are classified by composition. The most common are the stony. Next are the irons, composed mainly of iron-nickel alloy. A very small number are stony-irons. Their mix is similar to that which formed the rocky planets. Other kinds are fragments of asteroids that have differentiated into metal cores and rock surfaces.

6cm (2.5in) long, found in Antarctica

Iron

Stony-iron

Stony

The Night Sky

Star trails over Kitt Peak, Arizona
In this time-exposure photograph, stars circulate around the north pole star, Polaris (the bright star at top left of centre), while the shutter of a telescope dome stands open to observe the sky.

Observation

Astronomy is a science to which amateurs can still make worthwhile contributions. With only binoculars, or just the unaided eye, they can estimate the changing brightness of variable stars, count the meteors in the various annual showers, and record atmospheric phenomena such as aurorae. With modest telescopes, they can plot the rise and fall of sunspot numbers and monitor changes on the surfaces of planets. Advanced amateurs can undertake visual or photographic patrols to discover comets, asteroids, novae, and supernovae.

First, you have to learn to find your way around the night sky. To do that, all you need is a pair of binoculars and a guidebook such as this. Start by identifying the brightest stars and learning the most prominent star patterns, which you can find in the all-sky charts on pp.274–323. The stars and constellations on view change throughout the year as the Earth orbits the Sun, so the sky each month is slightly different. As familiar patterns slip below the horizon in the west, new ones rise in the east. Once you are able to recognize the main constellation shapes each month, you can navigate from these to the other, fainter parts of the sky, as needed.

Shifting skies

Although the stars on show each month remain the same from year to year, the positions of the Moon and planets are constantly changing. The information listed in the Almanac on pp.324–39 includes the dates of new and full Moon, appearances of the planets, and eclipses. Note that when the Moon is above the horizon, particularly around the time of full Moon, it will brighten the sky and faint objects, particularly nebulae and galaxies, will be difficult to see. Try to schedule your deep-sky observing sessions for times around the appearance of the new Moon, when the sky will be at its darkest.

Choosing equipment

This chapter on observing the night sky also tells you how to get started with choosing simple equipment. Binoculars should be first on the list, because of their portability and ease of use. For those wishing to go beyond the basics, telescopes are now more affordable than ever, and the amateur has a much wider range to choose from.

Amateur astronomy has entered the digital age, with computer controls that make telescopes much simpler to use, as well as electronic chips that can take images of the night sky that surpass those from professional observatories not many years ago. This is indeed a golden age for amateur astronomy.

Despite all these advances, most amateurs go stargazing to enjoy the sky for its own sake, without expecting to make scientific discoveries.

» Amateur telescope
Modern computer-controlled telescopes like this allow amateur astronomers to find their way around the sky more easily than ever before.

The appearance of the sky

It is convenient to think of the stars and other celestial bodies as fixed to the inside of a huge sphere – the celestial sphere. At the centre of this sphere is the Earth, which turns like a spinning top from west to east, completing one rotation each day. As a result, celestial objects appear to move slowly across the sky from east to west.

Where on Earth

How much of the celestial sphere you can see depends on your latitude on Earth – that is, how far north or south of the equator you are. If you were stationed exactly at the Earth's North Pole, with the north celestial pole directly above your head, you would only ever be able to see the northern hemisphere of sky. If, instead, you were at the equator (with the celestial equator above you), you would be able to see all the way from there to the celestial poles, which would lie on your northern and southern horizons. As the Earth turned, the objects on the celestial sphere would rise in the east and set in the west, and you would be able to see them all in the course of a year.

Of course, most people live at intermediate latitudes, so their view of the sky lies somewhere between these two extremes. They will be able to see all of one celestial hemisphere over a year; plus a part of the other. The closer you are to the equator, the greater the percentage of the celestial sphere you get to see.

⟫ The celestial sphere
Even though stars lie at widely differing distances from Earth, the celestial sphere, a relic of ancient Greek astronomy, remains a handy tool for mapping the skies as they appear from Earth. Astronomers use precisely defined points on its surface as references for describing the positions of stars.

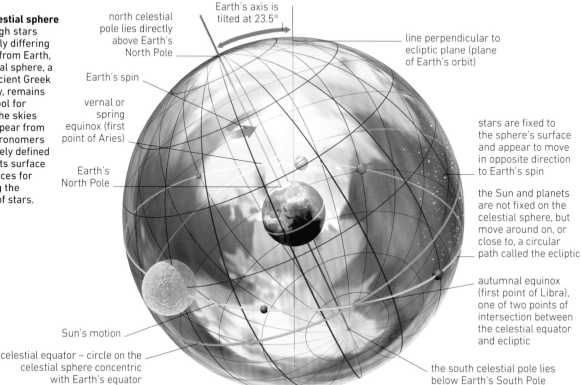

north celestial pole lies directly above Earth's North Pole

Earth's axis is tilted at 23.5°

line perpendicular to ecliptic plane (plane of Earth's orbit)

Earth's spin

vernal or spring equinox (first point of Aries)

Earth's North Pole

stars are fixed to the sphere's surface and appear to move in opposite direction to Earth's spin

the Sun and planets are not fixed on the celestial sphere, but move around on, or close to, a circular path called the ecliptic

autumnal equinox (first point of Libra), one of two points of intersection between the celestial equator and ecliptic

Sun's motion

celestial equator – circle on the celestial sphere concentric with Earth's equator

the south celestial pole lies below Earth's South Pole

Earth's merry-go-round

The Earth's motions define the two basic divisions of our timekeeping system: the year and the day. The Earth rotates once on its axis every day, while it takes a year to orbit the Sun. As the Earth moves along its orbit, the Sun appears to move against the distant stars. Hence there is a gradual change in the stars on show from night to night. For example, constellations such as Taurus and Gemini that are well seen in January and February will be behind the Sun six months later and lost in daylight. After another six months, the Earth is back to where it started and these constellations are visible again.

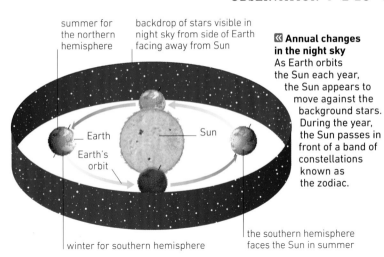

summer for the northern hemisphere

backdrop of stars visible in night sky from side of Earth facing away from Sun

Earth

Sun

Earth's orbit

winter for southern hemisphere

the southern hemisphere faces the Sun in summer

《 Annual changes in the night sky
As Earth orbits the Sun each year, the Sun appears to move against the background stars. During the year, the Sun passes in front of a band of constellations known as the zodiac.

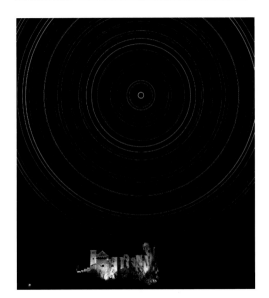

Around the pole

Unless you are very close to the equator, part of the sky around the celestial pole will always remain above the horizon as the Earth rotates. Stars in this area circle around the pole without setting – they are termed circumpolar. The amount of sky that is circumpolar increases with your distance from the equator. For example, as seen from latitude 30°, stars within 30° of the celestial pole are circumpolar, while from latitude 50°, stars within 50° of the celestial pole are circumpolar.

《 Circular sky over Vienna
Long-exposure photographs reveal the movement of the stars in the night sky. This image of the sky over Vienna, Austria, is the result of a relatively short exposure, in which the star trails have been extended digitally to create full circles. Polaris made the small bright circle near the centre.

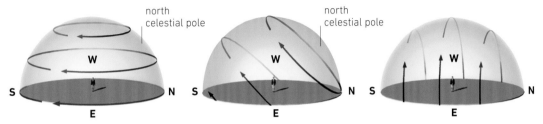

⬢ Motion at North Pole
All objects appear to circle the celestial pole without rising or setting. The motion is clockwise at the North Pole, anticlockwise at the South.

⬢ Motion at mid-latitude
Most objects rise in the east, cross the sky obliquely, then set in the west; others, however, are circumpolar and circle the celestial pole.

⬢ Motion at equator
All objects appear to rise vertically in the east, pass across the sky over the observer's head, then fall vertically to set in the west.

Getting coordinated

To pinpoint the positions of objects on the celestial sphere, astronomers use a system of coordinates similar to latitude and longitude. Right ascension (RA) is the equivalent of longitude, declination (dec.) the equivalent of latitude. Declination is the easier to understand, as it is graduated like terrestrial latitude in degrees, from 0° at the celestial equator to 90° at the celestial poles. Right ascension, by contrast, is divided into 24 hours, because the celestial sphere appears to rotate once in that time. RA runs from west to east, the direction in which the Earth turns.

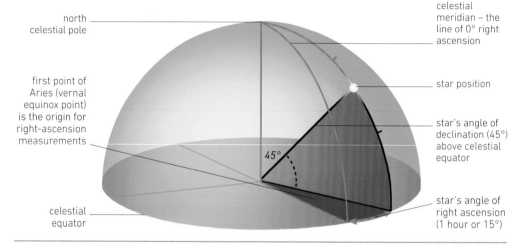

north celestial pole

celestial meridian – the line of 0° right ascension

» Recording a star's position
The star in this diagram has a declination of 45° and a right ascension of 1 hour, or 15°. Right ascension is the angle of the object measured from the celestial meridian, which intersects the celestial equator at the first point of Aries.

first point of Aries (vernal equinox point) is the origin for right-ascension measurements

star position

star's angle of declination (45°) above celestial equator

45°

celestial equator

star's angle of right ascension (1 hour or 15°)

Highway in the sky

The ecliptic is the highway that the Sun appears to follow against the background stars as the Earth orbits it every year. If the Earth's axis were perpendicular to its orbit, the ecliptic would coincide with the celestial equator. In practice, the Earth's axis is tilted at 23.5°, so the ecliptic is angled by 23.5° relative to the celestial equator. The ecliptic cuts the celestial equator at two points, termed the equinoxes. When the Sun is at these positions, in late March and late September, day and night are equal in length the world over. Right ascension begins from the position of the March equinox, in the same way that the Greenwich meridian marks the zero point of longitude on Earth.

All the planets orbit in a narrow plane remaining close to the ecliptic. The band of constellations through which the ecliptic passes is known as the zodiac.

» Seasonal differences
Earth's tilt causes seasonal differences. The Sun's paths across the sky are shown here as seen from mid-northern latitudes at the summer and winter solstices (top, bottom) and the equinoxes (middle).

⟫ MOTION OF THE PLANETS

The Moon and planets are constantly on the move. The inner planets, Mercury and Venus, stay close to the Sun. Mercury is particularly difficult to see, as it is usually low and lost in twilight, but Venus shines brilliantly as the Morning or Evening "Star", depending which side of the Sun it is on. The outer naked-eye planets, Mars, Jupiter, and Saturn, can be seen in any part of the sky as they orbit the Sun, appearing like bright stars among the familiar constellation shapes.

⏶ Evening Star
When setting after the Sun, brilliant Venus is known as the Evening "Star". Here it lies in evening twilight near the crescent Moon.

⏵ Star coordinates
Precession gradually changes the coordinates of stars, so for precise work, positions in catalogues are referred to a standard date known as an epoch. Current catalogues and charts are for Epoch 2000. These were exactly correct on 1 January 2000.

Wobbling Earth

Very slowly, the Earth is wobbling in space, so that its axis traces out a circle on the celestial sphere. This effect is termed "precession" and one complete wobble takes 25,800 years. Because of precession, the position of both celestial poles among the stars is slowly changing. In the year 2000, Polaris was ¾° (one and a half Moon diameters) from the exact celestial pole; in 2100 it will be just under ½° away, the closest it ever comes. The effects of precession are barely noticeable to the human eye.

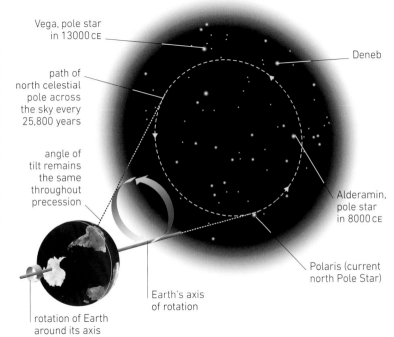

Vega, pole star in 13000 CE

Deneb

path of north celestial pole across the sky every 25,800 years

angle of tilt remains the same throughout precession

Alderamin, pole star in 8000 CE

Earth's axis of rotation

Polaris (current north Pole Star)

rotation of Earth around its axis

Summer solstice

Spring/ autumn equinox

Winter solstice

Finding your way

Travellers on land and sea have always used the celestial pole as a reference point. When learning your way around the sky, there are also handy techniques for measuring angles and apparent distances between objects in the sky without the need for any equipment.

Finding the North Pole

The key guide to Polaris is one of the most familiar patterns in the sky – the seven stars that make up the saucepan-shaped outline of Ursa Major, the Plough or Big Dipper, which is always visible from latitudes above 40° north. Two stars in the bowl of the saucepan, α (Alpha) and β (Beta) Ursae Majoris, are known as the Pointers because they indicate the position of the north celestial pole.

1. First locate the constellation of Ursa Major (commonly known as the Plough or Big Dipper), and then identify the Pointers, Alpha Ursae Majoris and Beta Ursae Majoris, on the right-hand side of the outline of the saucepan. Imagine that a straight line connects the two stars.

2. Keep in mind the distance between the two stars and then extend that line by five times, to arrive near the north Pole Star, Polaris. At second magnitude, the north celestial pole is the brightest star in Ursa Minor. Its brightness varies, but the variations are invisible to the naked eye.

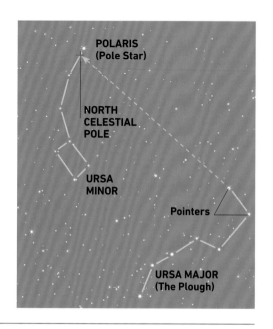

Finding the South Pole

In the southern sky, the Southern Cross acts as a pointer to the celestial pole. Extend the long axis of the Cross fivefold to reach the southern pole, an area bereft of any stars of note. As a further rough guide, the south celestial pole forms a triangle with the bright stars Canopus and Achernar.

1. First locate the familiar shape of Crux, the Southern Cross (highest on April and May evenings). It is the smallest of all the constellations, but one of the most distinctive. Then locate the two brightest stars in Centaurus, Alpha and Beta Centauri, known as the Southern Pointers.

2. Imagine a line connecting the top to the bottom of the Southern Cross. Now imagine that line extended by five times. Next imagine a straight line going from halfway between the Pointers to cross the line leading from the Southern Cross. The pole lies where the two lines intersect.

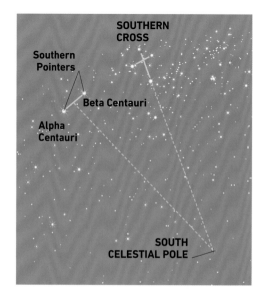

Angles in the sky

The position of an object in the sky can be described by its altitude and azimuth. Its altitude is its distance above the horizon in degrees, 0° being on the horizon and 90° directly overhead. Azimuth is the object's compass bearing from north. Due north is azimuth 0°, due east is 90°, due south is 180°, and so on back to north. An object's altitude and azimuth change as the Earth rotates. A planetarium-type computer program will tell you the altitude and azimuth of an object at any given time and from any desired location.

Horizon

Measuring altitude
You can learn to estimate the altitudes of celestial objects above the horizon with your arms. The altitude of the object shown here is 45°, halfway from horizon to zenith (the point directly above you).

Measuring azimuth
Azimuth is measured clockwise around the horizon from north, and can also be estimated with your arms. The object shown here is in the northeast at an azimuth of 45°. If it were in the northwest, its azimuth would be 315°.

The brightness of a star as it appears in the sky is termed its apparent magnitude, and this is shown on star charts by dots of different sizes. Star brightness can now be measured very precisely, but originally astronomers divided stars into six broad groups from 1st magnitude (the brightest) to 6th magnitude (the faintest visible to the naked eye).

Classification
Each step on the magnitude scale is equal to a brightness difference of about 2½ times, so a 1st-magnitude star is 100 times brighter than a 6th-magnitude one. Stars brighter than 1st magnitude are given zero or even negative values. Sirius, the brightest star, is of magnitude −1.46. A star's apparent brightness is affected both by its actual light output and by its distance from us.

Other objects
The magnitude scale is also applied to other celestial objects. For example, Venus, the brightest planet in the night sky, can appear as bright as magnitude −4.7, while the full Moon is magnitude −12.7.

The Moon and Venus

Sizing things up

The sizes of objects on the celestial sphere, and the distances between them, are measured in degrees and parts of a degree. A rough-and-ready ruler to help size things up is literally to hand. Your finger at arm's length is about 1° across, more than enough to cover the half-degree width of the Sun or Moon. A hand at arm's length has a width of about 10°, while a hand with splayed fingers is about 16° wide. Everyone's hand is different, so measure up the distances between bright stars and the sizes of various constellations for yourself.

Finger width
The Moon or Sun, both of which are only half a degree across, can easily be covered by an index finger at arm's length.

Half handspan
The back of a closed hand is about 10° across, the measure of the Plough or Big Dipper in the night sky.

Whole handspan
An outstretched hand with splayed fingers (here seen against the Square of Pegasus) is about 16° across.

Starting observing

One of the first things you need to know when going out to observe is what can be seen in the sky at any particular time. A useful portable device is a circular star chart termed a planisphere. The Monthly Sky Guide chapter in this book (pp.274–323) is another handy guide.

Going outside

Before going outside, ensure that you wrap up warmly, as even mild nights can feel chilly and damp after a while. Make yourself comfortable by sitting in a reclining chair. At night your eyes will become much more sensitive, a process known as dark adaptation, but the increased sensitivity builds up only slowly. Hence, when going outdoors from a brightly lit room, you should allow at least 10 minutes for your eyes to adjust before beginning to observe.

Planispheres

A planisphere is a simple portable device for studying the heavens. This consists of a base on which the stars visible from a given latitude are plotted, overlain by a rotating mask. The directions on the horizon are marked around the edge, making it easy to use. When facing south for example, you should hold the planisphere so that south is at the bottom, and so on.

Using a red torch
Green and blue light rapidly destroys dark adaptation, whereas red light does not, so when observers need to see to read or write at night they use a red light, such as a torch with a red filter.

>> **USING A PLANISPHERE**

To use a planisphere, turn the mask until the required time lines up with the date (remember to deduct an hour when Daylight Saving, or Summer Time, is in force). The stars above the horizon at that time and date will appear within the cut-out area of the mask. Rotating the mask will show how stars rise and set during the course of the night. North, south, east, and west are marked around the edge of the mask, corresponding to the directions on your horizon.

turn the map inside the casing

compass direction helps you relate the star map to the real sky

the edge of the window corresponds to your horizon

»» MOVING PLANETS

Planets are always on the move, and are easier to see at some times than at others. The inner (or inferior) planets, Mercury and Venus, never appear very far from the Sun. Their maximum separation from the Sun is termed greatest elongation, either east of the Sun (in the evening sky) or west (morning sky). They are invisible around the times of superior or inferior conjunction, although occasionally they cross in front of the Sun at inferior conjunction to cause a transit. The outer (or superior) planets can lie anywhere along the ecliptic, and are best seen when near opposition. They then lie directly opposite the Sun in the sky and are visible all night. At opposition the outer planets are closest to us and appear brightest. Conversely, at conjunction they are behind the Sun and invisible.

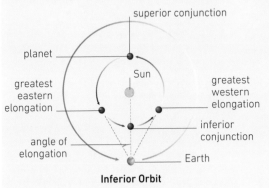

Inferior Orbit

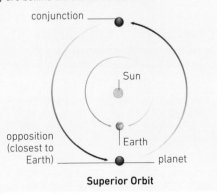

Superior Orbit

Planispheres are easy to use, cheap, portable, and never go out of date. Their disadvantages are that they do not show the positions of moving objects such as the Moon and planets, they work only for the latitude for which they are designed, and they depict only the brightest objects. A planisphere can be used up to about 5° from the latitude for which it is designed, after which a discrepancy with the sky will become noticeable.

Digital sky

Sophisticated planetarium-type computer programmes are commercially available for PC and Mac computers. With these you can see the sky as it appears from any place on Earth, at any time. You can travel at will into the future or the past, for instance, to see the sky as it would have looked on the day you were born.

Such programmes allow you to follow the movements of the Sun, Moon, and planets, zoom in on selected areas, and view simulations of eclipses. You can also add orbital information about newly discovered comets to the programme's database and the software will automatically calculate the object's movement across the sky.

«« Stars in detail
Star Walk is a useful app as it shows you helpful outlines, such as the Cancer crab shown here, to help you locate the constellations you are looking for.

«« Augmented stargazing apps
You can download apps such as Star Walk onto your phone or tablet, providing a portable planetarium that is readily available at your fingertips.

Binoculars

Binoculars are the ideal instrument for beginners. They are cheap, portable, easy to use, and will help you learn your way around the night sky before you move on to a telescope.

Aperture and magnification

Binoculars can be thought of as two small telescopes fixed together. Inside the binocular, two prisms fold up the light path, making them more compact than a normal telescope. Every binocular is marked with two numbers, such as 7 × 40 or 10 × 50. The first figure is the magnification, which is the enlargement relative to the naked-eye view. The second figure is the aperture (that is, the width) of the front lenses measured in millimetres. Wider apertures allow in more light, making fainter objects visible. However, larger-aperture binoculars are also inevitably heavier and more expensive than smaller ones. Binoculars with zoom magnification are available but generally produce poorer images.

Field of view

The field of view (also simply called the "field") of binoculars is typically 3° to 5° across, equivalent to six to ten Moon diameters. This is a good deal larger than the field of most telescopes, allowing you to scan large areas at a glance. Some objects are ideally suited to observation with binoculars, such as comets with long tails, scattered star clusters, and Milky Way star fields.

⌃ **Keeping binoculars steady**
For a steadier view when observing through binoculars, support your elbows on something solid such as the arms of a chair. Alternatively, sit with your elbows resting on your knees.

⟫ FOCUSING BINOCULARS

For best results, binoculars must be adjusted to suit the user's eyes. First, pivot each half of the binocular around the central bar until the distance between the eyepieces matches that of your eyes. For focusing, most binoculars have a knob on the central bar, plus one individually adjustable eyepiece. In some models, however, both eyepieces have to be adjusted individually.

 Focus left
focusing knob

 Focus right
eyepiece adjuster

1. Close the eye on the side of the adjustable eyepiece. Turn the central knob to focus the image you see with your other eye.

2. Now close the other eye and turn the individually adjustable eyepiece to bring that eye into focus.

3. From then on, you need only turn the central knob to bring both eyes into focus simultaneously.

Choosing binoculars

When buying binoculars, take into account the person who will be using them – a child, for example, will need a smaller, lighter pair. Do not be tempted by binoculars that promise high magnification combined with a small aperture, as the resulting images will be faint and indistinct.

Ideally, for astronomical use the aperture of a binocular in millimetres should be at least five times greater than the magnification figure. Binoculars with a smaller ratio are fine for daytime but will produce fainter images at night.

Compact binoculars

Roof prisms allow a "straight through" light path, resulting in a compact design with straight barrels. However, the lenses are relatively small, so they are not the best choice for astronomy.

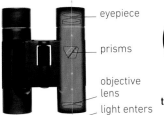

eyepiece

prisms

objective lens

light enters

The Moon through compact binoculars

rubber eye cups can be rolled back if you wear glasses

eyepiece

prism

objective lens

light enters

The Moon through standard binoculars

Standard binoculars

In these binoculars, the beam of light (yellow line) is folded by two prisms before it exits at the eyepiece, as shown here. The type of prisms in these binoculars are called Porro prisms. Some binoculars have rubber eye cups on the eyepieces to keep out stray light. Spectacle wearers can push these back to get their eyes closer to the eyepieces.

The Moon through large binoculars

Large binoculars

The more a pair of binoculars magnify, the more the image will seem to shake because of the movements of your hand. The solution is to mount the binoculars on a camera tripod. Better still, although more expensive, is a specialist binocular mount, available from telescope suppliers.

Newer, image-stabilizing binoculars sense movement and adjust the prisms accordingly These cut out hand shake very well, but not slower movements. Although these binoculars have the advantage of portability, they are heavier and more expensive than standard types.

>> **Binocular mounts**
When using large, heavy binoculars with magnifications greater than 10 times, a tripod mount is desirable. These are mounted 25 × 100 binoculars, used for deep-sky viewing.

Telescope astronomy

To see more than is possible with binoculars, you will need a telescope. There are several designs of amateur telescope, which can be supported on various types of mounting.

Gathering light

How much you can see with any telescope is governed by its aperture. Larger apertures collect more light, thereby showing fainter objects and finer detail. However, large lenses are expensive to make, are heavy, and require longer tubes. As a result, for telescopes with apertures greater than about 100mm (4in), amateur astronomers usually turn to reflectors, which use mirrors, and catadioptrics, a hybrid design. All large professional telescopes are reflectors.

Saturn through a 76mm (3in) aperture

Saturn through a 304mm (12in) aperture

⏶ **Zooming in**
Computer-controlled telescopes are now highly popular. This is a 200mm (8in) Schmidt-Cassegrain telescope on a GOTO mount, which automatically finds and tracks objects (see p.161).

⟫ EYEPIECES AND MAGNIFICATION

Telescopes have interchangeable eyepieces, and the magnification depends on the focal length of the eyepiece used. Shorter focal lengths produce greater magnification. Each eyepiece is marked with its focal length in millimetres and this must be divided into the telescope's focal length to find the resulting magnifying power. For example, on a telescope with a focal length of 1,200mm (47.2in), a 25mm (0.99in) focal-length eyepiece will produce a power of 48 times. On the same telescope, a 10mm (0.4in) eyepiece will give a magnification of 120 times. In practice, the highest magnification usable on a given telescope before the image becomes faint and indistinct is twice the aperture in millimetres.

Normal view

telescope end

eye end

Normal eyepiece

Wide-angle view

Wide-angle eyepiece

Types of eyepiece
Medium-power eyepieces are best for planetary and lunar studies, while the highest powers should be reserved for close double stars and planetary detail. Wide-angle eyepieces are useful for objects such as star clusters, nebulae, and galaxies as they give a wider field of view than a normal eyepiece.

Types of telescope

There are three main designs to choose from. Refracting telescopes collect and focus light with a lens. Reflecting telescopes use a mirror for the same purpose. Catadioptric telescopes use a combination of mirrors and lenses.

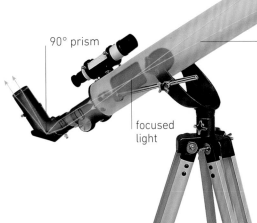

90° prism

objective lens

refracted light

focused light

Refracting telescopes

At the front end of a refractor's tube is the main lens, known as the objective lens or object glass. Light entering this lens is brought to a focus at the opposite end of the tube, where the eyepiece is placed. Virtually all small telescopes are refractors.

Reflecting telescopes

Reflecting telescopes employ mirrors to collect light and focus it into the eyepiece. A concave main mirror serves the same purpose as the objective lens in a refractor. The main mirror lies at the bottom end of the tube. Light falling on it is reflected back up the tube to a smaller mirror, known as the secondary, which diverts the beam into an eyepiece at the side of the tube. The design is termed a Newtonian, after Isaac Newton who first devised it, and is the type most commonly used.

light enters

eyepiece

objective lens

secondary mirror

lightweight tube

Catadioptric telescopes

These are reflectors with a thin lens, known as a corrector plate, across the front of the tube to increase the field of view. Light passes through the corrector plate and on to the main mirror, from where it is reflected to a convex secondary, which is usually attached to the rear of the corrector plate. The secondary reflects the light back down the tube into the eyepiece, which is positioned in the centre of the main mirror. Catadioptrics are popular because they are compact and portable, although they are more expensive than reflectors.

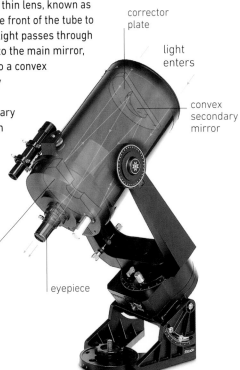

corrector plate

light enters

convex secondary mirror

concave primary mirror

eyepiece

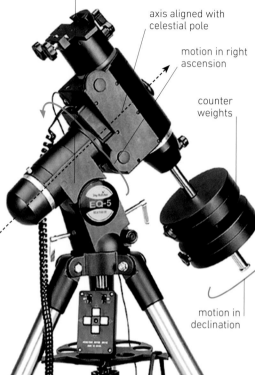

equatorial mounts are frequently motor-driven to allow hands-free observing

axis aligned with celestial pole

motion in right ascension

counter weights

motion in declination

Telescope mountings

There are two main types of mounting that can be used for any type of telescope. A simple "pan and tilt" mounting is known as an altazimuth. Such a mounting is convenient for reflectors because the eyepiece is at the upper end of the tube, unlike in refractors where the eyepiece is at the lower end and must be raised by a tripod so that it is high enough to see through. A prism can be used at the eyepiece end to improve the viewing position.

More sophisticated is the equatorial mounting, of which there are several varieties. However, they all have the common feature that the "pan" axis is aligned so that it points to the celestial pole. The telescope tube is attached to the other axis, known as the declination axis. Equatorial mounts are frequently motor-driven to allow hands-free observing, although a source of power, battery or mains, is required.

motion in altitude

motion in azimuth

⌂ Dobsonian mount
The Dobsonian is a cheap, popular form of altazimuth mounting, in which the lower end of the telescope tube rests in a box or fork.

» Altazimuth mount
A simple "pan and tilt" mounting allows the telescope to swivel freely from side to side (in azimuth) and tilt up and down (in altitude). It requires no special setting up or adjustment. Altazimuth mountings are commonly used for small refractors and reflectors.

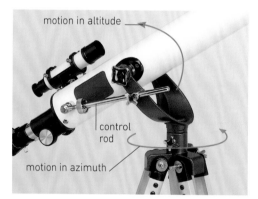

motion in altitude

control rod

motion in azimuth

⌂ Equatorial mount
In an equatorial mounting, the "pan" axis is aligned so that it points to the celestial pole. Once aligned, the telescope can be steered to follow a star simply by turning the polar-pointing axis as the Earth rotates. The type of equatorial mount most usually encountered with small refracting telescopes is known as the German mounting. This has a counterweight on the declination axis for balance.

» POLAR ALIGNMENT

For an equatorial mount to work successfully, it must be properly aligned. First the mounting must be set to your latitude and then levelled. The polar axis must then be aligned so that it points towards the celestial pole. This can be done with the aid of a compass or rough visual sighting towards the polar region of the sky at night. Approximate alignment is satisfactory for most normal observing requirements, but precise alignment is essential for accurate tracking during astrophotography. There are special alignment processes for computer-controlled GOTO telescopes.

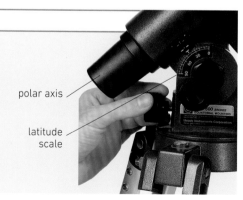

polar axis

latitude scale

Fork mountings

Many catadioptric telescopes use a fork mounting, either altazimuth or an equatorial. For smaller telescopes, such as this refractor, the fork may have only one arm, as shown here. The only difference between an altazimuth and an equatorial fork mounting is that the base is tilted at an angle equivalent to your latitude, so that the azimuth (or pan) axis becomes the polar axis.

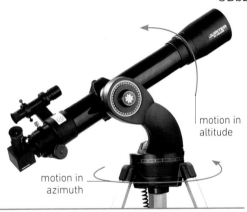

motion in altitude

motion in azimuth

≫ COMPUTERIZED TELESCOPES

Computers have revolutionized many aspects of life and astronomy is no exception. In recent years, computer-controlled telescopes that can automatically find and track objects have become popular and increasingly affordable. These are called GOTO telescopes and are controlled by a handset into which you can key the name or coordinates of the object you wish to find, or select from a built-in list. As well as the handset, most GOTO mounts can be controlled from PCs running a suitable software program, allowing access to an even wider database of objects. Unlike traditional altazimuth mountings, which are not motorized, GOTO telescopes on altazimuth mounts can automatically track objects as the Earth rotates. The accuracy is suitable for visual observation but for the smoothest tracking, particularly for long-exposure photography, an equatorial mount is still preferable. The instructions below explain how to set up a GOTO telescope.

1. A simple set-up procedure requires you to enter into the handset the date, time, and your location – although the most sophisticated mounts have a Global Positioning System (GPS) receiver that automatically gathers this information from satellites with high accuracy.

2. Then you need to align the telescope by pointing it in turn at two or more widely separated stars. These can be of your own choosing or, in some models, the telescope automatically slews to a bright star that you must centre in the field of view. Once calibrated in this way, the computer can find other objects with ease and accuracy.

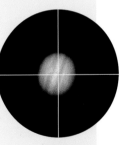

View through finder

simple low-power refractor acts as finder

slight adjustments made with handset

3. Use the keypad again to enter the details of what you would like to view. If necessary, the object can be centred in the field of view by using the hand controls.

⌃ Removing the eyepiece
Eyepieces push into place and are held by small screws. To remove the eyepiece, simply loosen the screws.

Astrophotography

Photography has two main advantages over visual observing – it keeps a permanent record of what is seen, and long exposures can build up images of objects far fainter than those visible to the naked eye. Astrophotography is a popular occupation for many amateurs.

Capturing images

Digital imaging has now completely supplanted film for astrophotography, as it has with normal photography. Digital cameras contain light-sensitive silicon chips known as charge-coupled devices (CCDs). CCDs have many advantages over film, including much greater sensitivity. Most astronomical objects are so faint that a long exposure is needed to see them. In a few seconds, a CCD can capture an image that might require an exposure of many minutes on film. When the exposure is finished, the image is simply read off the chip into a computer.

The CCD camera is attached to the telescope in place of the eyepiece. For short exposures, the body of a digital camera with a removable lens can be used, but for longer exposures a specialist CCD designed for astrophotography is necessary. These are cooled to reduce electronic noise in the chip. Several exposures of the same object can be added together to bring out faint features.

Colour pictures are created by taking three separate exposures through red, green, and blue filters and combining them in the computer. Further processing can be done to adjust the colour balance, brightness, and contrast, and to sharpen detail. The observer need not even be

⟫ USING FILTERS TO IMPROVE IMAGES

Special filters are available that reduce the effects of light pollution. These Light Pollution Reduction (LPR) filters make the sky darker so that faint nebulae and galaxies stand out better. Others, called nebula filters, allow through only the specific wavelengths of light emitted most strongly by nebulae. Ultra High Contrast (UHC) filters transmit the green light emitted

by hydrogen and oxygen. Most restrictive of all are the so-called OIII filters, which block all but the lines of ionized oxygen that are emitted strongly by planetary nebulae, but these are for specialized use only. All such filters screw into the barrel of the eyepiece.

use a solar filter for observing the Sun

⟨⟨ Coloured filters
Advanced observers use coloured filters to enhance details on the planets. For example, yellow and orange filters can emphasize dark markings on Mars, while light green and blue filters can bring out detail in the clouds of Jupiter and Saturn.

⟫ Solar filters
Solar filters consist of thin plastic, or sometimes glass, with a metallic coating that reduces the incoming light and heat from the Sun to safe levels. Such filters fit across the entire telescope tube, either refractor or

reflector, and with one of these in place you can look directly at the Sun to see sunspots and other features. Never use glass Sun filters that cover the eyepiece as these can crack suddenly under the concentrated light and heat, with disastrous consequences for your eyes.

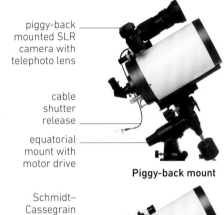

« High-quality amateur photograph

Modern telescopes, allied to CCDs and computer processing, have made it possible for dedicated amateurs to produce photographs that rival professional ones, such as this image of the nebula NGC 1977 in Orion.

at the telescope to make the exposures, but can sit comfortably indoors, operating the telescope and CCD camera remotely via computer control. Modified webcams are also used to take video sequences and stills through a telescope.

Simple views of the sky showing bright planets or the crescent Moon can be taken with an ordinary camera, or even a mobile phone, with an exposure of a few seconds. Pictures taken in twilight will record foreground details, adding visual interest. If the shutter can be kept open for long periods, star trails can be recorded and, with luck, a bright meteor.

piggy-back mounted SLR camera with telephoto lens

cable shutter release

equatorial mount with motor drive

Piggy-back mount

Schmidt–Cassegrain telescope

camera adaptor

camera body attached to eyepiece holder

Prime-focus mount

« Camera mountings

Exposures longer than a few seconds can be taken with the camera piggy-backed on a telescope with a motorized mount. The telescope tracks the stars as the Earth rotates. To record what the telescope sees, the lens can be removed and the body fixed to the eyepiece holder, with or without the eyepiece in place.

⌄ Post-processing

To produce a full-colour image, three exposures of the same object taken through red, green, and blue filters are combined in a computer. Using image-processing software, the resulting composite can be adjusted as required to enhance specific features.

red-dot finder for coarse pointing

standard low-power refractor finder

« CCD camera setups

Here, a CCD is attached at the eyepiece end of a motor-driven Schmidt–Cassegrain telescope. A cable connects the CCD to a computer, allowing the image to be downloaded once the exposure is complete. CCD chips are about the size of a postage stamp, and consist of millions of picture elements, or pixels.

CCD unit replaces eyepiece

CCD connected to computer via USB cable

three separate images taken through different coloured filters

Detail of a CCD

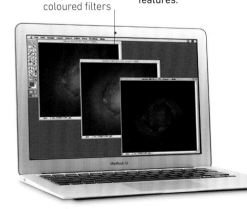

Solar eclipse sequence
Successive stages of an eclipse of the Sun, with the glowing corona visible during totality at the centre of the image, are seen in this multiple exposure taken from Svalbard, in the Arctic, on 20 March 2015. The individual exposures making up this composite were taken every three minutes. Totality itself lasted two and a half minutes.

Patterns in the sky
At least eight constellations can be seen in this image taken from Hakos, Namibia. Ancient peoples named the patterns of the stars after their mythical characters; these names are still in use.

The constellations

Astronomers divide the celestial sphere into sections called constellations. These originated with the star patterns imagined by ancient people to represent their mythical heroes, gods, and exotic beasts. Now, constellations are simply areas of sky with boundaries laid down by international agreement, although the ancient names such as Perseus, Andromeda, and Orion have been retained.

A catalogue of stars divided into 48 constellations was produced around 150 CE by a Greek astronomer and geographer called Ptolemy, and his catalogue has formed the basis of our system of constellations ever since. At the end of the 16th century, a Dutch cartographer, Petrus Plancius, and two Dutch navigators, Pieter Dirkszoon Keyser and Frederick de Houtman, added a number of new constellations, including a dozen in the far southern sky that had been below the horizon of the ancient Greeks. More were added at the end of the 17th century by Johannes Hevelius, a Polish astronomer, filling in the gaps between the Greek constellations.

An authoritative map

The picture was completed in the 1750s by a French astronomer, Nicolas Louis de Lacaille, who invented 14 new constellations representing devices from science and the arts in the southern sky. In all, 88 constellations now fill the sky, their names and boundaries defined by the International Astronomical Union, astronomy's governing body. All stars within the boundaries of a constellation area are considered to belong to that constellation, whether or not they form part of the constellation figure. Names are often shortened to a three-letter abbreviation – for example, Cassiopeia becomes Cas, and Canis Major, CMa.

The naming of stars

Stars are known by a confusing variety of designations, and any given star can have several aliases. Bright stars are labelled with Greek letters, a system devised in 1603 by a German astronomer, Johann Bayer. The letter is used with the genitive (possessive) form of the constellation name – for example, Alpha (α) Centauri. Some bright stars also have proper names, such as Sirius and Betelgeuse. With fainter stars, numbers are used, as in 61 Cygni.

A different system of naming is used for so-called deep-sky objects such as star clusters, nebulae, and galaxies. The first list of such objects was compiled by the French astronomer Charles Messier, and these are still known by their M, or Messier, numbers. Messier's final catalogue, which appeared in 1781, contained just over 100 objects, but the numbers known grew rapidly as telescopes improved. The New General Catalogue (NGC) of 1888 contained 7,840 objects and over 5,000 more were added in two supplements called the Index Catalogues (IC).

◄ The sky in a nutshell
This 18th-century hinged pocket globe depicts the celestial sphere (top) and the Earth (bottom).

Mapping the sky

Each of the 88 constellations is profiled on the following pages. A chart of the constellation is accompanied by an account of its origin and brief descriptions of the main objects of interest, with an emphasis on those within reach of amateur equipment.

Northern hemisphere

Polaris, the north Pole Star, lies almost centrally on the chart to the right, less than 1° from the exact north celestial pole. For observers in the Northern Hemisphere, the stars around the north pole never set – they are said to be circumpolar. The farther north the observer, the more of the sky will be circumpolar. The chart of the northern sky extends from declination 90° north to 30° south. The appearance of the constellations at the edge of each hemisphere is considerably distorted.

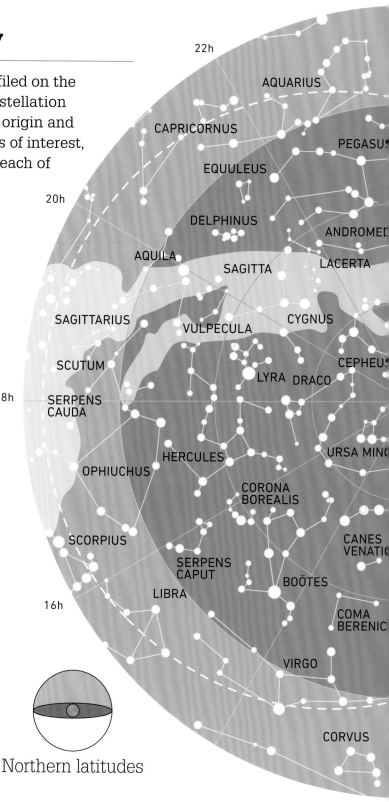

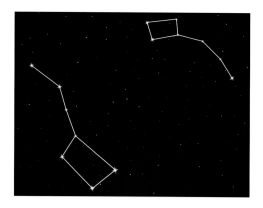

⊼ **Big and little bears**
Pictured here is the Ursa Minor (top right) constellation and a part of the Ursa Major (bottom left) constellation called the "Big Dipper". These constellations can also be located on the star map opposite.

Northern latitudes

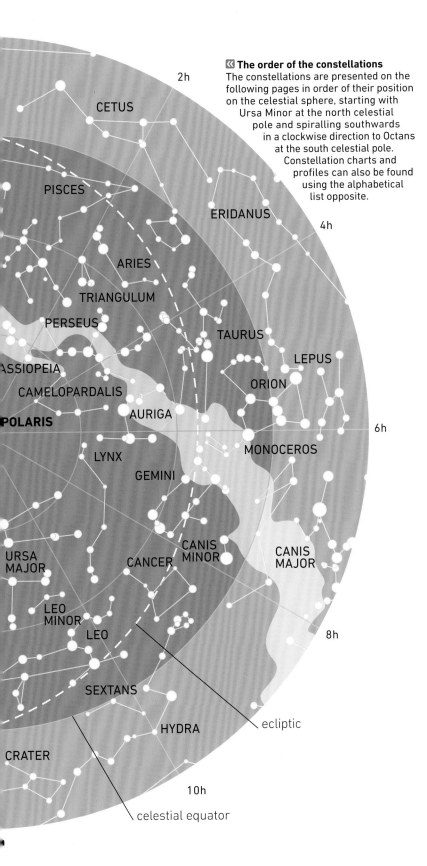

2h

CETUS

PISCES

ERIDANUS

4h

ARIES

TRIANGULUM

TAURUS

PERSEUS

LEPUS

CASSIOPEIA

ORION

CAMELOPARDALIS

AURIGA

POLARIS

6h

LYNX

MONOCEROS

GEMINI

URSA
MAJOR

CANIS
MINOR

CANCER

CANIS
MAJOR

LEO
MINOR

8h

LEO

SEXTANS

ecliptic

HYDRA

CRATER

10h

celestial equator

≪ The order of the constellations
The constellations are presented on the
following pages in order of their position
on the celestial sphere, starting with
Ursa Minor at the north celestial
pole and spiralling southwards
in a clockwise direction to Octans
at the south celestial pole.
Constellation charts and
profiles can also be found
using the alphabetical
list opposite.

≫ CONSTELLATIONS

Southern hemisphere

Unlike the northern hemisphere, which has the North Star, the southern hemisphere does not have a relatively bright star in its centre. The chart of the southern sky extends from declination 90° south to 30° north.

⌃ Mini constellations
Both the False Cross (right) and the Southern Cross (left) can be joined up using stars that comprise the much larger constellations of Carina and Vela, respectively. Stargazers can easily confuse the two due to their close proximity in the sky.

⌃ Clusters of galaxies
At the ALMA observatory in the Atacama Desert, Chile – when the sky is clear – you can get a beautiful view of the stars. The brighter clusters are not constellations of stars but galaxies that are over 150,000 light-years away.

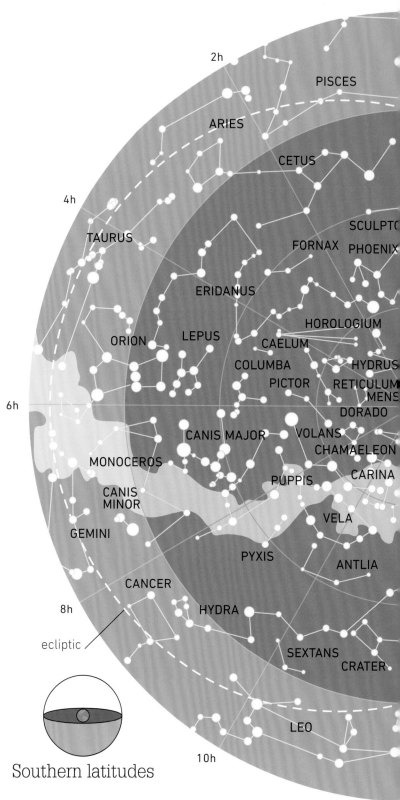

2h

PISCES

ARIES

CETUS

4h

SCULPTO

TAURUS

FORNAX PHOENIX

ERIDANUS

HOROLOGIUM

ORION LEPUS CAELUM

HYDRUS

COLUMBA

RETICULUM

PICTOR

MENS

6h

DORADO

CANIS MAJOR VOLANS

CHAMAELEON

MONOCEROS

CARINA

CANIS MINOR

PUPPIS

VELA

GEMINI

ANTLIA

PYXIS

CANCER

8h

HYDRA

ecliptic

SEXTANS

CRATER

LEO

10h

Southern latitudes

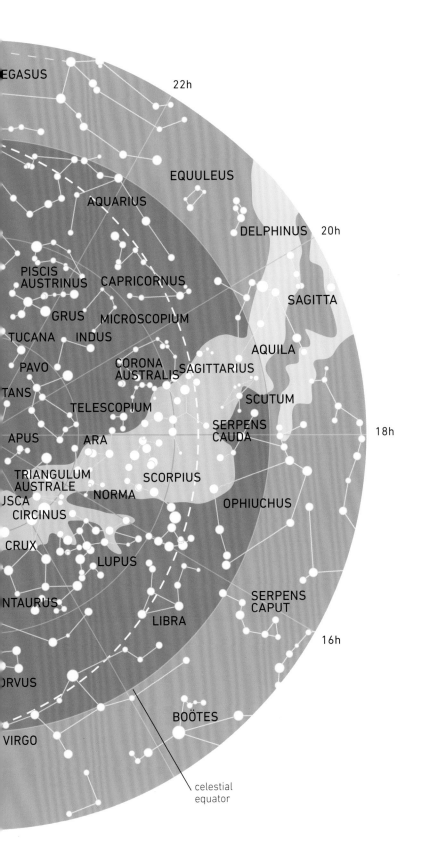

PEGASUS
22h
EQUULEUS
AQUARIUS
DELPHINUS   20h
PISCIS
AUSTRINUS   CAPRICORNUS
SAGITTA
GRUS   MICROSCOPIUM
TUCANA   INDUS
AQUILA
PAVO   CORONA
AUSTRALIS   SAGITTARIUS
TANS
TELESCOPIUM   SCUTUM
SERPENS
CAUDA   18h
APUS   ARA
TRIANGULUM
AUSTRALE   SCORPIUS
USCA   NORMA
CIRCINUS   OPHIUCHUS
CRUX
LUPUS
NTAURUS   SERPENS
CAPUT
LIBRA   16h
ORVUS
BOÖTES
VIRGO

celestial
equator

Locator maps
These show the position of each constellation in the northern or southern celestial hemisphere. Some, like Orion (above), span the celestial equator, so project from the edge of the map.

Understanding the charts

The constellation charts in this section are oriented with north at the top and south at the bottom. They are all reproduced to the same scale, to give an accurate reflection of the relative sizes of the constellations.

What the charts show

The charts show the constellation figure (the pattern of lines joining the bright stars) and the boundaries of the constellation, as defined by the International Astronomical Union, which are outlined in orange. Within each constellation, every star brighter than magnitude 5 is labelled. In addition, all stars brighter than magnitude 6.5 are shown, but not labelled. Deep-sky objects such as galaxies are represented by an icon, as shown in the key at right.

›› STARS AND DEEP-SKY OBJECTS

The charts show the major stars and deep-sky objects in each constellation. Individual stars are represented by dots that indicate their apparent magnitude (see scale at bottom left of page).

Deep-sky objects
Distant, nebulous objects, such as clusters of stars, nebulae, and galaxies are known as deep-sky objects. These are indicated on the constellation maps by the icons shown on the right. All are labelled with a catalogue number, which may be a Messier number or an NGC or IC number (see p.167).

Open cluster

Globular cluster

Diffuse nebula

Planetary nebula or supernova remnant

Galaxy

Black hole

Star names
The brightest stars in a constellation are labelled with Greek letters, usually, but not always, in order of magnitude. The Greek alphabet is given below. Close pairs or groups of stars may share a Greek letter and are distinguished by superscripts. In Orion, a chain of six stars bears the letter Pi (π) with a distinguishing superscript – Pi¹ (π^1), Pi² (π^2), and so on. Other stars are known simply by a number, for example, 15 Orionis. The stars are numbered in increasing order of right ascension (from right to left on the charts). These are known as Flamsteed numbers because the stars concerned were catalogued by the first British Astronomer Royal, John Flamsteed (1646–1719).

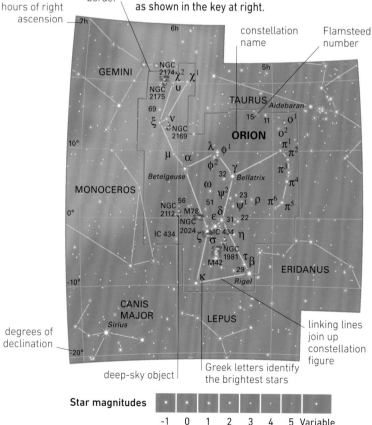

constellation border
hours of right ascension
constellation name
Flamsteed number
degrees of declination
deep-sky object
Greek letters identify the brightest stars
linking lines join up constellation figure

α Alpha	ι Iota	ρ Rho
β Beta	κ Kappa	σ Sigma
γ Gamma	λ Lambda	τ Tau
δ Delta	μ Mu	υ Upsilon
ε Epsilon	ν Nu	φ Phi
ζ Zeta	ξ Xi	χ Chi
η Eta	o Omicron	ψ Psi
θ Theta	π Pi	ω Omega

Star magnitudes

-1 0 1 2 3 4 5 Variable star

Ideal viewing

Which constellations you can see depends on your latitude on Earth. Far-southern constellations cannot be seen from far-northern latitudes, for example, because they never rise above the horizon. The bar at the top of each constellation entry specifies the latitudes on Earth from which each constellation is fully visible. Even so, when objects are close to the horizon, they will be dimmed by the atmosphere and not as easy to see.

partially visible (only part of constellation can be seen)   fully visible

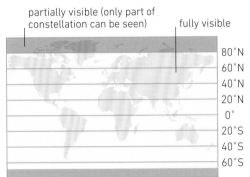

80°N
60°N
40°N
20°N
0°
20°S
40°S
60°S

⌃ The visibility of Orion
Orion is fully visible between latitudes 79°N and 67°S. As is clear from the map above, the whole constellation can be seen from virtually the entire inhabited world.

》 SIZING

Hand symbols are used to indicate how large a constellation appears in the sky. A splayed hand at arm's length spans about 16 degrees of sky, while a closed hand covers about 10 degrees. Combinations of these symbols are used to convey the full width and depth of each constellation.

16° of sky   10° of sky   42° of sky

⌃ Constellation figures
A small sketch shows how each constellation was originally imagined as a picture in the sky, as in this example of Orion the Hunter.

》 SEEING STARS

The icons below are used to indicate what equipment, if any, is needed to see each of the items of interest described in a constellation.

Visibility icons

👁 Naked eye   🖥 CCD

🔭 Binoculars   ⚲ Professional
                          observatory
🗡 Telescope
   (amateur)

《 Photographic images
Photographic or CCD images will show far more than can be seen with the eye alone.

Ursa Minor Ursae Minoris (UMi)

WIDTH 🖐 **DEPTH** 🖐🖐 **SIZE RANKING** 56th **FULLY VISIBLE** 90°N–0°

This ancient Greek constellation represents a little bear. Its main stars form a shape known as the Little Dipper, similar to the larger and brighter Big Dipper (Ursa Major), but with its handle curving in the opposite direction. Ursa Minor contains the north celestial pole. Less than 1° from the pole is Polaris, the north Pole Star, which has been used for navigation since the earliest times. Beta (β) and Gamma (γ) Ursae Minoris are known as the Guardians of the Pole.

Northern Hemisphere

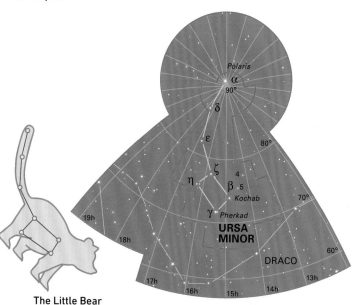

The Little Bear

⟫ FEATURES OF INTEREST

Alpha (α) Ursae Minoris (Polaris) 👁 ⚹
A creamy-white supergiant of magnitude 2.0, about 430 light-years away. It is a Cepheid variable, but the brightness changes are too small to be noticeable to the eye. A small telescope shows an 8th-magnitude star near to it, which is unrelated.

Gamma (γ) Ursae Minoris 👁 🔭 A wide double star that is part of the bowl of the Little Dipper. It is of magnitude 3.0 and has an unrelated 5th-magnitude companion that can be seen with good eyesight or through binoculars.

Eta (η) Ursae Minoris 🔭 Another wide double star in the bowl of the Little Dipper. Eta is of magnitude 5.0 and has a magnitude 5.5 companion, easily visible through binoculars. Like the companion of Gamma (Ursae Minoris), Eta's companion is an unrelated star lying in the same line of sight.

⟫ Ursa Minor in the night sky
The main stars of Ursa Minor form the Little Dipper, which curves away from the north Pole Star, Polaris, seen bottom centre.

Draco Draconis (Dra)

WIDTH 👐👐👐 **DEPTH** 👐👐👐 **SIZE RANKING** 8th **FULLY VISIBLE** 90°N–4°S

Draco, the Dragon, is a large constellation that winds nearly halfway around the north celestial pole. Despite its considerable size, Draco is not particularly easy to identify apart from a distorted diamond marking the dragon's head. The head contains the constellation's brightest member, Gamma (γ) Draconis, magnitude 2.2. In Greek mythology, Draco represents the dragon slain by Hercules as one of his 12 labours.

Northern Hemisphere

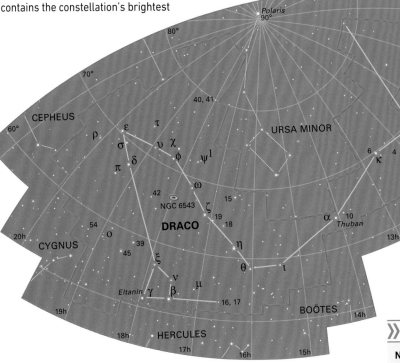

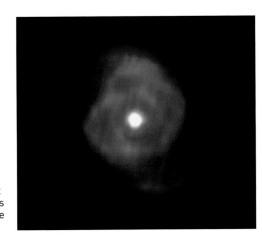

The Dragon

⟫ The Cat's Eye Nebula
Through a small telescope, the nebula appears blue-green, but CCD images such as this one bring out a red tinge in its outer regions.

⟫⟫ FEATURES OF INTEREST

Nu (ν) Draconis 🔭 The faintest of the four stars in the dragon's head. An easy double, it consists of identical white components of 5th magnitude and is regarded as one of the finest doubles for binoculars.

Psi (Ψ) Draconis 🔭 A closer pair of 5th- and 6th-magnitude stars, divisible with a small telescope.

16 and 17 Draconis 🔭🔭 A wide pair of stars, easily seen through binoculars. The brighter of the two, 17 Draconis, can be further divided by a small telescope with high magnification.

39 Draconis 🔭 A star seen as a double through a small telescope with low magnification. High powers divide the brighter component into a closer pair of magnitudes 5.0 and 8.0.

NGC 6543 (Cat's Eye Nebula) 🔭 A planetary nebula made famous by a Hubble Space Telescope image.

Northern Hemisphere

Cepheus Cephei (Cep)

WIDTH 🖐🖐 **DEPTH** 🖐🖐🖐 **SIZE RANKING** 27th **FULLY VISIBLE** 90°N–1°S

A constellation of the far northern sky, Cepheus lies between Cassiopeia and Draco, but is not particularly prominent. Its main stars form a shape resembling a square tower with a pointed steeple. It is an ancient Greek constellation that represents the mythical King Cepheus of Ethiopia, who was husband of Queen Cassiopeia and father of Andromeda.

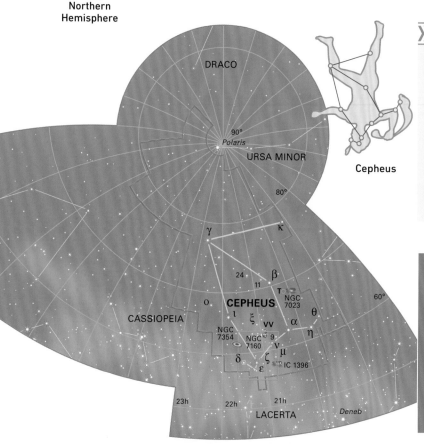

Cepheus

》 FEATURES OF INTEREST

Delta (δ) Cephei 👁 🔭 ✦ The star from which Cepheid variables take their name. A yellow supergiant around 865 light-years away, it varies from magnitude 3.5 to 4.4 and back every 5 days 9 hours. These changes can be followed with the naked eye (see chart, below). It is also a double star, with a blue-white companion of 6th magnitude visible through small telescopes.

Mu (μ) Cephei 👁 🔭 A red supergiant that varies semi-regularly between magnitudes 3.4 and 5.1 every 2 years or so. It is also known as the Garnet Star because of its strong coloration.

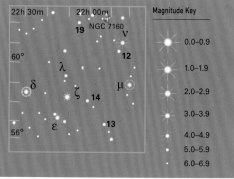

Variation in Delta and Mu Cephei
To gauge the magnitude of the variable stars Delta (δ) and Mu (μ) Cephei at any given time, their brightness can be compared to that of non-variable stars nearby. Useful yardsticks are Zeta (ζ) Cephei (magnitude 3.4), Epsilon (ε) Cephei (magnitude 4.2), and Lambda (λ) Cephei (magnitude 5.1).

Mu Cephei and IC1396
In this CCD image, the Garnet Star, Mu (μ) Cephei, appears above the nebula IC1396, which is centred on Struve 2816, a 6th-magnitude multiple star.

Cassiopeia Cassiopeiae (Cas)

WIDTH 〰〰 **DEPTH** 〰〰 **SIZE RANKING** 25th **FULLY VISIBLE** 90°N–12°S

A distinctive constellation of the northern sky, Cassiopeia lies in the Milky Way between Perseus and Cepheus. Its five main stars form a large, easily recognizable W shape. Cassiopeia is an ancient Greek constellation, representing a mythical Queen of Ethiopia, who was punished for her vanity by Poseidon. Her husband, Cepheus, and daughter, Andromeda, are represented by adjoining constellations.

Northern Hemisphere

Cassiopeia

≫ FEATURES OF INTEREST

Gamma (γ) Cassiopeiae 👁 A hot, rapidly rotating star in the middle of the W of Cassiopeia. It throws off rings of gas from its equator from time to time, causing unpredictable changes in its brightness. Currently it is of magnitude 2.2, but in the past has ranged between magnitudes 3.0 and 1.6.

Eta (η) Cassiopeiae 🔭 An attractive binary pair of yellow and red stars, magnitudes 3.5 and 7.5, easily divided by small telescopes. The fainter companion orbits the brighter star every 480 years.

Rho (ρ) Cassiopeiae 👁 🔭 An intensely luminous yellow-white supergiant that varies between 4th and 6th magnitudes every 27 months or so.

M52 🔭 🔭 An open cluster visible through binoculars, noticeably elongated and spanning about one-third the width of the full Moon.

M103 🔭 🔭 A small, elongated star cluster, better seen through small telescopes than binoculars.

≫ Elongated cluster
The main feature of the open star cluster M103 is a chain of three stars, which looks like a mini Orion's belt.

Camelopardalis Camelopardalis (Cam)

WIDTH 〰️〰️ **DEPTH** 〰️〰️ **SIZE RANKING** 18th **FULLY VISIBLE** 90°N–3°S

Northern Hemisphere

Camelopardalis is a large but dim constellation of the far northern sky, representing a giraffe. Its long neck stretches around the north celestial pole between Ursa Minor and the tail of Draco. Not one of the original Greek constellations, it was introduced in the early 17th century by the Dutch astronomer Petrus Plancius.

>> FEATURES OF INTEREST

Beta (β) Camelopardalis 🔭 The constellation's brightest star, at magnitude 4.0. It is a double with a wide 7th-magnitude companion that can be seen with small telescopes or even powerful binoculars.

NGC 1502 🔭 A small open star cluster visible through binoculars and small telescopes. It has a 7th-magnitude double star at its centre that is easily divided by small telescopes.

Kemble's Cascade 🔭 A chain of faint stars visible through binoculars that stretches for five Moon diameters from NGC 1502 towards Cassiopeia. This star chain is named after Lucian Kemble, a Canadian amateur astronomer who first drew attention to it.

The Giraffe

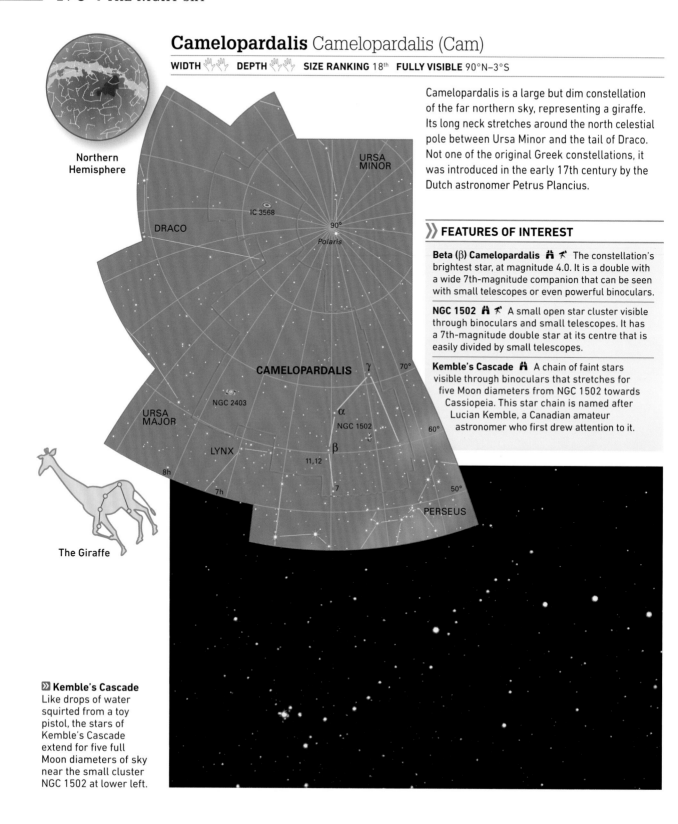

>> **Kemble's Cascade**
Like drops of water squirted from a toy pistol, the stars of Kemble's Cascade extend for five full Moon diameters of sky near the small cluster NGC 1502 at lower left.

Lynx Lyncis (Lyn)

WIDTH 〰🖐 **DEPTH** 〰🖐 **SIZE RANKING** 28th **FULLY VISIBLE** 90°N–28°S

Polish astronomer Johannes Hevelius introduced this constellation in the late 17th century to fill the gap in the northern sky between Ursa Major and Auriga. It gained its name because Hevelius felt that only the lynx-eyed would be able to see it.

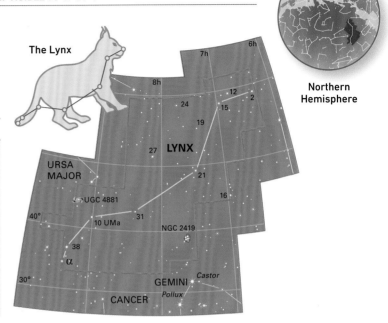

The Lynx

Northern Hemisphere

》 FEATURES OF INTEREST

12 Lyncis 🏹 A star that appears through a small telescope as a double of 5th and 7th magnitudes. Apertures of 75mm (3in) will divide the brighter star again. This pair, of magnitudes 5.4 and 6.0, forms a binary with an orbital period of about 900 years.

19 Lyncis 🏹 A pair of stars of 6th and 7th magnitudes with a wider 8th-magnitude companion, all visible with small telescopes.

38 Lyncis 🏹 A closer double, of 4th and 6th magnitudes. A telescope with a 75mm (3in) aperture is needed to divide it.

Auriga Aurigae (Aur)

WIDTH 〰🖐 **DEPTH** 〰🖐 **SIZE RANKING** 21st **FULLY VISIBLE** 90°N–34°S

A prominent constellation of the northern sky, Auriga contains the most northerly first-magnitude star, Capella. Lying in the Milky Way between Gemini and Perseus, to the north of Orion,

Auriga represents a charioteer, usually identified in Greek mythology as Erichthonius, a legendary king of Athens.

Northern Hemisphere

》 FEATURES OF INTEREST

Epsilon (ε) Aurigae 👁 👤 A luminous supergiant orbited by a mysterious dark companion that eclipses it every 27 years, the longest interval of any eclipsing binary. During the eclipses, it remains dimmed for over a year.

Zeta (ζ) Aurigae 👁 👤 An orange giant orbited by a smaller blue star, which eclipses it every 2.7 years, decreasing its brightness from magnitude 3.7 to 4.0, for six weeks.

M36, M37, M38 👤 🏹 Three large and bright open star clusters, embedded in a rich part of the Milky Way. All three will just fit within the same field of view in wide-angle binoculars.

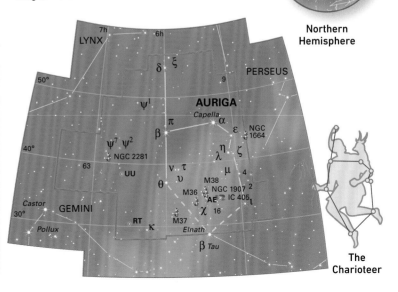

The Charioteer

Northern Hemisphere

Ursa Major Ursae Majoris (UMa)

WIDTH 〰〰〰〰 **DEPTH** 〰〰〰 **SIZE RANKING** 3rd **FULLY VISIBLE** 90°N–16°S

Ursa Major, the Great Bear, is a large and prominent constellation of the northern sky. Seven of its stars form the familiar shape known as the Plough or Big Dipper, but this is only part of the whole constellation. The two stars in the dipper's bowl furthest from the handle, Alpha (α) and Beta (β) Ursae Majoris, point towards the north Pole Star, Polaris. In Greek myth, the Great Bear is identified with two different characters. One is Callisto, a lover of Zeus, who was turned into a bear by Zeus's wife, Hera, in a fit of jealousy. The other is Adrasteia, one of two nymphs who nursed the infant Zeus, hiding him from his murderous father, Cronus. The second nymph, Ida, is represented by Ursa Minor, the Little Bear.

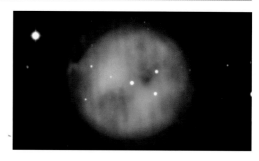

⌃ The Owl Nebula (M97)
This faint planetary nebula under the bowl of the Big Dipper gets its popular name from its dark, owl-like "eyes", although these are visible only through large telescopes or on photographs and CCD images such as this.

The Great Bear

Seeing Ursa Major in the night sky
The familiar saucepan shape of the Plough or Big Dipper is one of the most easily recognized patterns in the sky, but makes up only part of Ursa Major. The second star in the handle can be seen to be double with the unaided eye.

⟫ FEATURES OF INTEREST

Plough (Big Dipper) 👁 One of the best-known patterns in the sky, marked out by the stars Alpha (α), Beta (β), Gamma (γ), Delta (δ), Epsilon (ε), Zeta (ζ), and Eta (η) Ursae Majoris. With the exception of Alpha and Eta, these stars are at similar distances from us (about 80 light-years) and are travelling in the same direction through space, forming what is known as a moving cluster.

Zeta (ζ) Ursae Majoris (Mizar and Alcor) 👁 🔭 Mizar, the second star in the handle, has a fainter companion star called Alcor, which can be picked out with good eyesight and is easily seen with binoculars. A small telescope shows that Mizar also has a closer 4th-magnitude companion.

Xi (ξ) Ursae Majoris 🔭 A double star in the south of the constellation that can be separated through telescopes with small apertures. The two components, of 4th and 5th magnitudes, form a true binary, orbiting every 60 years, a relatively short period for a visual binary star.

M81 🔭 🔭 A spiral galaxy in northern Ursa Major, tilted at an angle to us.

M82 🔭 🖥 A spiral galaxy edge-on to us. It is thought to be undergoing a burst of star formation following a close encounter with the larger M81 some 300 million years ago.

M97 (the Owl Nebula) 🔭 A planetary nebula under the Big Dipper's bowl, one of the faintest objects in Charles Messier's catalogue.

M101 🔭 🔭 A spiral galaxy presented face-on to us near the end of the Big Dipper's handle.

⌃ **M81 and M82**
These two contrasting spiral galaxies are found in northern Ursa Major. The larger of them, M81, is visible on clear, dark nights as a slightly elongated patch of light. One full Moon diameter to the north of it is the smaller, fainter M82 (seen on the right here), which will require a telescope to be spotted.

Northern
Hemisphere

Canes Venatici Canum Venaticorum (CVn)

WIDTH 〰️🖐️ **DEPTH** 〰️🖐️ **SIZE RANKING** 38th **FULLY VISIBLE** 90°N–37°S

This constellation of the northern sky lies between Ursa Major and Boötes, south of the handle of the Plough (Big Dipper). It represents two hunting dogs held on a leash by Boötes, the herdsman.

Canes Venatici was formed at the end of the 17th century by the Polish astronomer Johannes Hevelius from stars that had previously been part of Ursa Major.

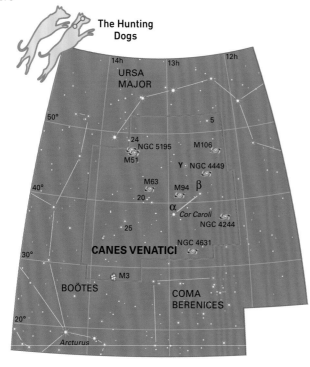

The Hunting
Dogs

▲ **The Whirlpool Galaxy**
The Whirlpool Galaxy M51 was the first galaxy in which spiral arms were detected. They were seen in 1845 by Lord Rosse in Ireland with his 1.8m (72in) reflector. At the end of one of the arms lies a smaller galaxy, NGC 5195.

▲ **Globular Cluster M3**
This spectacular globular cluster is easily found with binoculars, but a telescope with an aperture of at least 100mm (4in) is necessary in order to distinguish its individual stars.

》 FEATURES OF INTEREST

Alpha (α) Canum Venaticorum (Cor Caroli) 🔭
The constellation's brightest star, a wide double of magnitudes 2.9 and 5.6, easily separated by small telescopes. The name Cor Caroli (Charles's Heart) commemorates King Charles I of England.

Y Canum Venaticorum (La Superba) 🔭
A deep red supergiant that fluctuates between magnitudes 5.0 and 6.5 every 270 days or so.

M3 🔭🔭 An impressive 6th-magnitude globular cluster, easily found with binoculars as a rounded ball of light between Cor Caroli and Arcturus.

M51 (The Whirlpool Galaxy) 🔭 🔭 A famous spiral galaxy, visible through binoculars as a round patch of light. Moderate-sized telescopes are needed to make out the spiral arms.

M63 (The Sunflower Galaxy) 🔭 A spiral galaxy visible through small telescopes as a slightly elongated patch of light.

Boötes Boötis (Boo)

WIDTH 〰〰 **DEPTH** 〰〰〰 **SIZE RANKING** 13th **FULLY VISIBLE** 90°N–35°S

Boötes is a large and prominent constellation of the northern sky, extending from Draco and the handle of the Plough (Big Dipper) in the north to Virgo in the south. It contains the brightest star north of the celestial equator, Arcturus (Alpha Boötis), which is also the fourth-brightest star of all. Boötes represents a man herding a bear (Ursa Major). The name "Arcturus" means "bear guard" or "bear keeper" in Greek. The herdsman's two dogs are represented in the adjoining constellation, Canes Venatici. Some faint stars in the northern part of Boötes used to form a constellation called Quadrans Muralis. The name survives in the Quadrantids, a meteor shower originating in this part of the sky in January.

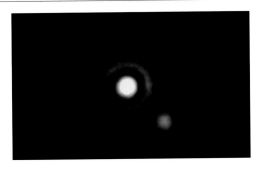

⌃ Epsilon Boötis
The orange star and its blue-green companion present one of the most beautiful contrasts of all doubles. As well as a high-magnification telescope, a night of steady air is needed to see this double at its best.

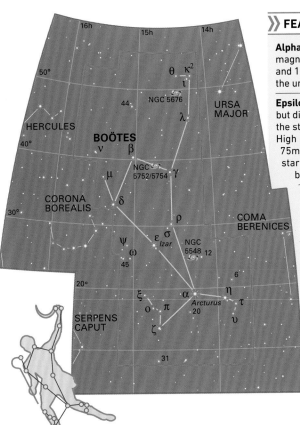

The Herdsman

》 FEATURES OF INTEREST

Alpha (α) Boötis (Arcturus) 👁 A red giant of magnitude −0.1, the fourth-brightest of all stars and 100 times more luminous than the Sun. To the unaided eye, it has a warm orange tint.

Epsilon (ε) Boötis 🏹 A celebrated double star but difficult to divide because of the closeness of the stars. To the eye, it appears of magnitude 2.4. High magnification on a telescope of at least 75mm (3in) aperture reveals an orange star accompanied by a 5th-magnitude blue-green companion.

Kappa (κ) Boötis 🏹 A double star with components of 5th and 7th magnitudes, divisible with a small telescope. The stars are unrelated.

Mu (μ) Boötis 🎏 A double star of 4th and 6th magnitudes, divisible with binoculars.

Xi (ξ) Boötis 🏹 Another double star of 5th and 7th magnitudes, divisible with small telescopes. The stars have warm yellow-orange hues and form a true binary with an orbital period of 150 years.

**Northern
Hemisphere**

Hercules Herculis (Her)

WIDTH 〰〰〰 **DEPTH** 〰〰〰 **SIZE RANKING** 5ᵗʰ **FULLY VISIBLE** 90°N–38°S

A large but not particularly prominent
constellation of the northern sky, Hercules
represents the strong man of Greek myth who
undertook 12 labours, clad in a lion's pelt and
brandishing a club. One of his tasks was to slay
a dragon, and in the sky Hercules kneels with
one foot on the head of Draco, to the north.

⟫ FEATURES OF INTEREST

The Keystone 👁 A distinctive quadrilateral
of stars that forms part of the body of Hercules.

Alpha (α) Herculis (Rasalgethi) 👁 📷 🔭 A red
giant that pulsates erratically, fluctuating between
3rd and 4th magnitudes. A small telescope reveals
a 5th-magnitude blue-green companion.

Rho (ρ) Herculis 🔭 A double star of 5th and 6th
magnitudes, divisible through small telescopes
with high magnification.

M13 👁 📷 A spectacular globular cluster, the
finest in northern skies, containing hundreds
of thousands of stars.

M92 🔭 A 7th-magnitude globular cluster.
Smaller and fainter than M13, it can be
mistaken for an ordinary star when seen
through binoculars, but a small telescope
quickly reveals its true nature.

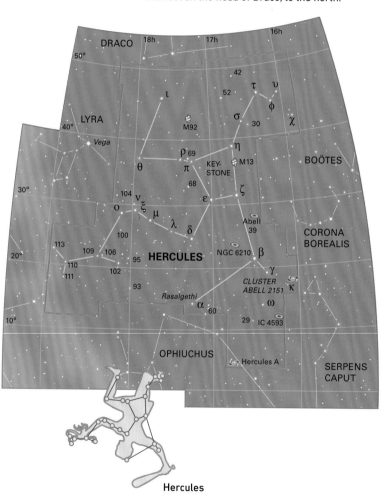

Hercules

⟫ Globular cluster M13
Under ideal conditions this magnificent globular
cluster can be glimpsed with the naked eye and is
easily found with binoculars, appearing like a hazy
star half the apparent width of the full Moon. It lies
on one side of the Keystone, about one-third of the
way from Eta (η) to Zeta (ζ) Herculis. M13 is 25,000
light-years away.

Lyra Lyrae (Lyr)

WIDTH 🖐 **DEPTH** 🖐🖐 **SIZE RANKING** 52ⁿᵈ **FULLY VISIBLE** 90°N–42°S

This compact but prominent constellation of the northern sky contains Vega, the fifth-brightest star in the sky. Blue-white Vega is part of a large triangle of stars in northern summer skies called the Summer Triangle (the other two members are Deneb, in Cygnus, and Altair, in Aquila). Lyra, which represents the instrument played by Orpheus, lies on the edge of the Milky Way next to Cygnus. The Lyrid meteors radiate from a point near Vega around 21–22 April every year.

The Lyre

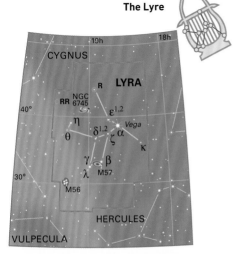

Northern Hemisphere

⟫ FEATURES OF INTEREST

Beta (β) Lyrae 🔭 A double star, easily resolved into cream and blue components by small telescopes. The brighter star (the cream one) is also an eclipsing binary that fluctuates between magnitudes 3.3 and 4.4 every 12.9 days. The two stars in this eclipsing binary are so close that gas from the larger one falls towards the smaller companion, and some of it spirals off into space.

Delta (δ) Lyrae 👁 🔭 A wide double, divisible with binoculars or good eyesight. It consists of a red giant of 4th magnitude and an unrelated blue-white star of 6th magnitude.

Epsilon (ε) Lyrae 🔭 🔭 The finest quadruple star in the sky. Binoculars show it as a neat pair of 5th-magnitude white stars. Each of these has a closer companion that is brought into view by telescopes of 60 or 75mm (2.5 or 3in) aperture with high magnification. All four stars are linked by gravity and are in long-term orbit around each other.

Zeta (ζ) Lyrae 🔭 🔭 A double star with components of 4th and 6th magnitudes, easily divided by binoculars or small telescopes.

M57 (The Ring Nebula) 🔭 💻 A planetary nebula that looks like a smoke ring. The Hubble Space Telescope has revealed that the "ring" is in fact a cylinder of gas thrown off from the central star.

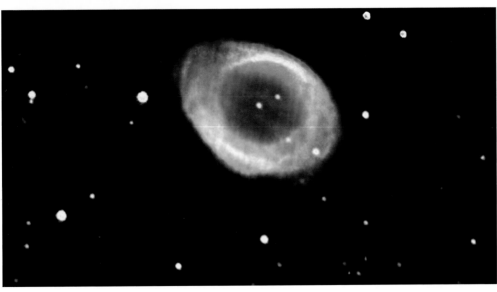

« The Ring Nebula
Through small telescopes, the ring appears as a disc larger than that of the planet Jupiter. Larger apertures are needed to make out the central hole. Photographs and CCD images like this emphasize colours that are not apparent visually.

Northern Hemisphere

Cygnus Cygni (Cyg)

WIDTH 〰〰 **DEPTH** 〰〰 **SIZE RANKING** 16th **FULLY VISIBLE** 90°N–28°S

One of the most prominent constellations of the northern sky, Cygnus contains numerous objects of interest. It is situated in a rich area of the Milky Way and represents a swan, the disguise adopted by the Greek god Zeus for one of his illicit seductions. Its main stars are arranged in the shape of a giant cross, hence its popular alternative name of the Northern Cross. Deneb, its brightest star, lies in the tail of the swan (or at the top of the cross, depending how the constellation is visualized). Deneb forms one corner of the northern Summer Triangle of stars, completed by Vega and Altair.

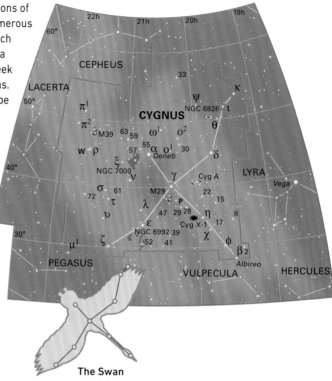

The Swan

⌃ Cygnus in the night sky
The main stars of Cygnus form a noticeable cross shape, and as a result the constellation is popularly known as the Northern Cross.

⌃ The North America Nebula
NGC 7000 is often called the North America Nebula, on account of its shape. Its full majesty is apparent only on long-exposure photographs.

》FEATURES OF INTEREST

Alpha (α) Cygni (Deneb) 👁 An immensely luminous supergiant. At magnitude 1.3 and about 1,400 light-years away, it is the most distant first-magnitude star.

Beta (β) Cygni (Albireo) ⚲ A beautiful coloured double star in the beak of the swan. The two stars can just be seen separately through binoculars, if steadily mounted, and are an easy target for a small telescope. The brighter star, magnitude 3.1, is orange and contrasts strikingly with the fainter one, magnitude 5.1, which is blue-green.

Omicron-1 (o¹) Cygni ⚹ ⚲ A wide double for binoculars consisting of a 4th-magnitude orange star with a bluish 5th-magnitude companion.

Chi (χ) Cygni 👁 ⚹ ⚲ A pulsating red giant of the same type as Mira. At its brightest, which it reaches every 13 months or so, it can appear of 3rd magnitude and is hence easily visible to the naked eye, but it fades to as faint as 14th magnitude so its full range can be followed only through telescopes.

61 Cygni ⚲ An easy double for small telescopes, consisting of two orange dwarfs of 5th and 6th magnitudes that orbit each other every 680 years.

M39 ⚹ ⚲ A large open cluster in northern Cygnus. It covers an area of similar size to the full Moon and is easily visible through binoculars. It is triangular in shape with a double star at the centre.

NGC 6826 (The Blinking Planetary) ⚲ A planetary nebula with a blue-green disc similar in size to the outline of Jupiter. It is popularly known as the Blinking Planetary, because it appears to blink on and off as the observer looks alternately at it and then to one side.

NGC 6992 (Veil Nebula) ⚲ 🖥 A huge loop of glowing gas, the remains of a star that exploded as a supernova thousands of years ago.

NGC 7000 ⚹ ⚲ 🖥 A large glowing gas cloud near Deneb. It can be glimpsed through binoculars on clear, dark nights.

Cygnus Rift (The Northern Coalsack) 👁 ⚹ A cloud of dust that divides the Milky Way in two.

Cygnus A and Cygnus X-1 🖥 Two objects that are beyond the reach of amateur observers but are of considerable astrophysical interest. Cygnus A is a powerful radio source, the result of two galaxies colliding billions of light-years away in distant space. Cygnus X-1 is an intense X-ray source near Eta (η) Cygni and is thought to be a black hole in orbit around a 9th-magnitude blue supergiant about 6,000 light-years away in our own galaxy.

⌃ The Veil Nebula
The remains of an exploded star form a wispy loop of glowing gas under the southern wing of Cygnus. Six Moon diameters wide, the nebula is best seen on photographs but, under ideal conditions, the brightest part, NGC 6992, can just be made out with binoculars and small telescopes.

Northern Hemisphere

Andromeda Andromedae (And)

WIDTH 〰〰 **DEPTH** 〰〰 **SIZE RANKING** 19th **FULLY VISIBLE** 90°N–37°S

A constellation of the northern sky, Andromeda adjoins Pegasus. One of the original Greek constellations, it depicts the daughter of the mythical Queen Cassiopeia, who is represented by the constellation to its north. Andromeda's head is marked by Alpha (α) Andromedae, which also forms one corner of the Square of Pegasus – long ago, this star was regarded as being shared with Pegasus.

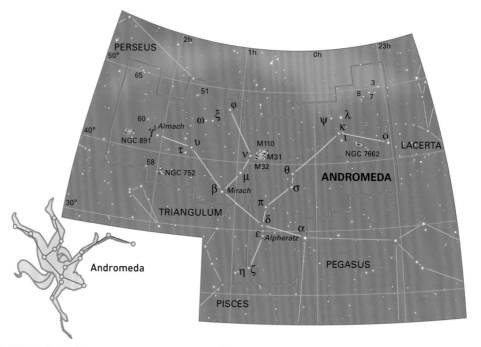

☑ The Andromeda Galaxy
The most distant object normally visible to the naked eye, this galaxy appears as a faint misty patch, elongated because it is tilted at an angle to us. Binoculars reveal more of its extent, and through telescopes signs of its spiral arms can be traced.

⟫ FEATURES OF INTEREST

Gamma (γ) Andromedae 🔭 A showpiece double star consisting of an orange giant star of magnitude 2.2 with a contrasting blue companion of magnitude 4.8, easily divided with small telescopes.

M31 (The Andromeda Galaxy) 👁 📷 🔭 A huge spiral galaxy about 2.5 million light-years away, similar in size and nature to our own galaxy.

NGC 752 📷 🔭 An open star cluster visible through binoculars and spread over an area larger than the full Moon. Small telescopes are needed to resolve its individual stars, which are of 9th magnitude and fainter.

NGC 7662 (The Blue Snowball) 🔭 One of the easiest planetary nebulae to see. Through a small telescope with low magnification it looks like a bright blue star, but higher powers clearly show its disc shape.

Lacerta Lacertae (Lac)

WIDTH 📏 **DEPTH** 📏 **SIZE RANKING** 68th **FULLY VISIBLE** 90°N–33°S

This small, unremarkable northern constellation consists of a zig-zag of faint stars that fills in the region between Andromeda and Cygnus, squeezed into a narrow space like a lizard between two rocks. Its brightest star is Alpha (α) Lacertae, magnitude 3.8. Lacerta is one of the constellations invented in the late 17th century by the Polish astronomer Johannes Hevelius. Although the constellation contains no objects of note for amateur astronomers, it does have one claim to fame, having given its name to a whole new class of galaxies, the so-called BL Lacertae or BL Lac objects.

Northern Hemisphere

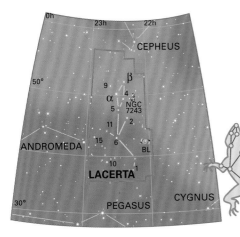

⟫ FEATURES OF INTEREST

BL Lacertae ⊟ The prototype of a distinctive class of galaxies that have bright, active nuclei, related to quasars. They are active galaxies from which a jet of gas shoots out directly towards Earth. Once the nucleus of BL Lacertae was thought to be a peculiar 14th-magnitude variable star and was classified as such.

The Lizard

Triangulum Trianguli (Tri)

WIDTH 🖐 **DEPTH** 📏 **SIZE RANKING** 78th **FULLY VISIBLE** 90°N–52°S

A small constellation lying between Andromeda and Aries, Triangulum consists of little more than a triangle of three stars. It is one of the constellations known to the ancient Greeks, who visualized it as either the Nile delta or the island of Sicily.

Northern Hemisphere

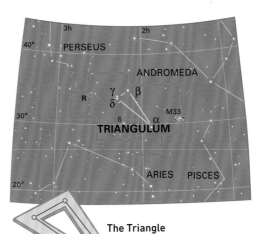

The Triangle

⟫ FEATURES OF INTEREST

M33 🔭 ✴ ⊟ The third-largest member of our Local Group of galaxies, about one-third the diameter of the Andromeda Galaxy. On dark, clear nights M33 can be seen through binoculars and telescopes as a large, pale patch similar in size to the full Moon. Larger amateur telescopes are needed to make out its spiral arms, where long-exposure photographs and CCD images show pink patches of gas. M33 lies about 2.7 million light-years away.

Bright neighbour
The brightness of the Andromeda Galaxy means that it has long been known to astronomers, the first recorded observations dating from the 10th century CE. It has been intensively studied ever since. In the 20th century, Edwin Hubble's studies of the galaxy led to greatly increased estimates of the Universe's overall size.

**Northern
Hemisphere**

Perseus Persei (Per)

WIDTH 〰〰〰 **DEPTH** 〰〰 **SIZE RANKING** 24th **FULLY VISIBLE** 90°N–31°S

A prominent northern constellation, Perseus lies in the Milky Way between Cassiopeia and Auriga. One of the original Greek constellations, it represents the character who was sent to slay Medusa, the Gorgon. In the sky he is depicted holding the Gorgon's head, which is marked by Algol, a famous variable star. Every August the Perseid meteors appear to radiate from the constellation.

The Victorious Hero

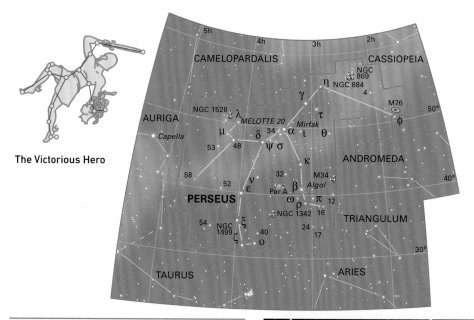

⟫ FEATURES OF INTEREST

Alpha (α) Persei ◉ ♐ The brightest star in Perseus, magnitude 1.8, at the centre of a group known as the Alpha Persei Cluster or Melotte 20.

Beta (β) Persei (Algol) ◉ ♐ The most famous eclipsing binary, and the first to be discovered. Algol fades by one-third in brightness, from magnitude 2.1 to 3.4, every 69 hours as one star eclipses the other, returning to normal after 10 hours. Predictions of Algol's eclipses can be found in astronomy yearbooks and magazines.

Rho (ρ) Persei ◉ ♐ A red giant that fluctuates by about 50 per cent in brightness, between magnitudes 3.3 and 4.0, every seven weeks or so.

M34 ♐ ✸ A scattered open cluster of several dozen stars near the border with Andromeda.

NGC 869 and NGC 884 (The Double Cluster) ◉ ♐ ✸ Two adjacent open clusters visible to the naked eye as a brighter patch in the Milky Way near the border with Cassiopeia. They are well seen through binoculars and small telescopes.

⌃ The Double Cluster
Each cluster contains hundreds of stars of 7th magnitude and fainter, covering a similar area of sky to the full Moon. Both clusters lie over 7,000 light-years away in the Perseus spiral arm of our galaxy.

Aries Arietis (Ari)

WIDTH 🖐🖐 **DEPTH** 🖐🖐 **SIZE RANKING** 39ᵗʰ **FULLY VISIBLE** 90°N–58°S

Aries is a constellation of the zodiac, lying between Pisces and Taurus, but is not particularly prominent. Its most recognizable feature is a crooked line of three stars, Alpha (α), Beta (β), and Gamma (γ) Arietis. In Greek mythology, it represents the ram whose golden fleece hung on an oak tree in Colchis on the eastern shore of the Black Sea. Jason and the Argonauts undertook their epic voyage to bring this fleece back to Greece.

Northern Hemisphere

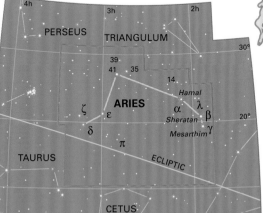

The Ram

☑ The crooked line of Aries
This line is formed by Aries's three brightest stars – Alpha (α), Beta (β), and Gamma (γ). Also visible in this photograph are Venus (below) and Mars to the left of Aries.

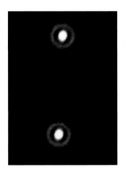

⌃ Gamma Arietis
To the naked eye, Gamma (γ) Arietis appears of magnitude 3.9, but through a small telescope, it is seen to be a striking double. This CCD image shows its two very similar components.

» FEATURES OF INTEREST

Gamma (γ) Arietis ⚹ A double star with nearly identical white components of magnitudes 4.6 and 4.7, easily divisible through small telescopes.

Lambda (λ) Arietis 🔭⚹ A 5th-magnitude star with a 7th-magnitude companion, visible through large binoculars or a small telescope.

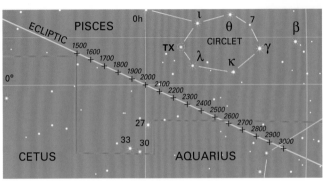

First point of Aries 1500–3000 CE

« The first point of Aries
In ancient Greek times, over 2,000 years ago, the vernal equinox – the point at which the ecliptic crosses the celestial equator – lay near the border of Aries and Pisces. The effect of precession (see p.151) has now moved the vernal equinox almost into Aquarius, but it is still called the first point of Aries. This chart shows its movement between the years 1500 and 3000 CE.

Taurus Tauri (Tau)

WIDTH 🖐🖐🖐 **DEPTH** 🖐🖐🖐 **SIZE RANKING** 17th **FULLY VISIBLE** 88°N–58°S

Northern Hemisphere

Representing a bull, Taurus is a large and prominent constellation of the zodiac that contains a wealth of objects for instruments of all sizes, most notably the Pleiades and Hyades star clusters and the Crab Nebula. The Hyades cluster outlines the bull's face while the constellation's brightest star, Aldebaran, marks its glinting eye. The Taurid meteors appear to radiate from a point south of the Pleiades in early November each year. In Greek mythology, Taurus was said to represent the disguise adopted by the god Zeus to carry off Princess Europa of Phoenicia to Crete, swimming across the Mediterranean with her on his back.

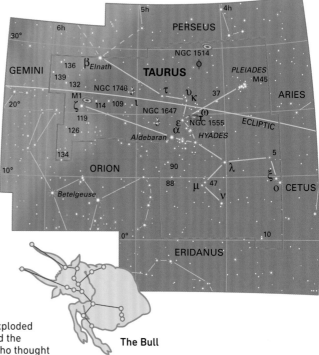

The Bull

⬒ The Crab Nebula
The beautiful remains of a star that exploded as a supernova, this object was named the Crab Nebula in 1844 by Lord Rosse, who thought the filaments of gas resembled the legs of a crab.

⬆ The Hyades cluster
The V-shaped Hyades star cluster forms the face of the bull. Its brightest stars are easily visible to the naked eye. Aldebaran (on the left) is not a member of the cluster but lies in the foreground.

FEATURES OF INTEREST

Alpha (α) Tauri (Aldebaran) 👁🔭 The brightest star in the constellation, magnitude 0.9. It is a red giant whose colour is clearly apparent to the eye. Although Aldebaran appears to be part of the Hyades star cluster, it is only 67 light-years away, less than half the cluster's distance, and so in fact is superimposed by chance.

Kappa (κ) Tauri 🔭 A wide double with components of 4th and 5th magnitudes, lying on the outskirts of the Hyades.

Lambda (λ) Tauri 👁 🔭 An eclipsing binary of the same type as Algol. It varies between magnitudes 3.4 and 3.9 in a cycle lasting just under four days.

Sigma (σ) Tauri 🔭 A wide double star in the Hyades, divisible with binoculars. Both components are of 5th magnitude.

M1 (The Crab Nebula) 🔭 ⬒ The remains of a star that exploded as a supernova. The event would have been witnessed from Earth in 1054 CE. Through small telescopes, it appears as a faint elliptical glow several times larger than the disc of Jupiter. However, large apertures are needed to make out the detail that the Irish astronomer Lord Rosse saw in 1844, when he gave the nebula its name.

M45 (The Pleiades) 👁 🔭 A large and prominent star cluster popularly known as the Seven Sisters. The brightest member is Alcyone, magnitude 2.9, near the centre. Six – rather than seven – members of the Pleiades can be detected with average eyesight, while through binoculars the cluster is a dazzling sight, with many members just beyond naked-eye visibility coming into view. In photographs and CCD images the stars appear in a bright blue haze of dust. In all, the Pleiades cluster spans an area of sky three times the apparent width of the full Moon. The cluster lies about 440 light-years from Earth.

The Hyades 👁 🔭 A large star cluster whose main stars form the shape of a V ten times the apparent width of the Moon. At 150 light-years away, it is the nearest major star cluster to us. Over a dozen members can be seen with the unaided eye, and binoculars bring dozens more into view. On the southern arm of the V is a wide double, Theta (θ) Tauri. The brighter of the pair of stars, magnitude 3.4, is the brightest member of the Hyades.

⌃ The Seven Sisters
Photographs and CCD images show that the Pleiades is immersed in a blue haze of dust that reflects light from the brightest stars.

**Northern
Hemisphere**

Gemini Geminorum (Gem)

WIDTH 🖐🖐 **DEPTH** 🖐🖐 **SIZE RANKING** 30th **FULLY VISIBLE** 90°N–55°S

This prominent constellation of the zodiac is easily identifiable by its two brightest stars, Castor and Pollux, named after the twins of Greek mythology whom the constellation represents. Castor and Pollux mark the heads of the twins but are far from identical. Pollux, the brighter of the pair at magnitude 1.1, is an orange giant 34 light-years away whereas Castor is blue-white,

magnitude 1.6, and 52 light-years from us. In mid-December each year, the Geminid meteors radiate from a point near Castor.

❯❯ FEATURES OF INTEREST

Alpha (α) Geminorum (Castor) 🏹 ⚖
A remarkable multiple star. To the eye it appears as a single star of magnitude 1.6, but a small telescope with high magnification divides it into a sparkling blue-white duo of 2nd and 3rd magnitudes. These form a genuine binary with an orbital period of 460 years. There is also a 9th-magnitude red dwarf companion. All three stars are spectroscopic binaries, making a total of six stars in the system.

Zeta (ζ) Geminorum 👁 📷 A Cepheid variable that ranges between magnitudes 3.6 and 4.2 in a cycle that lasts 10.2 days.

Eta (η) Geminorum 👁 📷 A red giant variable that ranges between magnitudes 3.1 and 3.9.

M35 📷 🏹 An open star cluster lying at the feet of the twins, easily found with binoculars.

NGC 2392 (The Eskimo Nebula) 🏹 🖥
A planetary nebula visible through small telescopes as a bluish disc similar in size to the globe of Saturn. Larger apertures, and CCD images, show a surrounding fringe of gas like the fur on an eskimo's parka.

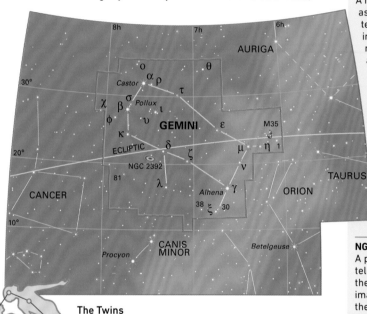

The Twins

❯❯ **Open cluster M35**
Under clear skies, the M35 star cluster can just be glimpsed with the naked eye. Through binoculars, it appears as an elongated patch of starlight of the same apparent width as the full Moon. When seen through small telescopes, its individual stars appear to form chains.

Cancer Cancri (Cnc)

WIDTH 🖑🖑 **DEPTH** 🖑🖑 **SIZE RANKING** 31st **FULLY VISIBLE** 90°N–57°S

Cancer, lying between Gemini and Leo, is the faintest of the 12 zodiacal constellations, but it includes a major star cluster, Praesepe, also known as the Beehive or Manger. The stars Gamma (γ) and Delta (δ) Cancri, north and south of the cluster, represent two donkeys feeding at the manger. In Greek mythology Cancer was the crab that attacked Hercules during his fight with the multi-headed Hydra but was crushed underfoot, a minor role that befits such a faint constellation.

Northern Hemisphere

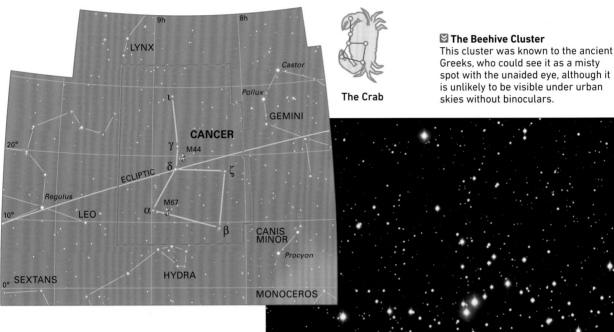

The Crab

☑ The Beehive Cluster
This cluster was known to the ancient Greeks, who could see it as a misty spot with the unaided eye, although it is unlikely to be visible under urban skies without binoculars.

》 FEATURES OF INTEREST

Zeta (ζ) Cancri 🔭 A double star for small telescopes. Its 5th- and 6th-magnitude stars form a binary with an orbital period of over 1,000 years.

Iota (ι) Cancri 🔭 🔭 A 4th-magnitude yellow giant with a nicely contrasting 7th-magnitude blue-white companion that is just detectable in 10 x 50 binoculars and easy to see in small telescopes.

M44 (Praesepe, The Beehive, The Manger) 🔭 🔭 A large open cluster at the heart of Cancer.

M67 🔭 🔭 An open cluster in southern Cancer, smaller and denser than M44.

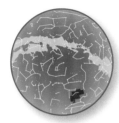

Leo Minor Leonis Minoris (LMi)

WIDTH 🖐🖐 **DEPTH** 🖐 **SIZE RANKING** 64th **FULLY VISIBLE** 90°N–48°S

This small and insignificant constellation north of Leo represents a lion cub, although such a shape is not suggested by its stars. Leo Minor was one of the constellations introduced at the end of the 17th century by the Polish astronomer Johannes Hevelius. It contains no objects of interest for users of small telescopes.

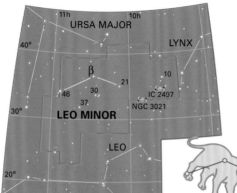

The Little Lion

❯❯ FEATURES OF INTEREST

Leo Minor has one unusual feature, the fact that the constellation has no star labelled Alpha (α), although its second-brightest star is labelled Beta (β). This is due to an error by the 19th-century English astronomer Francis Baily, who assigned Greek letters to Leo Minor's stars. In doing so, he overlooked 46 LMi, which should be Alpha.

❯❯ Distant spiral
NGC 3021 is a small spiral galaxy about 100 million light-years away in Leo Minor. It shows up well in this image from the Hubble Space Telescope, but is far too faint to be seen through most amateur instruments.

Coma Berenices Comae Berenices (Com)

WIDTH 🖑🖑 **DEPTH** 🖑🖑 **SIZE RANKING** 42nd **FULLY VISIBLE** 90°N–56°S

Coma Berenices is a faint but nonetheless interesting northern constellation, between Leo and Boötes. It represents the flowing locks of Queen Berenice of Egypt, which she cut off as a tribute to the gods for the safe return of her husband from battle. Coma Berenices was made into a separate constellation in the mid-16th century by the German cartographer Caspar Vopel. Before then, its stars were regarded as forming the tail of Leo. Numerous galaxies inhabit the southern part of the constellation, most of them members of the Virgo Cluster, such as M85, M88, M99, and M100.

Northern Hemisphere

》 FEATURES OF INTEREST

The Coma Star Cluster 👁 🔭 A large open cluster of faint stars to the south of Gamma (γ) Comae Berenices. Binoculars show the cluster to best advantage.

M64 (The Black Eye Galaxy) 🔭 A spiral galaxy, visible as an elliptical patch of light in small telescopes. A large dust cloud near the nucleus gives the galaxy its popular name, but an aperture of 150mm (6in) or more is needed to see it well.

NGC 4565 🔭 🖥 Another spiral galaxy, presented edge-on to us and so appearing long and thin. Telescopes with apertures of 100mm (4in) will show it.

Berenice's Hair

The Coma Star Cluster
Also known as Melotte 111, the Coma Star Cluster consists of a V-shaped grouping of several dozen stars. Although visible to the naked eye, it is best seen through binoculars.

Northern Hemisphere

Leo Leonis (Leo)

WIDTH 🖐🤏🤏 **DEPTH** 🖐🤏🤏 **SIZE RANKING** 12th **FULLY VISIBLE** 82°N–57°S

Leo is a large constellation of the zodiac and one of the easiest to recognize, because its outline really does bear a marked resemblance to a lion. The pattern of six stars that marks the lion's head and chest, shaped like a back-to-front question mark, is known as the Sickle. In Greek mythology, Leo represents the lion with the impenetrable hide that was slain by Hercules in the first of his 12 labours. The Leonid meteors radiate from the region of the Sickle each November.

The Sickle of Leo

The six stars – Epsilon (ε), Mu (μ), Zeta (ζ), Gamma (γ), Eta (η), and Alpha (α) Leonis (Regulus) – that form the Sickle of Leo are clearly visible at the right of this photograph. The brightest, Alpha (α) Leonis, marks the end of the handle of the Sickle.

The Lion

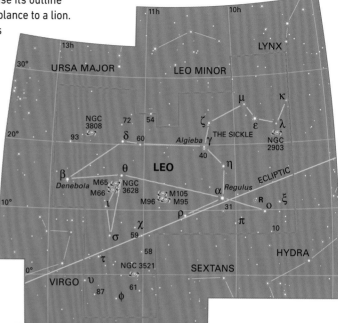

FEATURES OF INTEREST

Alpha (α) Leonis (Regulus) 👁 🔭 ✈ The brightest star in the constellation at magnitude 1.4. Small telescopes or binoculars show a wide companion of 8th magnitude.

Gamma (γ) Leonis (Algieba) ✈ A golden pair, of magnitudes 2.4 and 3.6, that can be divided by small telescopes with high magnification. Both are orange giants orbiting each other every 550 years or so.

Zeta (ζ) Leonis 🔭 A wide triple in the Sickle of Leo. Zeta is of 3rd magnitude, with unrelated 6th-magnitude stars to the north and south of it that are visible through binoculars.

M65 and M66 ✈ A pair of spiral galaxies lying beneath the hind quarters of Leo that can be glimpsed through small telescopes. They are tilted at steep angles to us and so appear elongated.

M95 and M96 ✈ A fainter pair of spiral galaxies, visible through moderate-sized telescopes.

Virgo Virginis (Vir)

WIDTH ༥ ༥ ༥ ༥ ༥   **DEPTH** ༥ ༥ ༥   **SIZE RANKING** 2nd   **FULLY VISIBLE** 67°N–75°S

The largest constellation of the zodiac and the second-largest overall, Virgo is shaped like a sloping Y with its brightest star, Spica, at the southern tip. On its northern border lies the nearest large cluster of galaxies to us, some 50 million light-years away. The Sun is within the boundaries of Virgo at the September equinox each year.

Northern Hemisphere

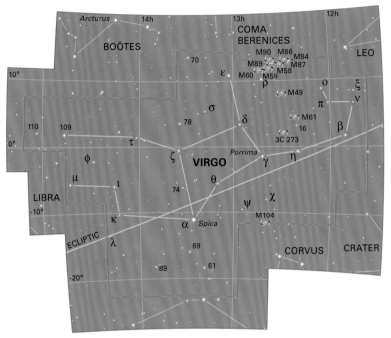

The Virgin

◀ Galaxy M87
A giant elliptical galaxy near the core of the Virgo cluster of galaxies, M87 has a highly active nucleus that is expelling a jet of gas, just visible here at the 2 o'clock position. M87 is a strong radio source.

▼ The Sombrero Galaxy
Virgo's best-known galaxy is not part of the Virgo cluster, lying at only about two-thirds of the distance from Earth to the cluster. The dark band across the central nucleus is created by dust in the spiral arms.

⟫ FEATURES OF INTEREST

Alpha (α) Virginis (Spica) 👁 The constellation's brightest member, magnitude 1.0. It is a blue-white star 260 light-years away.

Gamma (γ) Virginis 🔭 A 3rd-magnitude binary star whose companions orbit each other every 169 years. Currently moving apart, they are easily divisible in small telescopes.

The Virgo Cluster 🔭 Within the bowl of Virgo's Y lie numerous members of the Virgo cluster of galaxies. The brightest galaxies are giant ellipticals, notably M49, M60, M84, M86, and M87.

M104 (The Sombrero Galaxy) 🔭 A spectacular spiral galaxy oriented almost edge-on to us.

NGC 5793

Dark lanes of dust are seen crossing the bright nucleus of the spiral galaxy NGC 5793, in Libra, in this Hubble Space Telescope image. It is of the type known as a Seyfert Galaxy, which have highly luminous cores powered by supermassive black holes. NGC 5793 lies over 150 million light-years away.

Southern Hemisphere

Libra Librae (Lib)

WIDTH ✋ **DEPTH** ✋ **SIZE RANKING** 29th **FULLY VISIBLE** 60°N–90°S

A constellation of the zodiac, between Virgo and Scorpius, Libra represents the scales of justice held by Virgo, although the ancient Greeks visualized the constellation as the claws of the neighbouring scorpion, Scorpius. As a result, its two brightest stars have names that mean the northern and southern claw.

⟩⟩ FEATURES OF INTEREST

Alpha (α) Librae (Zubenelgenubi, The Southern Claw) 👁 📷 A wide double star with components of 3rd and 5th magnitudes, easily divided with binoculars or sharp eyesight.

Beta (β) Librae (Zubeneschamali, The Northern Claw) 📷 ✦ One of the few stars to display a greenish tinge, noticeable when viewed through binoculars and telescopes.

Iota (ι) Librae 📷 ✦ A binocular double with components of 5th and 6th magnitudes. Small telescopes show that the brighter star has a closer 9th-magnitude companion.

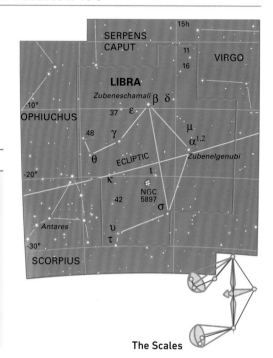

The Scales

Corona Borealis Coronae Borealis (CrB)

WIDTH 🖐 **DEPTH** ✋ **SIZE RANKING** 73rd **FULLY VISIBLE** 90°N–50°S

This small, distinctive constellation between Boötes and Hercules forms a horseshoe of seven stars. One of the constellations known to the ancient Greeks, it represents the crown worn by Princess Ariadne when she married the god Bacchus, who cast it into the sky in celebration.

Northern Hemisphere

⟫ FEATURES OF INTEREST

Zeta (ζ) Coronae Borealis 🔭 A pair of stars of 5th and 6th magnitudes, each blue-white in colour. They form an attractive sight when seen through small telescopes.

Nu (ν) Coronae Borealis 🔭 A wide double star consisting of a pair of 5th-magnitude red giants divisible in binoculars.

R Coronae Borealis 🔭 🔭 A highly luminous yellow supergiant that normally appears of 6th magnitude, but which suffers sudden dips in brightness due to a build-up of sooty particles in its atmosphere. The fades occur every few years and can last for months.

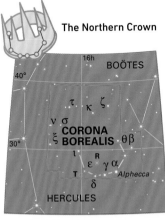

The Northern Crown

◩ Starry crown

The seven main stars of Corona Borealis form an arc like a celestial tiara between Boötes and Hercules. According to Greek myth, Dionysus threw Ariadne's jewelled crown into the sky, where it transformed into stars. The bright star at the bottom of the image is Arcturus in neighbouring Boötes.

Northern Hemisphere

Serpens Serpentis (Ser)

WIDTH 👐👐 **DEPTH** 👐👐 **SIZE RANKING** 23rd **FULLY VISIBLE** 74°N–64°S

Serpens is a unique constellation, for it is split into two – Serpens Caput, the head, and Serpens Cauda, the tail. Both halves count as a single constellation. It represents a huge snake coiled around Ophiuchus, who grasps the body in his left hand and the tail in his right. One of the original 48 Greek constellations, Serpens is linked in legend with the constellation Ophiuchus. The latter represents Asclepius, who was a great healer, reputedly able to revive the dead. In Greek myth, snakes were a symbol of rebirth, owing to the fact that they shed their skins.

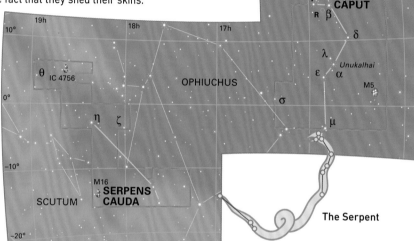

>> Star cluster M16 and nebula
The Eagle Nebula surrounding the star cluster M16 is well seen only through larger apertures or on photographs and CCD images.

>> FEATURES OF INTEREST

Theta (θ) Serpentis 🔭 A pair of 5th-magnitude white stars, easy to split with small telescopes.

M5 🔭 One of the finest globular clusters in the northern sky. Binoculars show it as a hazy patch about half the apparent size of the full Moon. Apertures of 100mm (4in) or so reveal curving chains of stars in its outskirts.

M16 🔭 A star cluster that can be seen easily through binoculars and small telescopes as a hazy patch covering a similar area of sky to the full Moon. It lies within the Eagle Nebula, which was made famous by a spectacular Hubble Space Telescope picture showing dark columns of dust within its glowing gas.

IC 4756 🔭 A good open cluster for binoculars, about twice the size of M16, lying near the tip of the serpent's tail.

Ophiuchus Ophiuchi (Oph)

WIDTH 🖐🖐🖐 **DEPTH** 🖐🖐🖐 **SIZE RANKING** 11th **FULLY VISIBLE** 59°N–75°S

Ophiuchus is a large constellation straddling the celestial equator, representing a man holding a snake. Ophiuchus's head adjoins Hercules in the north while his feet rest on Scorpius, the scorpion, in the south. The Sun passes through Ophiuchus in the first half of December, but despite this the constellation is not regarded as a member of the zodiac. In mythology, Ophiuchus is identified with Asclepius, the Greek god of medicine who had the power to revive the dead. Hades, god of the Underworld, feared that this ability endangered his trade in dead souls and so asked Zeus to strike Asclepius down with a thunderbolt. Zeus placed Asclepius among the stars, where he is seen holding a snake, the symbol of healing.

Southern Hemisphere

⟫ FEATURES OF INTEREST

Rho (ρ) Ophiuchi 🔭 ⚹ An outstanding multiple star. Binoculars will show it as a 5th-magnitude star with a 7th-magnitude companion on either side. Small telescopes with high magnification bring another 6th-magnitude companion into view much closer to the central star.

36 Ophiuchi ⚹ A neat pair of 5th-magnitude orange dwarfs divisible by small telescopes.

70 Ophiuchi ⚹ A beautiful double star, easy for small telescopes. It consists of yellow and orange dwarfs of 4th and 6th magnitudes.

M10 and M12 🔭 ⚹ Two globular clusters detectable with binoculars on a good night. These are the best of the seven globular clusters catalogued by Messier (see p.75) in Ophiuchus.

NGC 6633 🔭 An open cluster of similar apparent size to the full Moon, visible through binoculars.

IC 4665 🔭 A large and scattered open cluster visible with binoculars.

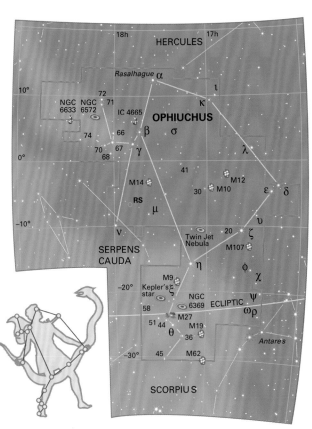

The Serpent Holder

◀◀ The Rho Ophiuchi Nebula
Photographs and CCD images show complex nebulosity in the area around Rho (ρ) Ophiuchi (top of picture) extending southwards to Antares, the bright star at bottom left. None of this can be detected with amateur telescopes, though.

Flying ducks
The open cluster M11 in Scutum contains hundreds of stars. When observed using amateur telescopes, the brightest of them form a fan shape like a flight of ducks, hence its popular name the Wild Duck Cluster. The stars in the cluster are estimated to be about 200 million years old.

Scutum Scuti (Sct)

WIDTH **DEPTH** **SIZE RANKING** 84th **FULLY VISIBLE** 74°N–90°S

Scutum is situated just south of the celestial equator in a rich area of the Milky Way, with Aquila to the north and Sagittarius to the south. It was introduced in the late 17th century by the Polish astronomer Johannes Hevelius under the name *Scutum Sobiescianum*, meaning Sobieski's Shield, to honour the king of Poland, Jan Sobieski. The Scutum star cloud in the north is one of the richest areas of the Milky Way.

Southern Hemisphere

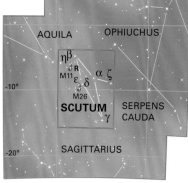

The Shield

》 FEATURES OF INTEREST

Delta (δ) Scuti 👁 📷 ⚊ A pulsating giant, the prototype of a class of stars that fluctuate in brightness very slightly over periods of a few hours.

R Scuti 📷 ✶ An orange supergiant that varies between magnitudes 4.2 and 8.6 every 20 weeks.

M11 (The Wild Duck Cluster) 📷 ✶ A beautiful open cluster, visible through binoculars as a smudgy glow half the apparent width of the full Moon. It is popularly known as the Wild Duck Cluster because, when seen through small telescopes, its stars form a fan shape, like a formation of ducks in flight.

M26 ✶ Another open cluster, fainter than M11 and best seen in small telescopes.

Sagitta Sagittae (Sge)

WIDTH 🖑🖑 **DEPTH** 🖑 **SIZE RANKING** 86ᵗʰ **FULLY VISIBLE** 90°N–69°S

Faint and easily overlooked (it is the third-smallest constellation in the sky), Sagitta lies in the Milky Way south of Vulpecula and north of Aquila. It was known to the ancient Greeks, who said that it represented an arrow shot by either Apollo, Hercules, or Eros. It is distinctly arrow-shaped, with its brightest star, Gamma (γ) Sagittae, magnitude 3.5, marking the arrow's tip. Alpha (α) Sagittae is only of magnitude 4.4, the same brightness as Beta (β) Sagittae.

》 FEATURES OF INTEREST

Zeta (ζ) Sagittae 🏹 A 5th-magnitude star with a 9th-magnitude companion visible in small telescopes.

S Sagittae 🏹🏹 A Cepheid variable that halves in brightness from magnitude 5.2 to 6.0 and then recovers again every 8.4 days.

M71 🏹🏹 A modest globular cluster, detectable in binoculars but better seen with a telescope. It lacks the central condensation typical of most globulars and so looks more like a dense open cluster.

Northern Hemisphere

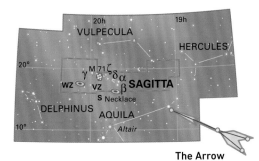

The Arrow

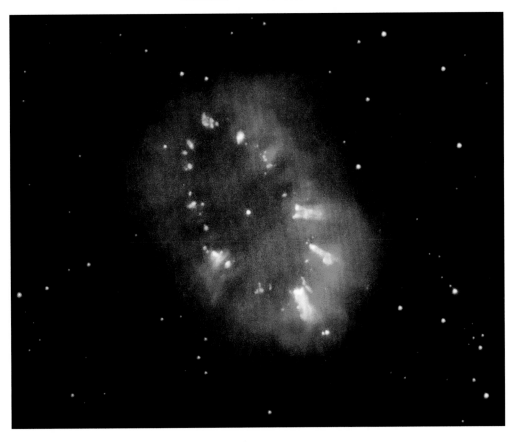

《 Gaseous necklace
Knots of glowing gas are strung along the ring-shaped Necklace Nebula, which is located some 15,000 light-years away in Sagitta. The loop of gas, approximately 2 light-years in diameter, was shed by a close pair of stars, which are seen as a dot at the centre of this image. The nebula is very faint and can be seen only through large professional telescopes. This view was taken through the Hubble Space Telescope.

Northern Hemisphere

Aquila Aquilae (Aql)

WIDTH ✋✋ **DEPTH** ✋✋ **SIZE RANKING** 22nd **FULLY VISIBLE** 78°N–71°S

Aquila lies on the celestial equator in a rich area of the Milky Way, with Cygnus to the north and Scutum and Sagittarius to the south. It represents a flying eagle. Aquila's brightest star, Altair, magnitude 0.8, lies in the eagle's neck and forms one corner of the northern Summer Triangle of stars, completed by Vega and Deneb. Altair is flanked by Beta (β) Aquilae, or Alshain, of 4th magnitude, and Gamma (γ) Aquilae, Tarazed, of 3rd magnitude, forming a distinctive trio. In Greek mythology, the eagle was the bird that carried the thunderbolts that the god Zeus hurled at his enemies. One story says that Zeus either sent an eagle, or turned himself into an eagle, to carry the shepherd boy Ganymede up to Mount Olympus to serve the gods. Ganymede is represented by the adjoining constellation Aquarius.

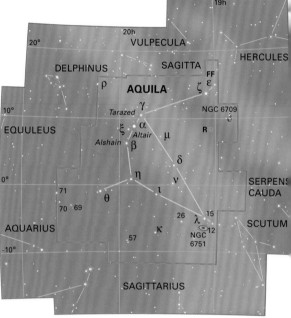

The Eagle

》 Altair, Tarazed, and Alshain
Altair, the brightest star in Aquila, is flanked by the stars Tarazed (Gamma Aquilae) to its north and Alshain (Beta Aquilae) to the south. Tarazed, of 3rd magnitude, is an orange giant and has a noticeable orange colour.

》 FEATURES OF INTEREST

Eta (η) Aquilae 👁 📷 One of the brightest Cepheid variables. It cycles between magnitudes 3.5 and 4.3 every 7.2 days. Its distance is estimated at 1,400 light-years.

15 and 57 Aquilae 🔭 Two easy double stars for small telescopes. 15 Aquilae has components of 5th and 7th magnitudes. In 57 Aquilae, both components are of 6th magnitude.

FF Aquilae 👁 📷 A Cepheid variable. Its variations, between magnitudes 5.2 and 5.5 every 4.5 days, can easily be followed through binoculars.

R Aquilae 📷 🔭 A variable red giant of the same type as Mira. At its brightest, which it reaches every nine months, it is visible in binoculars.

NGC 6709 📷 🔭 A modest open cluster, irregular in shape, containing stars of magnitude 9 and fainter.

Vulpecula Vulpeculae (Vul)

WIDTH 〰〰〰 **DEPTH** 〰 **SIZE RANKING** 55th **FULLY VISIBLE** 90°N–61°S

Vulpecula is a small and faint northern constellation lying in the Milky Way at the head of Cygnus. It was introduced in the late 17th century by the Polish astronomer Johannes Hevelius under the name Vulpecula cum Anser, the Fox with Goose, which has since been simplified to just the Fox. Despite its obscurity, Vulpecula contains two unmissable, distinctively shaped objects for binocular users: the Dumbbell Nebula and Brocchi's Cluster (also known as the Coathanger).

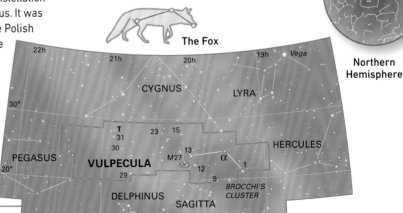

The Fox

Northern Hemisphere

》 FEATURES OF INTEREST

Alpha (α) Vulpeculae 📷 The constellation's brightest star, a 4th-magnitude red giant. Through binoculars, a 6th-magnitude orange star is visible nearby, but the two lie at different distances and are hence unrelated.

M27 (The Dumbbell Nebula) 📷 ✈ 🖥 Widely regarded as the easiest planetary nebula to see. Through binoculars, M27 becomes visible as a rounded patch about one-third the apparent size of the full Moon. Larger instruments and long-exposure photographs show the twin-lobed shape that gives rise to its popular name.

Brocchi's Cluster (The Coathanger) 📷 One of the binocular treasures of the sky, a grouping of 10 stars from 5th to 7th magnitudes, nicknamed the Coathanger because of its shape. A line of six stars forms the bar of the hanger while the remaining four are the hook. The stars are all unrelated and so do not form a true cluster. The Coathanger is therefore the delightful product of a chance alignment.

The Dumbbell Nebula
Through binoculars, the Dumbbell Nebula is visible as a misty patch. Colours in the gases are apparent only on photographs and CCD images such as this.

Brocchi's Cluster
The 10 stars that form the distinctive upside-down-coathanger shape of Brocchi's Cluster are clearly visible at the centre of this photograph. It lies in the southern part of Vulpecula.

Northern Hemisphere

Delphinus Delphini (Del)

WIDTH 🖑 **DEPTH** 🖑🖑 **SIZE RANKING** 69th **FULLY VISIBLE** 90°N–69°S

Delphinus, a small but distinctive constellation shaped like a flag on a stick, is tucked between Aquila and Pegasus, just north of the celestial equator. It is one of the original Greek constellations and bears a fair resemblance to a dolphin leaping from the waves. According to myth, it represents either the dolphin that saved Arion, a celebrated poet and musician, from drowning when he was attacked by robbers on a ship or the dolphin sent by Poseidon to bring the sea nymph Amphitrite for him to marry. Four stars, all of 4th magnitude, form a distinctive diamond shape at the head of the dolphin. This used to be popularly known as Job's Coffin.

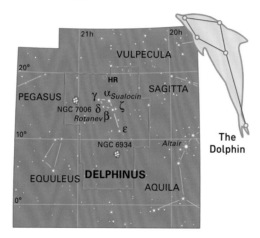

The Dolphin

⟫ FEATURES OF INTEREST

Gamma (γ) Delphini 🏹 An attractive orange and yellow double with components of 4th and 5th magnitudes, divisible with small telescopes. Both stars lie approximately 125 light-years from Earth. Adding to the interest of this part of the sky, a fainter and closer double star, known as Struve 2725, can be seen in the same field of view. Its components, both of 8th magnitude, are also divisible with small telescopes.

⟫ NGC 6934
NGC 6934, a globular cluster in Delphinus, is visible through amateur telescopes as a hazy patch, but seen through the Hubble Space Telescope it breaks up into a ball of glittering stars. It lies in the outer reaches of the Milky Way, some 50,000 light-years from Earth.

Equuleus Equulei (Equ)

WIDTH 🖐 **DEPTH** 🖐 **SIZE RANKING** 87th **FULLY VISIBLE** 90°N–77°S

Equuleus is the second-smallest constellation. It represents the head of a small horse or foal, lying next to the larger celestial horse, Pegasus. No legends are associated with Equuleus, which is thought to have been added to the sky in the 2nd century CE by Ptolemy, the ancient Greek astronomer who wrote the *Almagest*. Ptolemy's *Almagest* was a compendium of astronomy that included a catalogue of the original Greek constellations.

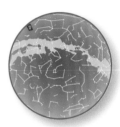

Northern Hemisphere

》 FEATURES OF INTEREST

Gamma (γ) Equulei 🔭 A wide double star with components of 5th and 6th magnitudes, easily divisible with binoculars. The two stars are not related.

1 Equulei 🔭 A 5th-magnitude star with a 7th-magnitude companion visible through small telescopes. On some maps, this star is also labelled as Epsilon (ε) Equulei. The brighter star is, in fact, a true binary, having a faint second companion with an orbital period of 100 years. The stars are too close to be separated with a small aperture.

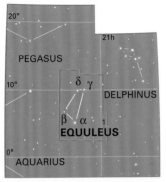

The Foal

☑ Head of the foal

One of the smallest and faintest constellations, Equuleus is easily overlooked. Its four main stars form a quadrilateral near the head of Pegasus, the flying horse. In this image, north is to the left and the distinctive shape of Delphinus is visible above it.

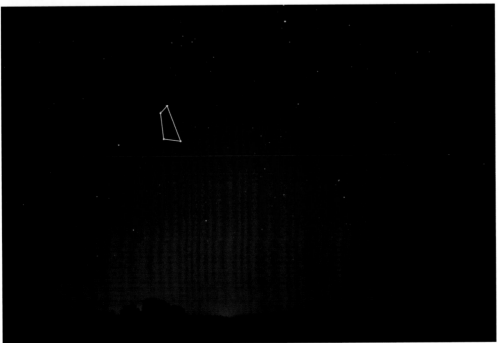

**Northern
Hemisphere**

Pegasus Pegasi (Peg)

WIDTH 〰〰〰 **DEPTH** 〰〰 **SIZE RANKING** 7th **FULLY VISIBLE** 90°N–53°S

Pegasus adjoins Andromeda, north of the zodiacal constellations Aquarius and Pisces. Its most notable feature is the Great Square formed by four stars, although one of these actually belongs to Andromeda. Only the forequarters of the horse are shown in the sky, but even so it is still the seventh-largest constellation. One of the original 48 Greek constellations, Pegasus was the flying horse ridden by the Greek hero Bellerophon. Pegasus was born from the body of Medusa the Gorgon when she was decapitated by Perseus. Sometimes Pegasus is wrongly identified as the steed of Perseus.

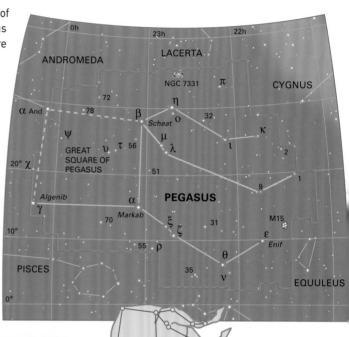

The Winged
Horse

☑ **Globular
Cluster M15**
M15 lies just at the limit of naked-eye visibility under clear conditions. Although the cluster is some 30,000 light-years from Earth, it is easy to locate with binoculars or small telescopes, even from towns, appearing like a hazy star.

〉〉 FEATURES OF INTEREST

The Great Square of Pegasus 👁 📷 The box shape formed by Alpha (α), Beta (β), and Gamma (γ) Pegasi, plus Alpha (α) Andromedae. The large area within the box is surprisingly devoid of stars, the brightest being Upsilon (υ) Pegasi, magnitude 4.4.

Beta (β) Pegasi (Scheat) 👁 📷 A red giant that varies irregularly between magnitudes 2.3 and 2.7.

Epsilon (ε) Pegasi 🔭 Jointly the brightest star in the constellation. This yellow star, of magnitude 2.4, has a wide 8th-magnitude companion.

51 Pegasi 👁 📷 A 5th-magnitude star, just outside the Square of Pegasus. It was the first star beyond the Sun confirmed to have a planet in orbit around it. The planet, discovered in 1995, has a mass about half that of Jupiter.

M15 📷 🔭 One of the finest globular clusters in the northern skies, easily found with binoculars.

Aquarius Aquarii (Aqr)

WIDTH 〰〰〰 **DEPTH** 〰〰 **SIZE RANKING** 10th **FULLY VISIBLE** 65°N–86°S

Aquarius is a large constellation of the zodiac, between Capricornus and Pisces. It is visualized as a youth (or, sometimes, an older man) pouring water from a jar. The stars Gamma (γ), Zeta (ζ), Eta (η), and Pi (π) Aquarii make up the water jar, from which a stream of water, represented by more stars, flows southwards to Piscis Austrinus. In Greek mythology, Aquarius represented a beautiful shepherd boy, Ganymede, to whom the god Zeus took a fancy. Zeus sent down his eagle (or, in some stories, turned himself into an eagle) to carry the boy to Mount Olympus, where he became a waiter to the gods. The eagle is represented by nearby Aquila. In early May each year, the Eta Aquariid meteor shower radiates from the area of the jar.

Southern Hemisphere

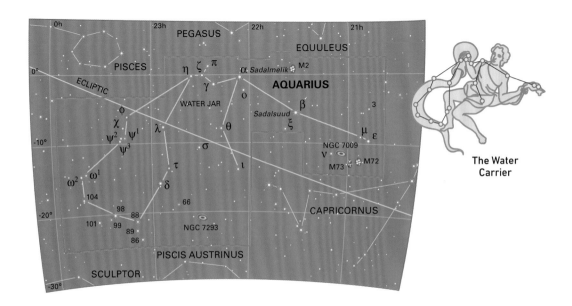

The Water Carrier

⟫ FEATURES OF INTEREST

Zeta (ζ) Aquarii ⚹ A close binary of 4th-magnitude stars just at the limit of resolution in telescopes of 60mm (2.4in) aperture.

M2 📷 ⚹ A globular cluster that appears as a fuzzy star when viewed through binoculars and small telescopes.

NGC 7009 (The Saturn Nebula) 📷 ⚹ 💻 A planetary nebula appearing of similar size to the disc of Saturn through small telescopes. Larger telescopes reveal faint extensions either side, rather like Saturn's rings.

NGC 7293 (The Helix Nebula) 📷 ⚹ 💻 This nebula is estimated to be the closest planetary nebula to us, about 700 light-years away.

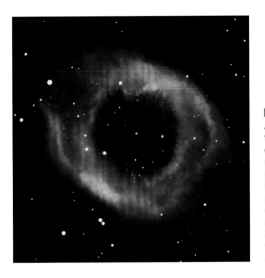

≪ The Helix Nebula
Almost half the width of the full Moon, this is one of the largest planetary nebulae as measured in apparent size. As its light is spread over such a large area, it requires clear, dark skies to be seen. Visually it appears as a pale grey patch, with none of the colours seen on photographs.

**Northern
Hemisphere**

Pisces Piscium (Psc)

WIDTH 🖐🖐🖐🖐 **DEPTH** 🖐🖐🖐 **SIZE RANKING** 14th **FULLY VISIBLE** 83°N–56°S

Pisces is a constellation of the zodiac, but not a particularly prominent one. Its main claim to fame is that it contains the point where the Sun crosses the celestial equator into the northern hemisphere each year. This point, known as the March equinox or the vernal equinox, is where the 0 hours line of right ascension intersects 0° declination (the celestial equator). Because of the slow wobble of the Earth known as precession, this point is gradually moving along the celestial equator and will enter neighbouring Aquarius in about 2600 CE. In Greek myth, Pisces represents Aphrodite and her son Eros, who transformed themselves into fish and plunged into the Euphrates to escape a monster called Typhon.

》 FEATURES OF INTEREST

The Circlet 👁 🔭 The ring of seven stars that describes the body of one of the fish.

Alpha (α) Piscium (Alrescha) 🏃 A close pair of stars of 4th and 5th magnitudes that can be divided with an aperture of 100mm (4in) or more. They form a true binary with a period of over 3,000 years.

Zeta (ζ) Piscium 🏃 A wide double consisting of stars of 5th and 6th magnitudes, divisible with small telescopes.

Psi-1 (Ψ¹) Piscium 🏃 Another double of 5th and 6th magnitudes, divisible with small telescopes.

M74 🏃 🖥 A beautiful face-on spiral galaxy. Small telescopes show it as a rounded glow with a bright centre. The spiral arms show up well only on larger apertures and long-exposure photographs.

The Fishes

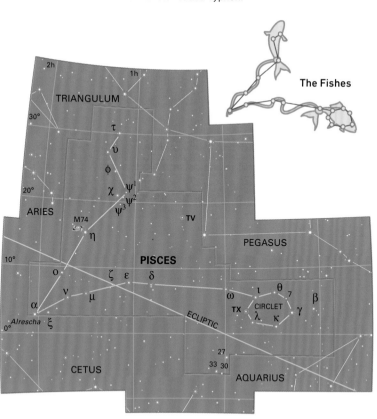

⌃ The Circlet
The seven stars in the Circlet are of 4th and 5th magnitudes. However, TX Piscium (also known as 19 Piscium) is a red giant that varies irregularly between magnitude 4.8 and 5.2. In this photograph, it is on the far left of the ring of stars.

Cetus Ceti (Cet)

WIDTH 〰️〰️〰️ **DEPTH** 〰️🤚🤚 **SIZE RANKING** 4ᵗʰ **FULLY VISIBLE** 65°N–79°S

Cetus is a large but not particularly prominent constellation in the equatorial region of the sky, south of the zodiacal constellations Pisces and Aries. It is home to a famous variable star, Mira, and a peculiar galaxy, M77. Cetus is one of the original 48 Greek constellations listed by Ptolemy in his *Almagest*. In the famous story of Perseus and Andromeda, Cetus was the sea monster which was about to devour Andromeda before it was killed by Andromeda's rescuer, Perseus.

Southern Hemisphere

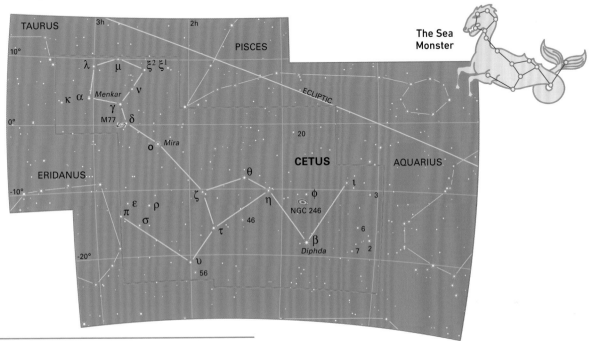

The Sea Monster

》 FEATURES OF INTEREST

Alpha (α) Ceti (Menkar) 🔭 A red giant of magnitude 2.5 with a wide and unrelated 6th-magnitude companion visible with binoculars.

Gamma (γ) Ceti 🔭 A challenging close double. Apertures of at least 60mm (2.4in) with high power will reveal two stars of 4th and 6th magnitudes.

Omicron (o) Ceti (Mira) 👁 🔭 🔭 The brightest and best-known example of a common class of pulsating red giants that fluctuate in size over months or years. Mira has an 11-month cycle and can reach second magnitude at its brightest. At its faintest, it drops to 10th magnitude.

Tau (τ) Ceti 🔭 A star of magnitude 3.5 lying 11.9 light-years away, whose temperature and brightness make it the most Sun-like of the nearby stars.

M77 🔭 🖥 🔭 A Seyfert spiral galaxy oriented face-on to us, just under 50 million light-years away.

⌃ Galaxy M77
This is the brightest example of a Seyfert galaxy, a class of galaxies, related to quasars, which have extremely bright centres. In small telescopes only these central regions will show up, making it appear like a fuzzy star.

Northern Hemisphere

Orion Orionis (Ori)

WIDTH 🖐🖐 **DEPTH** 🖐🖐 **SIZE RANKING** 26th **FULLY VISIBLE** 79°N–67°S

Orion is one of the most glorious constellations, representing a giant hunter or warrior followed by his dogs (the constellations Canis Major and Canis Minor). Its most distinctive feature is Orion's belt, a line of three second-magnitude stars. In Greek mythology, Orion was stung to death by a scorpion. He is placed in the sky opposite Scorpius, so that he sets as the scorpion rises. In late October each year, the Orionid meteors appear to radiate from a point near the border with Gemini.

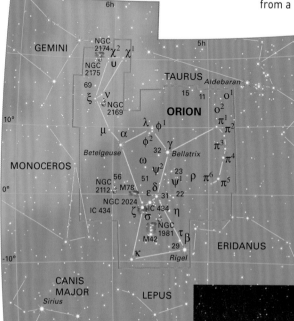

The Hunter

☑ **Armed Hunter**
The most interesting and rewarding area of Orion lies around the sword that hangs from his distinctive belt.

⏩ **The Orion Nebula**
On photographs and CCD images, the nebula appears multi-coloured. Visually, it appears grey-green because the eye is not sensitive to colours in faint objects. On clear nights, it is visible to the naked eye as a hazy patch of light.

≫ FEATURES OF INTEREST

Alpha (α) Orionis (Betelgeuse) 👁 A red supergiant hundreds of times larger than the Sun. It varies irregularly in brightness between magnitudes 0 and 1.3, with an average value around magnitude 0.5. Betelgeuse is about 500 light-years away, closer to us than the other bright stars in Orion.

Beta (β) Orionis (Rigel) 👁 🔭 A luminous blue supergiant of magnitude 0.1. Apart from the rare times when Betelgeuse is at its maximum, Rigel is the brightest star in the constellation. Small telescopes will just pick out a 6th-magnitude companion star from Rigel's surrounding glare.

Delta (δ) Orionis (Mintaka) 🔭 🔭 The star at the northern end of the belt. It has a 7th-magnitude companion that is visible in small telescopes or even binoculars.

Zeta (ζ) Orionis (Alnitak) 🔭 The southernmost belt star. Telescopes of at least 75mm (3in) aperture will reveal its close 4th-magnitude companion.

Theta-1 (θ¹) Orionis (The Trapezium) 🔭 A multiple star at the centre of the Orion Nebula. Through small telescopes, it appears as a group of four stars of 5th to 8th magnitudes. To one side of the nebula lies Theta-2 (θ²) Orionis, a binocular double with components of 5th and 6th magnitudes.

Iota (ι) Orionis 🔭 A double, with stars of 3rd and 7th magnitudes, at the tip of Orion's sword, divisible with small telescopes. A wider double nearby, of 5th and 6th magnitudes, is called Struve 747.

Sigma (σ) Orionis 🔭 An impressive multiple star. A small telescope shows that the main 4th-magnitude star has two 7th-magnitude companions on one side and a closer 9th-magnitude companion on the other. A fainter triple star, Struve 761, is also visible in the same telescopic field of view.

M42 (The Orion Nebula) 👁 🔭 🔭 🖥 An enormous star-forming cloud of gas 1,500 light-years away and covering over two Moon diameters of sky. A northern extension of it bears a separate number, M43, but both are part of the same cloud.

NGC 1977 🔭 An elongated patch of nebulosity surrounding the stars 42 and 45 Orionis.

NGC 1981 🔭 A large, scattered cluster of stars south of Orion's belt. Its brightest members are of 6th magnitude.

The Horsehead Nebula 🖥 🖥 Probably the best-known dark nebula in the sky. It appears silhouetted against IC 434, an area of brighter nebulosity that extends southwards from Zeta (ζ) Orionis.

⌂ **The Horsehead Nebula**
The nebula's silhouette appears like the knight of a chess set against a strip of bright nebulosity, IC 434. Photographs show it well, but to see it visually requires a large telescope and a dark site.

Southern
Hemisphere

Canis Major Canis Majoris (CMa)

WIDTH · **DEPTH** · **SIZE RANKING** 43rd · **FULLY VISIBLE** 56°N–90°S

This prominent constellation contains the brightest star in the entire sky, Sirius, which forms a sparkling triangle with two other first-magnitude stars, Procyon (in Canis Minor) and Betelgeuse (in Orion). Canis Major was known to the ancient Greeks as one of the two dogs following Orion, the hunter.

⟫ FEATURES OF INTEREST

Alpha (α) Canis Majoris (Sirius) 👁 ♨ The brightest star in the sky, at magnitude -1.5. Sirius is among the closest stars to us, 8.6 light-years away. A faint white dwarf, Sirius B, orbits it every 50 years, but this can be seen only with a large telescope.

M41 👁 🔭 🏹 A large open cluster, bright enough to be visible as a hazy patch to the naked eye. Binoculars show its stars scattered over an area about the size of the full Moon. Through telescopes, chains of stars can be seen radiating from its centre.

NGC 2362 🏹 A tight cluster of stars around the 4th-magnitude blue giant Tau (τ) Canis Majoris, best seen through telescopes.

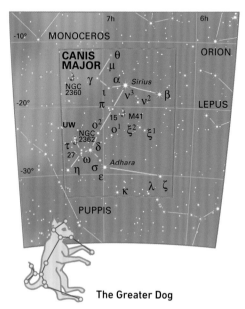

The Greater Dog

Canis Minor Canis Minoris (CMi)

WIDTH · **DEPTH** · **SIZE RANKING** 71st · **FULLY VISIBLE** 89°N–77°S

The smaller of the two dogs of Orion, Canis Minor is easily found from its brightest star, Procyon, which forms a large triangle with Betelgeuse in Orion and Sirius in Canis Major. One of the original Greek constellations, Canis Minor is usually identified with one of the two dogs of Orion. Procyon is a Greek word meaning "before the dog". The name is applied to this star because it rises earlier than the other celestial dog, Canis Major.

Northern
Hemisphere

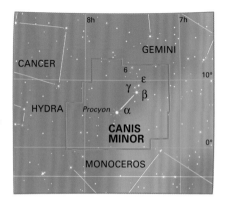

The Little Dog

⟫ FEATURES OF INTEREST

Alpha (α) Canis Minoris (Procyon) 👁 ♨ The eighth-brightest star in the sky with a magnitude of 0.4. At a distance of 11.5 light-years, it is slightly further away than the other dog star, Sirius in Canis Major. Like Sirius, it has a white dwarf partner, Procyon B, but this is so faint and so close to Procyon that it can be seen only with very large telescopes.

Monoceros Monocerotis (Mon)

WIDTH 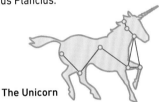 **DEPTH** **SIZE RANKING** 35TH **FULLY VISIBLE** 78°N–78°S

Monoceros is often overlooked, being overshadowed by neighbouring Orion, Gemini, and Canis Major. It is easy to find, lying in the middle of the large triangle formed by brilliant Betelgeuse, Procyon, and Sirius. Although none of the stars of Monoceros are bright, it lies in the Milky Way and contains numerous deep-sky objects of interest. The constellation was introduced in the early 17th century by the Dutch astronomer and cartographer Petrus Plancius.

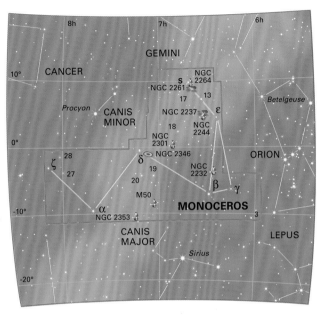

Southern Hemisphere

The Unicorn

The Rosette Nebula
Surrounding the NGC 2244 cluster is a flower-like cloud of gas known as the Rosette Nebula, although this is faint and can be seen well only on photographs and CCD images.

❯❯ FEATURES OF INTEREST

Beta (β) Monocerotis ✶ One of the finest triple stars in the sky for small telescopes, consisting of an arc of three 5th-magnitude stars.

8 Monocerotis ✶ A double star – on some charts labelled Epsilon (ε) Monocerotis – with components of 4th and 7th magnitudes.

M50 ♓ ✶ An open cluster about half the apparent size of the full Moon, visible in binoculars but requiring telescopes to resolve individual stars.

NGC 2244 ♓ ✶ 🖥 A group of stars of 6th magnitude and fainter, visible through binoculars. The cluster is surrounded by the Rosette Nebula.

NGC 2264 ♓ ✶ 🖥 Another combination of open cluster and nebula. Photographs and CCD images show a surrounding nebulosity into which protrudes a dark wedge known as the Cone Nebula.

Butterfly nebula
The butterfly-shaped planetary nebula NGC 2346 spreads its wings in the constellation Monoceros, captured here by the Hubble Space Telescope. At the centre of the nebula lies a close pair of stars, one of which evolved into a red giant and lost its outer atmosphere, thereby creating the nebula. NGC 2346 lies about 2,000 light-years away.

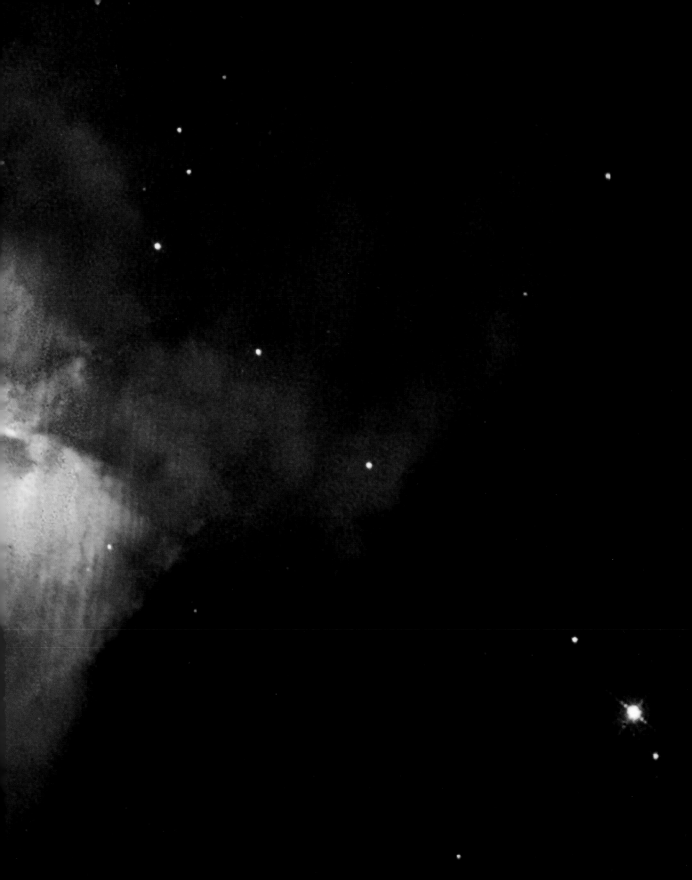

Southern Hemisphere

Hydra Hydrae (Hya)

WIDTH **DEPTH** **SIZE RANKING** 1st **FULLY VISIBLE** 54°N–83°S

Hydra is the largest of the constellations, stretching more than a quarter of the way around the sky from the top of its head, south of the constellation Cancer, to the tip of its tail, between Libra and Centaurus. For all its size, there is little within Hydra to catch the eye other than a group of six stars of modest brightness south of Cancer that forms its head. Its brightest star is Alphard, magnitude 2.0, whose name, coined by Arab astronomers, means "the solitary one" in reference to its position in an area of sky with no other prominent stars. Hydra represents the multi-headed monster fought and killed by Hercules in the second of his labours, although it is depicted as a single-headed water snake.

The Water Snake

The Head of Hydra
The most easily recognizable part of Hydra is its head, formed by six stars. The brightest of these are Epsilon (ε) Hydrae (top centre) and Zeta (ζ) Hydrae (top left), both of third magnitude.

LEO

0° SEXT

CRATER

CORVUS

LIBRA

CENTAURUS

ANTL

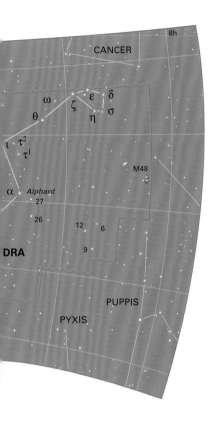

⟫ FEATURES OF INTEREST

Alpha (α) Hydrae (Alphard) 👁 An orange-coloured giant, the brightest star in Hydra.

Epsilon (ε) Hydrae 🔭 A close binary star with components of contrasting colours, requiring apertures of at least 75mm (3in) and high magnification to be separated. The yellow and blue component stars are of 3rd and 7th magnitudes and have an orbital period of nearly 600 years.

R Hydrae 👁 🔭 🔭 A red giant variable of the same type as Mira that ranges in brightness between 4th and 10th magnitudes every 13 months or so.

M48 🔭 🔭 An open star cluster larger than the apparent size of the full Moon, well seen through binoculars and small telescopes.

M83 🔭 🖥 An impressive face-on spiral galaxy. Through small telescopes, it appears as an elongated glow, but larger apertures reveal its spiral structure and a noticeable central bar, possibly similar to the bar that is thought to lie across the centre of our own Milky Way Galaxy. M83 lies about 15 million light-years away.

NGC 3242 (The Ghost of Jupiter) 🔭 🖥 A relatively prominent planetary nebula. It shows a disc of similar apparent size to the planet Jupiter when seen through small telescopes, hence its popular name, the Ghost of Jupiter.

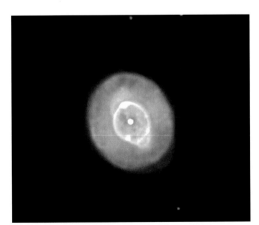

⌃ The Ghost of Jupiter
As its name suggests, this planetary nebula is a much fainter object than Jupiter. Small telescopes show it as a blue-green ellipse, but larger apertures are needed to see the inner ring and the central white dwarf.

⟪ M83 Spiral Galaxy
M83, on the border of Hydra and Centaurus, is visible in small telescopes. Amateur CCD images, such as this one, bring out the pink clouds of gas dotted along its spiral arms where stars are forming.

Spiral Galaxy NGC 2997
This galaxy is inclined at an angle of about 45° to our line of sight. Photographs and CCD images show pinkish clouds of hydrogen along its spiral arms.

Antlia Antliae (Ant)

WIDTH 🖐🖐 **DEPTH** 🖐 **SIZE RANKING** 62nd **FULLY VISIBLE** 49°N–90°S

Southern Hemisphere

This faint constellation of the southern sky consists of a handful of stars between Vela and Hydra. It was one of the constellations introduced in the mid-18th century by the French astronomer Nicolas Louis de Lacaille to commemorate scientific and technical inventions, in this case an air pump used for experiments on gases.

≫ FEATURES OF INTEREST

Zeta (ζ) Antliae 👥 ✺ A multiple star that appears as a wide pair of 6th-magnitude stars when viewed through binoculars. A small telescope reveals that the brighter of the pair has a 7th-magnitude companion.

NGC 2997 ✺ 🖥 This elegant spiral galaxy is just too faint to be seen well in small telescopes but is captured beautifully on photographs and CCD images.

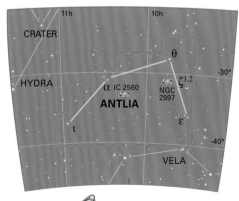

The Air Pump

Sextans Sextantis (Sex)

WIDTH 〰 **DEPTH** 〰 **SIZE RANKING** 47ᵗʰ **FULLY VISIBLE** 78°N–83°S

This faint and unremarkable constellation lies on the celestial equator, south of Leo. It was introduced in the late 17th century by the Polish astronomer Johannes Hevelius. It represents a sextant, the kind of instrument used by Hevelius himself for measuring and cataloguing the positions of stars in the sky.

》 FEATURES OF INTEREST

17 and 18 Sextantis 🔭 A line-of-sight double formed by two unrelated stars of 6th magnitude, shown neatly by binoculars.

NGC 3115 (The Spindle Galaxy) 🔭 ✈ A highly elongated lenticular galaxy that has acquired its popular name on account of its shape. Lying about 30 million light-years away from the Earth, it is detectable through small- to medium-sized telescopes.

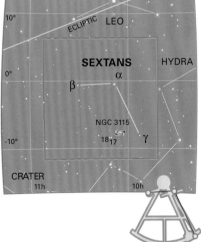

Southern Hemisphere

The Sextant

《 Cosmic spindle
NGC 3115 in Sextans, also known as the Spindle Galaxy, is a so-called lenticular galaxy, meaning that it has a disk of stars and a central bulge but no detectable spiral pattern. It is seen here in a composite of X-ray (blue) and optical (gold) images. The X-ray image reveals hot gas falling towards a supermassive black hole at the galaxy's centre. NGC 3115 lies about 30 million light-years away.

**Southern
Hemisphere**

Crater Crateris (Crt)

WIDTH 〜 **DEPTH** 〜 **SIZE RANKING** 53rd **FULLY VISIBLE** 65°N–90°S

This faint constellation lies next to Corvus on the back of Hydra, the water snake. It represents a goblet or chalice. Crater and adjacent Corvus feature together in a Greek myth in which the god Apollo sent the crow (Corvus) to fetch water in a cup (Crater). On the way, the greedy crow stopped to eat figs. As an alibi, the crow snatched up a water snake (Hydra) and blamed it for delaying him, but Apollo saw through the deception and banished the trio to the skies. Crater is larger than Corvus, but contains no objects of interest to users of small telescopes.

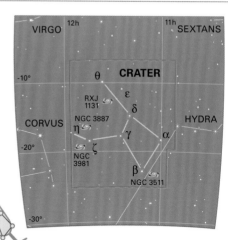

The Cup

» Galactic lens
Multiple images of the quasar RX J1131, shown here in pink, are caused by an intervening giant elliptical galaxy, which is coloured orange in this composite view made from observations by the Chandra X-ray Observatory and the Hubble Space Telescope. Three of the quasar images lie to the left of the galaxy, and one to the right. RX J1131 lies about 6 billion light-years away in Crater.

Corvus Corvi (Crv)

WIDTH ✋ **DEPTH** 🤚 **SIZE RANKING** 70th **FULLY VISIBLE** 65°N–90°S

Corvus is a small constellation south of Virgo. Its four brightest stars – Beta (β), Gamma (γ), Delta (δ), and Epsilon (ε) Corvi – form a distinctive keystone shape. Oddly, the star labelled Alpha (α) Corvi, at magnitude 4.0, is fainter than these four stars. Corvus is one of the original 48 Greek constellations and represents a crow, the sacred bird of Apollo. It is linked in legend with the neighbouring constellation Crater, the cup.

Southern Hemisphere

》 FEATURES OF INTEREST

Delta (δ) Corvi 🎇 A double star with components of unequal magnitudes, 3rd and 8th, divisible with small telescopes.

NGC 4038 and NGC 4039 (The Antennae)
🎇 🖥 ⚖ A remarkable pair of interacting galaxies. A major "traffic accident" is happening, 65 million light-years away, as the galaxies NGC 4038 and 4039 collide with each other. At 10th magnitude, the galaxies are too faint to see with small telescopes, but photographs reveal their true structure. Stretching away from them on either side are two antenna-like arcs, consisting of plumes of gas and millions of stars that have been flung into intergalactic space as a result of this collision.

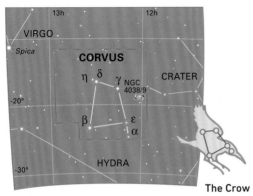

The Crow

☑ The Antennae
As the galaxies NGC 4038 and NGC 4039 sweep past each other, gravity draws out long streams of stars and gas, seen extending like the antennae of an insect, hence the object's popular name.

Centaurus Centauri (Cen)

WIDTH 🖐🖐🖐 **DEPTH** 🖐🖐 **SIZE RANKING** 9th **FULLY VISIBLE** 25°N–90°S

**Southern
Hemisphere**

Centaurus is one of the dominant constellations of the southern skies, containing a variety of notable objects. These include the closest star to the Sun, the brightest globular cluster, and a peculiar galaxy. The constellation represents a centaur, a mythical creature with the torso of a man and the legs of a horse. Its two brightest stars, Alpha (α) and Beta (β) Centauri, point towards the Southern Cross.

⟩⟩ FEATURES OF INTEREST

Alpha (α) Centauri (Rigil Kentaurus) 👁 ⚡
A celebrated multiple star. To the naked eye, it appears of magnitude -0.3, making it the third-brightest star in the sky, but a small telescope shows it to be a double of two yellow stars that orbit each other every 80 years. They appear so bright because they are a mere 4.3 light-years away. Only one star is nearer to us – the third member of the system, Proxima Centauri, a faint 11th-magnitude red dwarf.

Omega (ω) Centauri (NGC 5139) 👁 🏃 ⚡
The largest and brightest globular cluster. To the eye, it appears like a large, hazy star.

NGC 3918 (The Blue Planetary) ⚡ A planetary nebula easily visible through small telescopes, appearing like a larger version of the disc of Uranus.

NGC 5128 (Centaurus A) 🏃 ⚡ 🖥 🖥 A peculiar galaxy and a strong radio source, thought to be a merging giant elliptical and spiral galaxy.

The Centaur

⟩⟩ **Omega Centauri**
Through binoculars, the cluster appears larger than the full Moon, while small telescopes resolve the brightest individual members. Omega Centauri lies about 17,000 light-years away.

Lupus Lupi (Lup)

WIDTH 🖐🖐 **DEPTH** 🖐🖐 **SIZE RANKING** 46th **FULLY VISIBLE** 34°N–90°S

This is a southern constellation lying on the edge of the Milky Way between the better-known figures of Centaurus and Scorpius. Lupus was one of the original 48 constellations known to the ancient Greeks, who visualized it as a wild animal speared by Centaurus. It contains numerous double stars of interest to amateur observers.

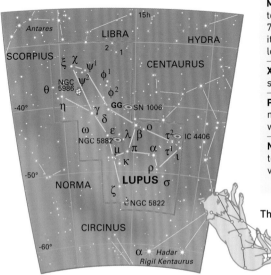

The Wolf

➤➤ FEATURES OF INTEREST

Epsilon (ε) Lupi 🔭 A 3rd-magnitude star with a companion of 9th magnitude that becomes visible with small telescopes.

Kappa (κ) Lupi 🔭 A double star with components of 4th and 6th magnitudes that are easily divided with small telescopes.

Mu (μ) Lupi 🔭 An interesting multiple star. Small telescopes show a 4th-magnitude star with a wide 7th-magnitude companion. The primary star is itself a close double, requiring apertures of at least 100mm (4in) to separate.

Xi (ξ) Lupi 🔭 A double of 5th and 6th magnitudes, separable with small telescopes.

Pi (π) Lupi 🔭 A double that can be divided into matching blue-white 5th-magnitude components with apertures of 75mm (3in).

NGC 5822 🔭 🔭 A large open cluster close to the southern boundary of the constellation, visible through binoculars and small telescopes.

Southern Hemisphere

⏶ Mu Lupi
The two principal components of Mu (μ) Lupi – the primary star of magnitude 4.3 (centre) and its wide companion star – are clearly visible in this photograph. But the magnification is not great enough to divide the primary star and its closer companion.

⏴ Open cluster NGC 5822
This scattered cluster can be made out with binoculars and small telescopes against the background of the Milky Way. However, its brightest stars are of only 9th magnitude, so it is not particularly prominent. This photograph shows it as it appears through a small telescope.

Southern Hemisphere

Sagittarius Sagittarii (Sgr)

WIDTH 🖐🖐🖐 **DEPTH** 🖐🖐 **SIZE RANKING** 15th **FULLY VISIBLE** 44°N–90°S

Sagittarius is a large and prominent constellation of the zodiac, between Scorpius and Capricornus. Its most recognizable feature is a star pattern that resembles a teapot. The handle of the teapot is sometimes also called the Milk Dipper, imagined as scooping into the Milky Way. The exact centre of our galaxy is thought to coincide with a radio source known as Sagittarius A*, near where the borders of Sagittarius, Ophiuchus, and Scorpius meet. In Greek mythology, Sagittarius was said to represent Crotus, son of Pan, who invented archery and went hunting on horseback.

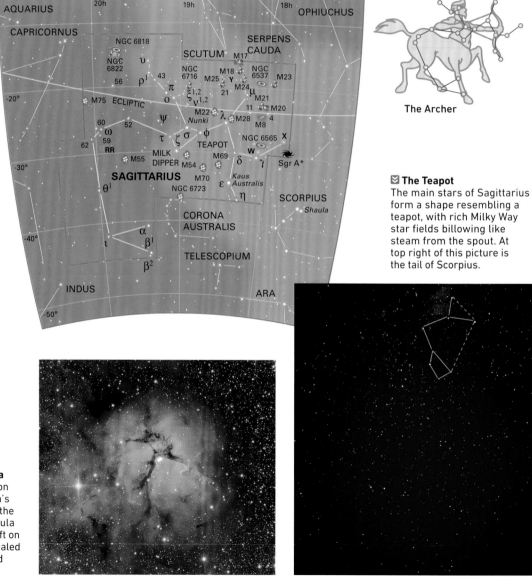

The Archer

☑ **The Teapot**
The main stars of Sagittarius form a shape resembling a teapot, with rich Milky Way star fields billowing like steam from the spout. At top right of this picture is the tail of Scorpius.

⏩ **The Trifid Nebula**
The pinkish emission of the Trifid Nebula's gas contrasts with the blue reflection nebula lying to its north (left on this image), as revealed on photographs and CCD images.

⟫ FEATURES OF INTEREST

Beta (β) Sagittarii 👁 🏹 A pair of 4th-magnitude stars, visible separately to the naked eye. A small telescope shows that the more northerly (and slightly brighter) of the two has a 7th-magnitude companion. All three stars are unrelated.

W Sagittarii 👁 📷 A Cepheid variable that ranges between magnitudes 4.3 and 5.1 every 7.6 days.

X Sagittarii 👁 📷 A Cepheid variable that ranges between magnitudes 4.2 and 4.9 every 7.0 days.

M8 (The Lagoon Nebula) 👁 📷 🏹 🖥 A patch of glowing gas that extends for three times the apparent width of the full Moon, bright enough to be visible to the naked eye and well seen through binoculars. One half of the nebula contains the open cluster NGC 6530, with stars of 7th magnitude and fainter, while in the other half lies the 6th-magnitude blue supergiant 9 Sagittarii.

M17 (The Omega Nebula) 🏹 🖥 A gaseous nebula that takes its name from its supposed resemblance to the Greek letter Omega. It is also known as the

Swan Nebula from an alternative interpretation of its shape. It can be glimpsed through binoculars, as can the loose cluster of stars within it.

M20 (The Trifid Nebula) 🏹 🖥 A spectacular emission nebula that gets its popular name because it is trisected by dark lanes of dust.

M22 👁 📷 🏹 One of the finest globular clusters in the entire sky. M22 is visible to the naked eye under good conditions and is an easy object for binoculars, appearing as a woolly ball about two-thirds the apparent diameter of the Moon. Apertures of 75mm (3in) will resolve its brightest stars.

M23 📷 🏹 A large open cluster visible through binoculars near the border with Ophiuchus but requiring a telescope to resolve its individual stars.

M24 👁 📷 Not a star cluster as such, but a bright Milky Way star field. Four apparent Moon diameters long, it is best seen through binoculars.

⌂ The Lagoon Nebula
This nebula is one of the largest in the sky, visible as an elongated, milky cloud. It is a good subject for viewing through binoculars. As with all nebulae, the red colour revealed by photographs is not apparent visually.

Towards the Galactic Centre
The centre of our home galaxy lies about 30,000 light-years away in the direction of the brighest part of the band of the Milky Way, seen here in the summer sky over Arizona, USA. Scorpius is at the right of the image, while Sagittarius is just below and to the left of centre.

Southern Hemisphere

Scorpius Scorpii (Sco)

WIDTH ✋✋ **DEPTH** ✋✋✋ **SIZE RANKING** 33rd **FULLY VISIBLE** 44°N–90°S

This beautiful and easily recognizable constellation of the zodiac lies in the southern sky between Libra and Sagittarius. In Greek mythology, Scorpius represents the scorpion that stung Orion to death. In the sky, the scorpion's heart is marked by the red star Antares, while a distinctive curve of stars marks the scorpion's tail, raised ready to strike. The tail extends into a rich area of the Milky Way towards the centre of our galaxy.

The Scorpion

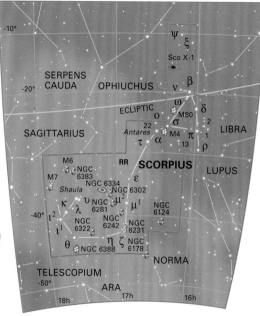

⟫ FEATURES OF INTEREST

Alpha (α) Scorpii (Antares) 👁 The constellation's brightest star, a red supergiant hundreds of times larger than the Sun. It fluctuates in brightness from about magnitude 0.8 to 1.2 every six years or so.

Delta (δ) Scorpii 👁 A star that unexpectedly began to brighten in the year 2000. Normally of magnitude 2.3, it rose by over 50 per cent as it ejected a shell of gas. It has since faded.

Xi (ξ) Scorpii 🔭 A complex multiple star. Through small telescopes, it appears as a white and orange pair of 4th and 7th magnitudes. In the same field of view, a fainter and wider pair can also be seen. All four are gravitationally linked, making this a genuine quadruple.

M4 🔭 🔭 A large, loosely scattered globular cluster near Antares, 7,000 light-years away.

M6 (The Butterfly Cluster) 🔭 🔭 An open cluster about twice as distant from us as M7 and hence appearing somewhat smaller.

M7 👁 🔭 🔭 An open cluster visible to the eye as a hazy patch against the Milky Way.

▽ The Butterfly Cluster
An open cluster near the tail of Scorpius, M6 gained its popular name because of its butterfly-like shape as seen through binoculars and small telescopes. On one wing lies its brightest star, the orange giant BM Scorpii.

⌃ Open cluster M7
M7 is the most prominent of three open clusters near the sting of the scorpion's tail. Binoculars show dozens of stars of 6th magnitude and fainter.

Capricornus Capricorni (Cap)

WIDTH 🖑🖑 **DEPTH** 🖑🖑 **SIZE RANKING** 40th **FULLY VISIBLE** 62°N–90°S

Capricornus, the smallest constellation of the zodiac and not at all prominent, lies in the southern sky between Sagittarius and Aquarius. In Greek myth, Capricornus represents the goat-like god Pan. In one tale of his exploits, he jumped into a river and turned himself into a creature that was part fish in order to escape from the sea monster Typhon. Hence the constellation is depicted as a goat with the tail of a fish.

Southern Hemisphere

》 FEATURES OF INTEREST

Alpha (α) Capricorni 👁 🔭 A wide pairing of unrelated 4th-magnitude stars, visible separately through binoculars or even with sharp eyesight. Alpha-1 (α^1) Capricorni is a yellow supergiant about 570 light-years away, while Alpha-2 (α^2) is a yellow giant that lies at less than one-fifth of that distance.

Beta (β) Capricorni 🔭 🔭 A wide double star, comprising a 3rd-magnitude yellow giant with a 6th-magnitude blue-white companion visible through small telescopes or even good binoculars.

M30 🔭 A modest globular cluster visible as a hazy patch through small telescopes. Larger apertures show chains of stars extending from it like fingers.

The Sea Goat

《 Compact group of galaxies
Hickson Compact Group 87 (HCG 87) is a family of contrasting galaxies some 400 million light-years away in Capricornus. The largest member, at the bottom, is a spiral galaxy seen edge-on. At the top is another spiral, with an elliptical galaxy on the right. The small spiral near the centre is either a fourth member or an unrelated background object. This image was taken with the Hubble Space Telescope.

Tangled galaxies
Millions of years ago, two galaxies in Microscopium passed so close to each other that they became entangled, creating this jellyfish-like hybrid. The colliding pair, called ESO 286-19, lies some 600 million light-years away and are seen here through the Hubble Space Telescope.

Microscopium Microscopii (Mic)

WIDTH 🖐 **DEPTH** 🖐 **SIZE RANKING** 66th **FULLY VISIBLE** 45°N–90°S

Southern Hemisphere

Microscopium is a faint and obscure southern constellation lying between Sagittarius and Piscis Austrinus. Representing an early form of compound microscope, it was one of the constellations invented in the 18th century by the French astronomer Nicolas Louis de Lacaille. Its brightest stars, Gamma (γ) and Epsilon (ε) Microscopii, are both of magnitude 4.7.

》 FEATURES OF INTEREST

Alpha (α) Microscopii 🏹 A 5th-magnitude star with a 10th-magnitude companion visible through amateur telescopes.

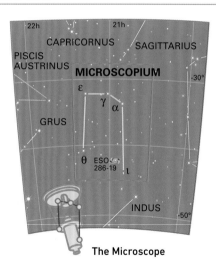

The Microscope

Piscis Austrinus Piscis Austrini (PsA)

WIDTH 𝄔𝄔 **DEPTH** 𝄔 **SIZE RANKING** 60th **FULLY VISIBLE** 53°N–90°S

This constellation south of Aquarius is made prominent by the presence of the star Fomalhaut, magnitude 1.2, although there is little else of note. To the ancient Greeks, Piscis Austrinus was the parent of the two fish of the zodiacal constellation Pisces. In the sky, the stream of water from the jar of Aquarius flows towards the mouth of the fish, marked by Fomalhaut, an Arabic name meaning "fish's mouth".

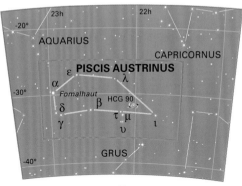

Southern Hemisphere

》 FEATURES OF INTEREST

Beta (β) Piscis Austrini 🏹 A wide double with components of 4th and 7th magnitudes, divisible by small telescopes.

Gamma (γ) Piscis Austrini 🏹 A closer pair, of 5th and 8th magnitudes, which is rather more difficult to divide than Beta.

The Southern Fish

《 Galactic tug of war In this image from the Hubble Space Telescope, a spiral galaxy with a distorted dust lane is being torn apart by the gravitational pulls of the two elliptical galaxies below and to the left of it. All three will one day merge to create a single super-galaxy, bigger than our own Milky Way. This trio is part of a cluster called Hickson Compact Group 90 (HCG 90) that lies about 100 million light-years away in Piscis Austrinus.

**Southern
Hemisphere**

Sculptor Sculptoris (Scl)

WIDTH ꙮꙮ **DEPTH** ꙮ **SIZE RANKING** 36th **FULLY VISIBLE** 50°N–90°S

This faint southern constellation adjoining Piscis Austrinus was introduced in the 18th century by the French astronomer Nicolas Louis de Lacaille, who described it as representing a sculptor's studio, although the name has since been shortened. Sculptor contains the south pole of our galaxy, that is, the point 90° south of the plane of the Milky Way. This enables us to see numerous far-off galaxies in this direction, because they are not obscured by intervening stars.

⟩⟩ FEATURES OF INTEREST

Epsilon (ε) Sculptoris ⚹ A binary that can be separated with a small telescope. Its components, of 5th and 9th magnitudes, have an orbital period of over 1,000 years.

NGC 55 ⚹ An edge-on spiral galaxy, similar in size and shape to NGC 253 but not quite as easy to see.

NGC 253 ⚇ ⚹ A spiral galaxy seen nearly edge-on so that it appears highly elongated. Under good sky conditions, it can be picked up with binoculars and small telescopes. Nearby lies the fainter and smaller globular cluster NGC 288.

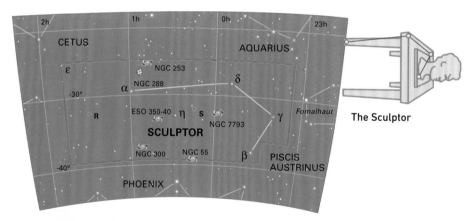

The Sculptor

⟩⟩ **Sculptor spiral**
The central regions of NGC 7793, a spiral galaxy some 13 million light-years away in Sculptor, are seen in this image from the Hubble Space Telescope. Unlike many other spiral galaxies, such as our own Milky Way, NGC 7793 does not have a very large central bulge or well-developed arms.

Fornax Fornacis (For)

WIDTH 〰️〰️ **DEPTH** 〰️ **SIZE RANKING** 41st **FULLY VISIBLE** 50°N–90°S

This undistinguished constellation of the southern sky is made up of faint stars tucked into a bend in the river Eridanus, south of Cetus. It was introduced in the 18th century by the French astronomer Nicolas Louis de Lacaille, originally under the name *Fornax Chemica*, referring to the chemical furnace of the kind used for distillation. In the southern part of Fornax lies a cluster of galaxies including the radio source Fornax A.

Southern
Hemisphere

》 FEATURES OF INTEREST

Alpha (α) Fornacis ✦ The brightest star in the constellation at magnitude 3.9, with a yellow 7th-magnitude companion visible through small telescopes.

Fornax Cluster ✦ 🖵 ⬒ A small cluster of galaxies about 60 million light-years away in the southern part of the constellation. The brightest member of the group is the peculiar spiral NGC 1316, a radio source also known as Fornax A. Another prominent member of the cluster is the beautiful barred spiral galaxy NGC 1365.

⬓ Central ring

NGC 1097 is a barred spiral galaxy about 50 million light-years away, in Fornax. Surrounding its nucleus is a bright ring of newly formed stars, captured in this close-up from the Hubble Space Telescope.

**Southern
Hemisphere**

Caelum Caeli (Cae)

WIDTH ⁽ᵐ⁾ **DEPTH** ⁽ᵐ⁾⁽ᵐ⁾ **SIZE RANKING** 81st **FULLY VISIBLE** 41°N–90°S

Caelum is a small and faint southern constellation, sandwiched between Eridanus and Columba, introduced in the 18th century by the French astronomer Nicolas Louis de Lacaille. Lacaille originally described the constellation as representing a pair of sharp engraving tools called burins, but now it is simply depicted as a stonemason's chisel. There is virtually nothing here of interest to users of small telescopes. Its brightest star, Alpha Caeli, is of only magnitude 4.4.

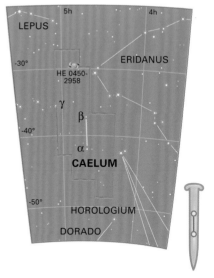

≫ FEATURES OF INTEREST

Gamma (γ) Caeli ✶ A 5th-magnitude orange giant with an 8th-magnitude companion. A modest-sized telescope is required to separate them because of the closeness of the pair.

The Chisel

≫ **Quasar without a home**
Quasars normally reside at the heart of galaxies, but in the case of HE0450-2958 in Caelum, no surrounding galaxy has been detected. This view of it is a combination of images from the European Southern Observatory and the Hubble Space Telescope. The object at upper right is a foreground star.

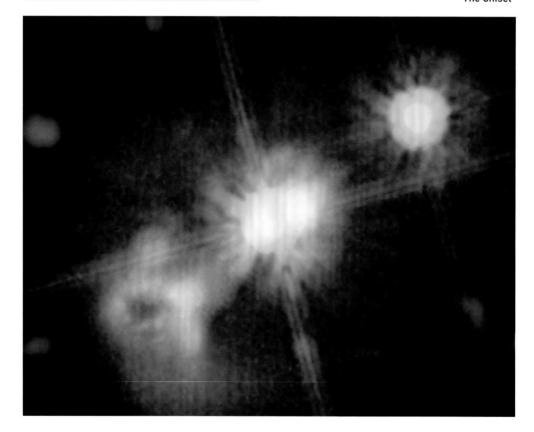

Eridanus Eridani (Eri)

WIDTH 🖑🖑🖑 **DEPTH** 🖑🖑🖑🖑🖑 **SIZE RANKING** 6th **FULLY VISIBLE** 32°N–89°S

Eridanus represents a river, meandering from the foot of Taurus in the north to Hydrus in the south, giving it a range in declination of 58°, the greatest of any constellation. Its only star of any note, Achernar, magnitude 0.5, lies at its southern tip. Eridanus features in the story of Phaethon, son of the Sun-god Helios, who tried to drive his father's chariot across the sky but lost control and fell like a meteor into the river below.

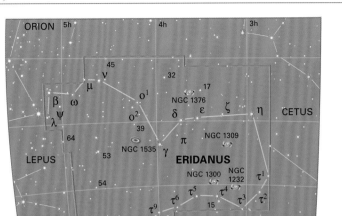

Southern Hemisphere

The River

⟫ FEATURES OF INTEREST

Theta (θ) Eridani 🏹 A double star consisting of white stars of 3rd and 4th magnitudes.

Omicron-2 (o²) Eridani (40 Eridani) 🏹 A multiple star that appears to the eye as a 4th-magnitude orange star. A small telescope reveals a 10th-magnitude companion – an easily seen white dwarf. This forms a binary with a fainter red dwarf.

32 Eridani 🏹 A pair of orange and blue stars of 5th and 6th magnitudes visible with small telescopes.

NGC 1300 🏹 🖵 A barred spiral galaxy, some 70 million light-years away. Its central bar is longer than the diameter of our own galaxy.

⟫ **Galaxy NGC 1300**
Too faint for easy viewing through small telescopes, this classic barred spiral galaxy shows up beautifully on photographs.

Southern Hemisphere

Lepus Leporis (Lep)

WIDTH 🖑 **DEPTH** 🖑 **SIZE RANKING** 51st **FULLY VISIBLE** 62°N–90°S

One of the constellations known to the ancient Greeks, Lepus lies under the feet of Orion, the hunter. It represents a hare, pursued across the sky by one of Orion's dogs, Canis Major. Surrounded by the sparkling stars of Orion and Canis Major, it is often overlooked, but is well worth attention. Its brightest star, Alpha Leporis, magnitude 2.6, is known as Arneb, from the Arabic meaning "hare". According to one legend, the constellation commemorates a plague of hares that overran the Greek island of Leros when a breeding programme got out of hand.

>> **FEATURES OF INTEREST**

Gamma (γ) Leporis ✸ A 4th-magnitude yellow star with a 6th-magnitude orange companion visible through binoculars.

Kappa (κ) Leporis ✸ A 4th-magnitude star with a close 7th-magnitude companion, difficult to separate with the smallest apertures.

R Leporis 🔭 ✸ A variable star of the same type as Mira, noted for its intensely red colour. Its brightness ranges from 6th to 12th magnitudes every 14 months or so.

M79 ✸ A modest globular cluster visible with a small telescope, over 40,000 light-years away and so not easy to resolve into individual stars. Herschel 3752, a triple star with components of 5th, 7th, and 9th magnitudes, lies in the same field of view.

NGC 2017 ✸ A compact group of stars, consisting of a 6th-magnitude star with four companions of 8th to 10th magnitudes visible through small telescopes. Larger apertures reveal three more stars. However, all these stars seem to be a chance line-of-sight grouping, so NGC 2017 is not a true cluster at all.

The Hare

⌂ Globular cluster M79
This somewhat sparse 8th-magnitude globular cluster, 42,000 light-years away, has long chains of stars which give it the appearance of a starfish when seen through small telescopes.

>> **NGC 2017**
This chance grouping of stars resembles a small cluster. The discs and spikes are effects of the large professional telescope through which the picture was taken.

Columba Columbae (Col)

WIDTH 〰️ **DEPTH** 〰️ **SIZE RANKING** 54th **FULLY VISIBLE** 46°N–90°S

Columba is a constellation of the southern sky formed in the late 16th century by the Dutch theologian and astronomer Petrus Plancius from stars south of Lepus and Canis Major that had not previously been allocated to any constellation. It represents the dove released by Noah from his Ark to try to find dry land in the Biblical story. The dove returned with a twig from an olive tree in its beak, indicating that the waters of the Great Flood were at last receding.

Southern Hemisphere

⟩⟩ FEATURES OF INTEREST

Mu (μ) Columbae 🏃 A 5th-magnitude star whose rapid movement suggests it was thrown out from the area of the Orion Nebula, south of Orion's belt, about 2.5 million years ago. Astronomers think it was once a member of a binary system that was disrupted by a close encounter with another star. The star thought to have been the other member of the binary is 6th-magnitude AE Aurigae, which is moving away from Orion in the opposite direction.

NGC 1851 🏹 A modest globular cluster visible as a hazy patch through small telescopes.

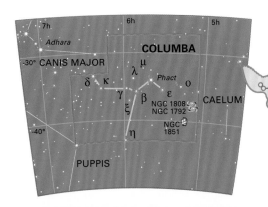

The Dove

✉️ **NGC 1851**
NGC 1851 is a rich globular cluster in Columba that is visible with small telescopes. It has a noticeably condensed core. The cluster lies 39,000 light-years away. This view in ultraviolet light from NASA's GALEX satellite picks out the hottest stars.

Southern Hemisphere

Pyxis Pyxidis (Pyx)

WIDTH 🖐 **DEPTH** 🖐🖐 **SIZE RANKING** 65th **FULLY VISIBLE** 52°N–90°S

Pyxis is a faint, unremarkable southern constellation lying next to Puppis on the edge of the Milky Way. It represents a ship's magnetic compass and was introduced in the 18th century by the French astronomer Nicolas Louis de Lacaille. Pyxis lies in the area that was once occupied by the mast of Argo Navis, the ship of the Argonauts, an ancient Greek constellation now divided into Carina, Puppis, and Vela. The brightest star, Alpha (α) Pyxidis, is of magnitude 3.7.

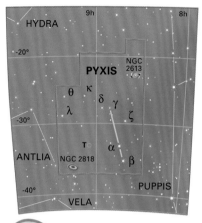

The Compass

❱❱ FEATURES OF INTEREST

T Pyxidis 🔭 🏃 A so-called recurrent nova – that is, one that has undergone several recorded outbursts. Six eruptions have been seen since 1890, the last being in 2011. During these outbursts, it has brightened from 15th magnitude to 6th or 7th magnitude. It is likely to brighten again at any time and so become visible through binoculars.

❱❱ NGC 2818
A dying central star lights up surrounding clouds of gas in the planetary nebula NGC 2818 in Pyxis, seen here in false colour by the Hubble Space Telescope. NGC 2818 lies in the same line of sight as a more distant star cluster, NGC 2818A, but is not connected with the cluster.

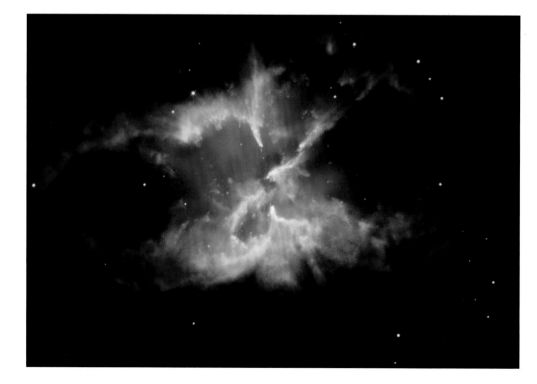

Puppis Puppis (Pup)

WIDTH 〰🖐 **DEPTH** 〰🖐🖐 **SIZE RANKING** 20th **FULLY VISIBLE** 39°N–90°S

Puppis is a rich southern constellation straddling the Milky Way. It was originally part of the ancient Greek figure of Argo Navis, the ship of Jason and the Argonauts, until it was made into a separate constellation in the 18th century by the Frenchman Nicolas Louis de Lacaille. Puppis represents the ship's stern and is the largest of the three parts into which Argo was divided. Lacaille labelled the stars of Argo with Greek letters, and in the case of Puppis the lettering started at Zeta (ζ).

Southern Hemisphere

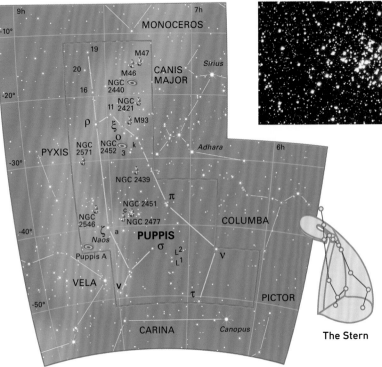

The Stern

⌃ **Open cluster M93**
Shaped roughly like an arrowhead, the cluster appears at the centre of this photograph with two orange giants clearly visible near the tip.

⟫ FEATURES OF INTEREST

Xi (ξ) Puppis 🔭 A 3rd-magnitude star with a wide and unrelated 5th-magnitude companion visible in binoculars.

k Puppis 🔭 A pair of nearly identical 5th-magnitude stars divisible through small telescopes.

L Puppis 👁 🔭 A wide naked-eye and binocular pair. The more northerly of them, L² Puppis, is a variable red giant that ranges between 3rd and 8th magnitudes every 5 months or so.

M46 and M47 🔭 🔭 A pair of open clusters that together create a brighter patch in the Milky Way. Both appear of similar size to the full Moon. M46 is

the richer of the two, while M47 is more scattered. M47 is also the closer, about 1,300 light-years away, just over a quarter the distance of M46.

M93 🔭 🔭 An attractive open cluster for binoculars and small telescopes. It appears triangular or like an arrowhead in shape and has two orange giants near its apex. It lies about 3,500 light-years away.

NGC 2451 👁 🔭 A more scattered open cluster than M93. Its brightest star is c Puppis, a 4th-magnitude orange giant, near its centre.

NGC 2477 🔭 🔭 One of the richest open clusters, containing an estimated 2,000 stars.

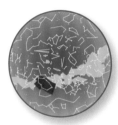

**Southern
Hemisphere**

Vela Velorum (Vel)

WIDTH ⊀⊀⊀ **DEPTH** ⊀⊀ **SIZE RANKING** 32ⁿᵈ **FULLY VISIBLE** 32°N–90°S

Vela is one of the three parts into which the ancient Greek constellation Argo Navis (depicting the ship of Jason and the Argonauts) was divided in the 18th century. Vela represents the ship's sails. The labelling of the stars in Vela starts with Gamma (γ) because Argo's stars Alpha (α) and Beta (β) are now in neighbouring Carina.

❯❯ FEATURES OF INTEREST

False Cross A pseudo "Southern Cross" formed by the stars Delta (δ) and Kappa (κ) Velorum combined with Epsilon (ε) and Iota (ι) Carinae.

Gamma (γ) Velorum 👁 📷 🏃 The constellation's leading star at magnitude 1.8. It is the brightest example of a Wolf-Rayet star, a rare type that has lost its outer layers, thereby exposing its ultra-hot interior. A 4th-magnitude companion is visible through small telescopes or even good binoculars. Two wider companions, of 7th and 9th magnitudes, can be seen with telescopes.

NGC 2547 📷 🏃 An open cluster half the apparent size of the full Moon.

NGC 3132 (The Eight-Burst Nebula) 🏃 🖥 🔭 A notable planetary nebula that gets its popular name from complex loops of gas that give it the appearance of intertwining figures of eight.

IC 2391 👁 📷 A group of several dozen stars covering an area greater than the apparent size of the full Moon. The brightest is 4th-magnitude Omicron (ο) Velorum. To the north of it lies another binocular cluster, IC 2395.

The Vela Supernova Remnant 🖥 🔭 The gaseous remains of a supernova created by a star that exploded around 11,000 years ago. It lies between Gamma (γ) and Lambda (λ) Velorum.

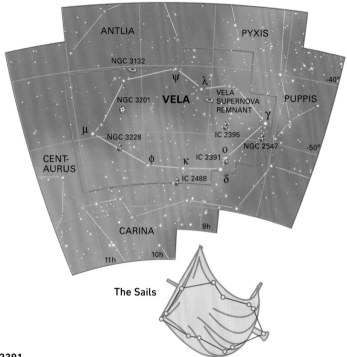

The Sails

🖼 The Eight-Burst Nebula
The nebula's loops of gas are only visible on photographs such as this one taken with large instruments. Small telescopes will show the nebula's disc, of similar apparent size to Jupiter, and the 10th-magnitude star at its centre.

🖼 IC 2391
This large open cluster lies 500 light-years away. It is visible to the naked eye and makes an excellent sight through binoculars.

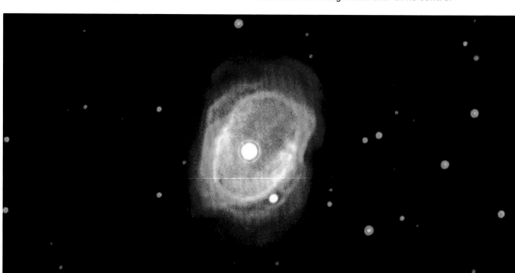

Carina Carinae (Car)

WIDTH 🖐🖐🖐 **DEPTH** 🖐🖐 **SIZE RANKING** 34th **FULLY VISIBLE** 14°N–90°S

Carina, a major southern constellation, was originally part of the larger figure of Argo Navis, representing the ship of the Argonauts, until that constellation was split into three in the 18th century. In Greek mythology, the Argo was a mighty 50-oared galley in which Jason and 50 of the greatest Greek heroes, called the Argonauts, sailed to Colchis, on the eastern shore of the Black Sea, to fetch the golden fleece of a ram. Their journey there and back is one of the epic stories of Greek myth. Carina, which represents the ship's keel, inherited many of the best objects from Argo Navis, including its most prominent star, Canopus.

Southern Hemisphere

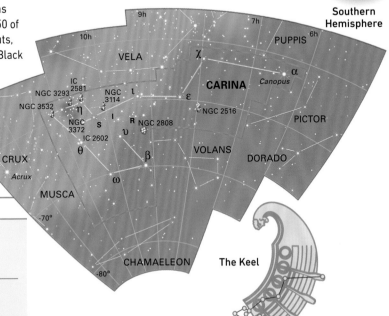

The Keel

》 FEATURES OF INTEREST

Alpha (α) Carinae (Canopus) 👁 A white supergiant of magnitude –0.7, the second-brightest star in the entire sky. It lies 310 light-years away.

NGC 2516 👁 🔭 A large open cluster visible to the naked eye. It appears cross-shaped through binoculars.

NGC 3372 (The Eta Carinae Nebula) 👁 🔭 🏃 🖥 🖥 A patch of glowing gas, four apparent Moon diameters wide, visible to the eye against the background of the Milky Way and well seen with binoculars. The brightest part of the nebula is around the peculiar variable star Eta (η) Carinae. During the 19th century, Eta Carinae flared up temporarily to become brighter than Canopus, although it has now subsided to around 5th magnitude.

NGC 3532 👁 🔭 An elongated cluster that makes an excellent sight through binoculars.

IC 2602 (The Southern Pleiades) 👁 🔭 An open cluster with several stars visible to the naked eye, the brightest of them being 3rd-magnitude Theta (θ) Carinae.

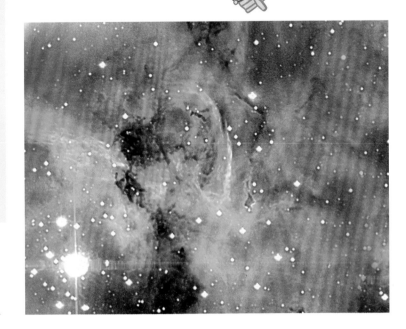

》 The Eta Carinae Nebula
Near Eta Carinae, the bright star at bottom left, telescopes show a dark and bulbous cloud of dust called the Keyhole, silhouetted against the nebula.

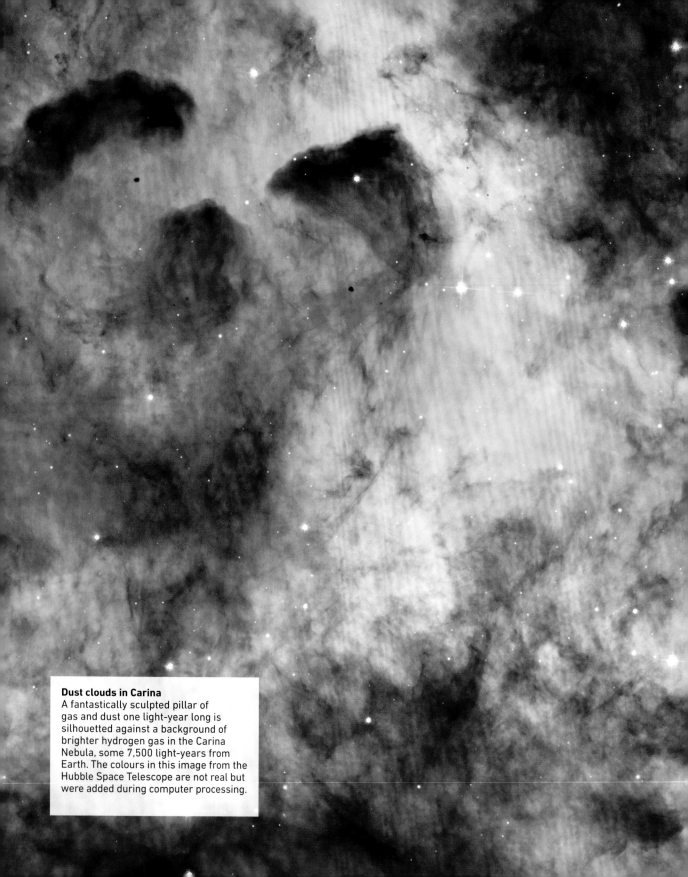

Dust clouds in Carina
A fantastically sculpted pillar of
gas and dust one light-year long is
silhouetted against a background of
brighter hydrogen gas in the Carina
Nebula, some 7,500 light-years from
Earth. The colours in this image from the
Hubble Space Telescope are not real but
were added during computer processing.

**Southern
Hemisphere**

Crux Crucis (Cru)

WIDTH 🖑 **DEPTH** 🖑 **SIZE RANKING** 88th **FULLY VISIBLE** 25°N–90°S

The smallest constellation, but instantly recognizable, Crux is squeezed between the legs of the centaur, Centaurus. The ancient Greeks regarded its stars as part of Centaurus, but it was made into a separate constellation by European seafarers on voyages of exploration in the 16th century, who dubbed it the Southern Cross. They would set their course using the fact that the longer axis of the cross points to the south celestial pole. Crux lies in a rich area of the Milky Way, which is here interrupted by the dark Coalsack Nebula.

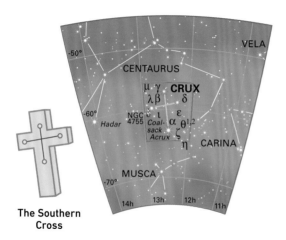

**The Southern
Cross**

» **FEATURES OF INTEREST**

Alpha (α) Crucis 👁 🔭 The most southerly first-magnitude star. To the eye, it appears of magnitude 0.8, but small telescopes divide it into a sparkling blue-white double of magnitudes 1.3 and 1.8.

Gamma (γ) Crucis 🔭 A 2nd-magnitude red giant with an unrelated 6th-magnitude companion visible through binoculars.

Mu (μ) Crucis 🔭 🔭 A wide pair of 4th- and 5th-magnitude stars, easily divisible by small telescopes or even good binoculars.

NGC 4755 (The Jewel Box Cluster) 👁 🔭 🔭 An open cluster that is one of the gems of the southern sky, visible to the naked eye as a brighter patch in the Milky Way. Binoculars and small telescopes show its individual stars covering about one-third the apparent width of the full Moon. A ruby-coloured supergiant near the centre contrasts with the other stars, most of which are blue-white supergiants, giving the impression of a collection of colourful jewels, hence its popular name.

The Coalsack Nebula 👁 🔭 A dark cloud of dust that blocks light from the stars of the Milky Way behind. Prominent to the naked eye and in binoculars, the Coalsack spans the length of 12 full Moons and extends into the constellations of Centaurus and Musca.

**»The Coalsack
Nebula**
This nebula appears as a dark patch in the bright Milky Way next to the stars of the Southern Cross, here at centre left. Alpha and Beta Centauri are far left while the Eta Carinae Nebula is the pink patch just right of centre.

Musca Muscae (Mus)

WIDTH ✋ **DEPTH** ✋ **SIZE RANKING** 77ᵗʰ **FULLY VISIBLE** 14°N–90°S

Musca lies in the Milky Way south of Crux and Centaurus. It is one of the southern constellations introduced at the end of the 16th century by the Dutch navigator–astronomers Pieter Dirkszoon Keyser and Frederick de Houtman, and represents a fly. The southern tip of the dark Coalsack Nebula extends into it from neighbouring Crux, but otherwise the constellation contains little of note.

≫ FEATURES OF INTEREST

Theta (θ) Muscae 🔭 A double star with components of 6th and 8th magnitudes divisible in small telescopes. The brighter star is a blue supergiant, while its fainter component is an example of a Wolf–Rayet star, a type of hot star that has lost its outer layers.

NGC 4833 🔭 🔭 A globular cluster visible with binoculars and small telescopes.

Southern Hemisphere

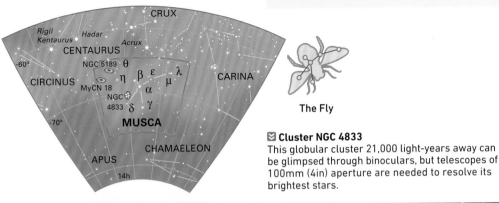

The Fly

☑ Cluster NGC 4833
This globular cluster 21,000 light-years away can be glimpsed through binoculars, but telescopes of 100mm (4in) aperture are needed to resolve its brightest stars.

Southern Hemisphere

Circinus Circini (Cir)

WIDTH 🖑 **DEPTH** 🖐 **SIZE RANKING** 85th **FULLY VISIBLE** 19°N–90°S

This is a small southern constellation, squeezed awkwardly in between Centaurus and Triangulum Australe. It is not difficult to find, lying next to Alpha (α) Centauri, but contains little of note. In shape, Circinus is a long, slim isosceles triangle, representing a pair of dividing compasses, as used by surveyors and navigators. It is one of a number of constellations based on the scientific instruments of the day that were introduced in the 18th century by the French astronomer Nicolas Louis de Lacaille.

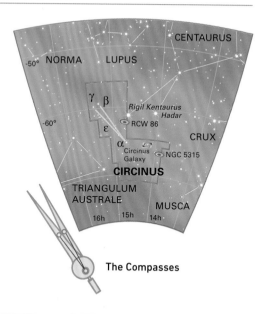

The Compasses

》 FEATURES OF INTEREST

Alpha (α) Circini 🏹 The brightest star of the constellation. It is an easy double of 3rd and 9th magnitudes divisible through small telescopes.

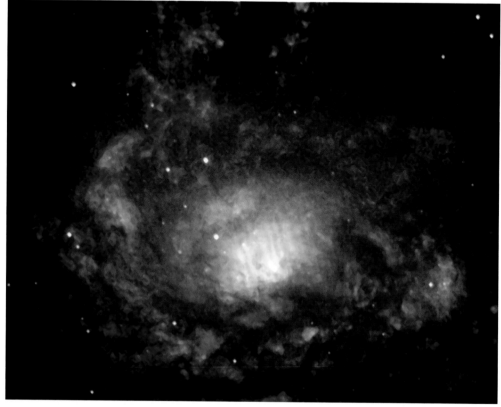

》 Circinus Galaxy
Hot gas, coloured deep-red at the top of this image, which was taken by the Hubble Space Telescope, is being ejected from the centre of this galaxy some 13 million light-years away in Circinus. The galaxy has a bright, active core, which is believed to harbour a supermassive black hole.

Norma Normae (Nor)

WIDTH **DEPTH** **SIZE RANKING** 74ᵗʰ **FULLY VISIBLE** 29°N–90°S

Norma is an unremarkable southern constellation lying in the Milky Way between Lupus and Scorpius. It was introduced in the 1750s by the Frenchman Nicolas Louis de Lacaille, although at first it was known as Norma et Regula, the Set Square and Rule. The stars that Lacaille originally designated Alpha (α) and Beta (β) have since been incorporated into neighbouring Scorpius.

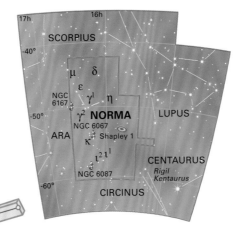

Southern Hemisphere

The Set Square

⟫ FEATURES OF INTEREST

Gamma-2 (γ²) Normae 👁 The constellation's brightest star, magnitude 4.0. It is one half of a naked-eye double with 5th-magnitude Gamma-1 (γ¹) Normae. The two stars lie at widely different distances and hence are unrelated.

Epsilon (ε) Normae 🏹 An easy double of 5th and 6th magnitudes for small telescopes.

Iota-1 (ι¹) Normae 🏹 An easy double with components of 5th and 8th magnitudes for small telescopes. Iota-2 (ι²) Normae is an unrelated star, some distance away.

NGC 6087 🔭 A large open cluster with radiating chains of stars, visible through binoculars. Its brightest star, near its centre, is S Normae, a Cepheid variable that ranges from 6th to 7th magnitudes every 9.8 days.

⟫ Smoke ring

Shapley 1, seen here in an image taken through the New Technology Telescope at the European Southern Observatory in Chile, is a beautifully symmetrical planetary nebula located some 2,500 light-years away in Norma. At the centre is a close binary star, which has spun off the ring of gas.

Interacting pair of galaxies
The shapes of this interacting pair of galaxies, jointly known as ESO 69-6, resemble musical notes. Long tidal tails of stars and gas were torn away from their outer regions as they brushed by each other in the distant past. The galaxies lie about 650 million light-years away in Triangulum Australe.

Triangulum Australe Trianguli Australis (TrA)

WIDTH 🖑 **DEPTH** 🖑 **SIZE RANKING** 83rd **FULLY VISIBLE** 19°N–90°S

Triangulum Australe is one of the southern constellations introduced in the late 16th century by the Dutch navigators Pieter Dirkszoon Keyser and Frederick de Houtman. It lies in the Milky Way not far from Alpha (α) Centauri. Although smaller than the northern triangle, Triangulum, its stars are brighter and so it is more prominent. Alpha (α) Trianguli Australis is magnitude 1.9, Beta (β) is 2.8, and Gamma (γ) is 2.9. There is little in this constellation to attract users of small telescopes.

Southern Hemisphere

》 FEATURES OF INTEREST

NGC 6025 🏃 ⚹ An open cluster visible with binoculars, noticeably elongated in shape and about one-third the apparent diameter of the full Moon.

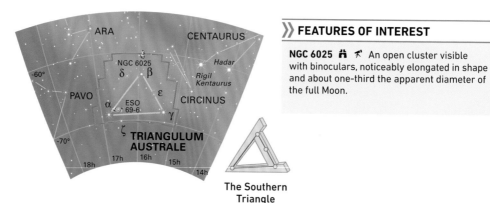

The Southern Triangle

Ara Arae (Ara)

WIDTH 〰 **DEPTH** 🖐🖐 **SIZE RANKING** 63rd **FULLY VISIBLE** 22°N–90°S

Ara is a southern constellation in the Milky Way, south of Scorpius. It was visualized by the ancient Greeks as the altar on which the gods of Olympus swore an oath of allegiance before their battle with the Titans for control of the universe. The altar's top faces south, and the Milky Way can be imagined to be smoke rising from the incense.

Southern Hemisphere

» FEATURES OF INTEREST

NGC 6193 🔭 An attractive open cluster for binoculars. It consists of about 30 stars of 6th magnitude and fainter covering an area half the apparent width of the full Moon.

NGC 6397 🔭 ✴ One of the closest globular clusters to us, around 7,500 light-years away and well seen in binoculars and small telescopes. It appears relatively large, over half the apparent size of the full Moon. The stars in its outer regions are more widely scattered than in many globulars.

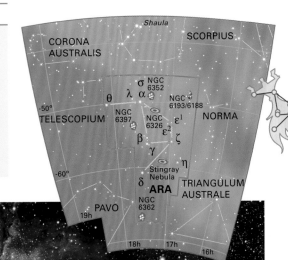

The Altar

《 **Cluster and nebula**
Stars in the naked-eye cluster NGC 6193, in Ara, illuminate the faint gas in the surrounding nebula NGC 6188. Dark arms of gas tens of light-years long are silhouetted against a brighter red background in this infrared image. Cluster and nebula lie nearly 4,000 light-years away.

Corona Australis Coronae Australis (CrA)

WIDTH  **DEPTH** **SIZE RANKING** 80th **FULLY VISIBLE** 44°N–90°S

Southern
Hemisphere

This small but attractive southern constellation lies on the edge of the Milky Way, under the feet of Sagittarius and next to the tail of Scorpius. It consists of an arc of stars of 4th magnitude and fainter, representing a crown or laurel wreath, and was one of the 48 constellations recognized by the ancient Greek astronomer Ptolemy in the 2nd century CE.

≫ FEATURES OF INTEREST

Gamma (γ) Coronae Australis ✴ A challenging binary. An aperture of 100mm (4in) is needed to separate the two 5th-magnitude stars. Their orbital period is 122 years. Currently they are slowly moving apart as seen from Earth, making them easier to see separately.

Kappa (κ) Coronae Australis ✴ An unrelated pair of 6th-magnitude stars easily separable with small telescopes.

NGC 6541 🔭 ✴ A modest globular cluster visible in binoculars and small telescopes, about one-third the apparent width of the full Moon.

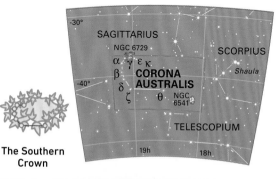

The Southern
Crown

◻ Stellar coronet

Glittering stars in the Coronet Cluster lie at the heart of a star-forming region some 500 light-years away in Corona Australis. Here the cluster is seen in a composite of invisible X-ray and infrared images from space satellites. The individual stars and surrounding nebulosity are too faint to see without telescopes.

◁ Dusty spiral
Located in Telescopium, NGC 6861, a faint spiral galaxy tilted at a steep angle to us, displays dark bands of dust in its spiral arms in this Hubble Space Telescope view. The galaxy is part of a small cluster of a dozen or so galaxies some 90 million light-years away that are named the NGC 6868 group after its brightest member.

Telescopium Telescopii (Tel)

WIDTH 🖑🖑 **DEPTH** 🖑 **SIZE RANKING** 57th **FULLY VISIBLE** 33°N–90°S

Telescopium is an unremarkable constellation south of Sagittarius and Corona Australis. It was invented in the 18th century by the French astronomer Nicolas Louis de Lacaille to commemorate the telescope, although its pattern of stars cannot be said to bear any resemblance to that instrument.

❯❯ FEATURES OF INTEREST

Delta (δ) Telescopii 👁 🔭 An unrelated pair of 5th-magnitude stars divisible with binoculars or even good eyesight.

Southern Hemisphere

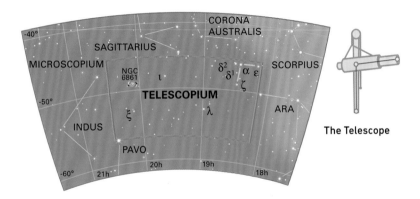

The Telescope

Southern Hemisphere

Indus Indi (Ind)

WIDTH 🖐 **DEPTH** 🖐🖐 **SIZE RANKING** 49th **FULLY VISIBLE** 15°N–90°S

This is one of the southern constellations introduced in the late 16th century by the Dutch navigator-astronomers Pieter Dirkszoon Keyser and Frederick de Houtman. It represents a man carrying a spear and arrows, although whether he is supposed to be a native of the East Indies, as encountered by the Dutch explorers during their expeditions, or a native of the Americas is not known. It is hard to trace the figure of a human from these stars.

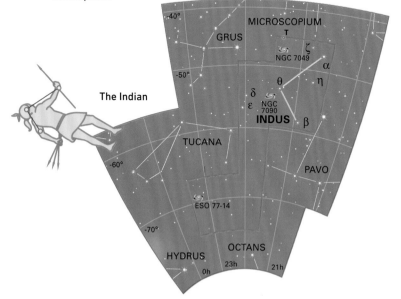

The Indian

≫ FEATURES OF INTEREST

Epsilon (ε) Indi 👁 🔭 One of the closest stars to our own Solar System, just 11.8 light-years away. Somewhat smaller and cooler than the Sun, it appears of magnitude 4.7 and is pale orange in colour.

Theta (θ) Indi 🔭 A double star with components of magnitudes 4.5 and 6.9. The companion star can be identified through small telescopes.

✉ Spiral galaxy NGC 7090
The spiral galaxy NGC 7090 in Indus appears elliptical because it is oriented side-on to us, however it is actually a spiral. Pinkish regions dotted over the galaxy in this view, taken from the Hubble Space Telescope, are glowing gas clouds, while darker patches are areas of dust. NGC 7090 lies nearly 30 million light-years away.

Grus Gruis (Gru)

WIDTH 🤚🤚 **DEPTH** 🤚🤚 **SIZE RANKING** 45ᵗʰ **FULLY VISIBLE** 33°N–90°S

Grus, a constellation of the southern sky between Piscis Austrinus and Tucana, was introduced at the end of the 17th century by the Dutch navigator-astronomers Pieter Dirkszoon Keyser and Frederick de Houtman. It represents a long-necked wading bird, the crane.

Southern Hemisphere

⟫ FEATURES OF INTEREST

Beta (β) Gruis 👁 📷 A variable red giant whose brightness ranges from magnitude 2.0 to 2.3 with no set period.

Delta (δ) Gruis 👁 📷 One of two notable doubles in the constellation that are wide enough to be divisible with the unaided eye. It consists of a pair of 4th-magnitude giants, yellow and red.

Mu (μ) Gruis 👁 📷 A pair of 5th-magnitude yellow giants divisible with the unaided eye. Like Delta (δ) Gruis (above), it is a chance alignment, not a true binary.

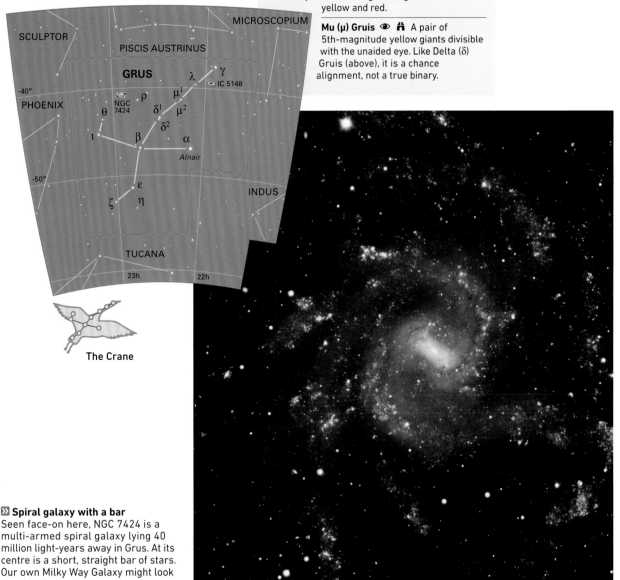

The Crane

⟫ **Spiral galaxy with a bar**
Seen face-on here, NGC 7424 is a multi-armed spiral galaxy lying 40 million light-years away in Grus. At its centre is a short, straight bar of stars. Our own Milky Way Galaxy might look like this if seen from outside.

Southern Hemisphere

Phoenix Phoenicis (Phe)

WIDTH 🖑🖑 **DEPTH** 🖑 **SIZE RANKING** 37th **FULLY VISIBLE** 32°N–90°S

Phoenix, which lies at the southern end of Eridanus, is the largest of the 12 constellations introduced by the Dutch explorers Pieter Dirkszoon Keyser and Frederick de Houtman. It represents the mythical bird that was reborn from the ashes of its predecessor.

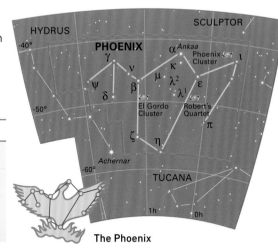

The Phoenix

》 FEATURES OF INTEREST

Zeta (ζ) Phoenicis 🏹 Both a double and a variable star. Small telescopes show it as a 4th-magnitude star with an 8th-magnitude companion. The brighter of the two stars is an eclipsing binary of the same type as Algol, varying between magnitudes 3.9 and 4.4 every 1.7 days.

》 X-ray cluster
Hot gas, shown here in blue, fills the space between galaxies in this massive cluster of galaxies over 5 billion light-years away in Phoenix. It is the most powerful producer of X-rays of any known cluster. This image of the Phoenix Cluster comes from the Chandra X-ray Observatory.

Tucana Tucanae (Tuc)

WIDTH 🖑🖑 **DEPTH** 🖑 **SIZE RANKING** 48th **FULLY VISIBLE** 14°N–90°S

This far southern constellation lies at the end of the celestial river, Eridanus. It now represents the large-beaked bird of South and Central America. When it was originally introduced in the late 16th century by the Dutch navigator-astronomers Pieter Dirkszoon Keyser and Frederick de Houtman, it was a bird of the East Indies. Its brightest star is Alpha (α) Tucanae, magnitude 2.8.

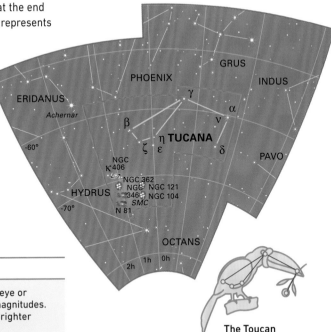

Southern Hemisphere

The Toucan

A telescope reveals that the globular cluster has an extremely bright, condensed centre with less dense outer regions.

47 Tucanae

⟩⟩ FEATURES OF INTEREST

Beta (β) Tucanae ◉ 👥 ⚲ A naked-eye or binocular double star of 4th and 5th magnitudes. Small telescopes further divide the brighter component into two.

Kappa (κ) Tucanae ⚲ A double star of 5th and 7th magnitudes, divisible through a small telescope.

The Small Magellanic Cloud (SMC) ◉ 👥 ⚲ The lesser of the two satellite galaxies that accompany our own home galaxy. It lies about 190,000 light-years away. To the naked eye, the Cloud looks like a detached patch of the Milky Way, seven times the apparent width of the full Moon. Binoculars and small telescopes show star fields and clusters within the Small Magellanic Cloud.

47 Tucanae (NGC 104) ◉ 👥 ⚲ A prominent globular cluster near the Small Magellanic Cloud. Actually it is a foreground object in our own galaxy and not associated with the Cloud. To the naked eye, it appears like a hazy 4th-magnitude star. Through binoculars and small telescopes, 47 Tucanae appears to cover the same area of sky as the full Moon. It is regarded as the second-best globular cluster in the entire sky, bettered only by Omega (ω) Centauri.

NGC 362 👥 ⚲ A smaller and fainter globular cluster than 47 Tucanae near the northern tip of the Small Magellanic Cloud, which will need binoculars or a small telescope to be seen. Like 47 Tucanae, it is a foreground object in our own galaxy.

Galaxy and cluster
The Small Magellanic Cloud (left) is an irregular galaxy in orbit around the Milky Way, whereas the globular cluster 47 Tucanae (right) is 175,000 light-years closer to Earth and part of our own galaxy.

**Southern
Hemisphere**

Hydrus Hydri (Hyi)

WIDTH 🖐🖐 **DEPTH** 🖐🖐 **SIZE RANKING** 61st **FULLY VISIBLE** 8°N–90°S

This constellation of the far southern sky lies between the two Magellanic Clouds and was introduced in the late 16th century by the Dutch navigator-astronomers Pieter Dirkszoon Keyser and Frederick de Houtman. It is not to be confused with Hydra, the large water snake that was one of the constellations known to the ancient Greeks. Its brightest star is Beta (β) Hydri, magnitude 2.8.

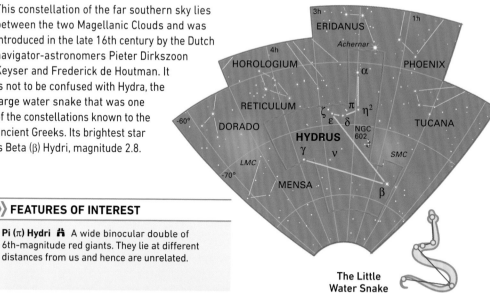

**The Little
Water Snake**

⟫ FEATURES OF INTEREST

Pi (π) Hydri 🔭 A wide binocular double of 6th-magnitude red giants. They lie at different distances from us and hence are unrelated.

⟫ Young star cluster
NGC 602 is a cluster of hot young stars in the outer reaches of the Small Magellanic Cloud, some 200,000 light-years away. Fierce radiation from the stars is eating away the remains of the gas cloud from which they formed a few million years ago. Denser parts of the nebula resist the erosion better, leaving long pillars pointing towards the central stars.

Horologium Horologii (Hor)

WIDTH 🖑🖑 **DEPTH** 🖑🖑 **SIZE RANKING** 58th **FULLY VISIBLE** 23°N–90°S

Horologium is a faint and unremarkable constellation of the southern sky near the foot of Eridanus. It represents a pendulum clock of the kind once used in observatories. It is one of the group of scientific constellations introduced in the 18th century by the French astronomer Nicolas Louis de Lacaille.

⟫ FEATURES OF INTEREST

R Horologii 🔭 ✴ A red giant variable star of the same type as Mira, ranging between 5th and 14th magnitude every 13 months or so.

NGC 1261 ✴ A compact globular cluster, more than 50,000 light-years from us. It is dimly visible through small telescopes.

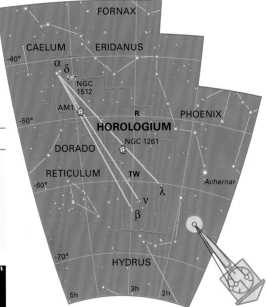

Southern Hemisphere

The Pendulum Clock

⌃ Heart of a globular
NGC 1261 is a globular cluster in Horologium visible through small telescopes. This exposure at infrared wavelengths, through the European Southern Observatory's New Technology Telescope, penetrates to the heart of the cluster.

⟫ Ring of stars
The Hubble Space Telescope reveals a ring of young star clusters surrounding the core of NGC 1512, a barred spiral galaxy some 30 million light-years away in Horologium. The bar in the galaxy, which is not visible in this image, funnels gas into the ring, causing bursts of star formation.

Southern Hemisphere

Reticulum Reticuli (Ret)

WIDTH ⟨🖐⟩ **DEPTH** ⟨🖐⟩ **SIZE RANKING** 82nd **FULLY VISIBLE** 23°N–90°S

This small constellation in the southern sky, near the Large Magellanic Cloud, was introduced in the 18th century by the French astronomer Nicolas Louis de Lacaille. It represents the reticule, a grid of fine lines, in the eyepiece of his telescope. He used this for measuring star positions when cataloguing the southern stars. The main stars of Reticulum form a diamond shape, which is reminiscent of the shape of the reticule in Lacaille's eyepiece.

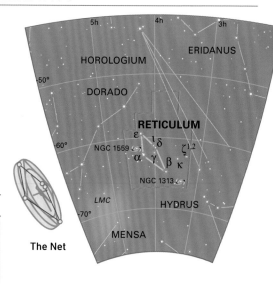

》 FEATURES OF INTEREST

Zeta (ζ) Reticuli 🔭 A pair of 5th-magnitude yellow stars similar to the Sun, divisible with binoculars or even good eyesight. Both are 39 light-years away.

The Net

Southern Hemisphere

Pictor Pictoris (Pic)

WIDTH ⟨🖐🖐⟩ **DEPTH** ⟨🖐🖐⟩ **SIZE RANKING** 59th **FULLY VISIBLE** 26°N–90°S

This faint constellation of the southern sky adjoining Carina and Puppis was invented in the 18th century by the French astronomer Nicolas Louis de Lacaille. He imagined it as an artist's easel and originally called it *Equuleus Pictoris*, although that name has since been shortened.

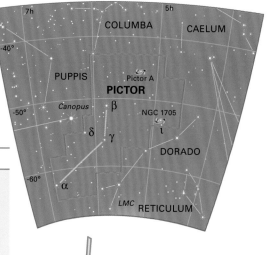

》 FEATURES OF INTEREST

Beta (β) Pictoris 🖥 🔭 A 4th-magnitude blue-white star encircled by a disc of dust and gas. This is thought to be a planetary system in the process of formation, in the same way that our own Solar System was born around the Sun. The disc can be seen only with special equipment on professional telescopes. Beta Pictoris is 63 light-years away.

Iota (ι) Pictoris 🔭 A double star consisting of 6th-magnitude components easily divisible with small telescopes.

The Painter's Easel

Dorado Doradus (Dor)

WIDTH 🖑🖑 **DEPTH** 🖑🖑 **SIZE RANKING** 72nd **FULLY VISIBLE** 20°N–90°S

Dorado contains most of the Large Magellanic Cloud (a nearby mini-galaxy), although some of it extends over the constellation's southern border into Mensa. Dorado is one of the constellations introduced in the late 16th century by the Dutch navigator–astronomers Pieter Dirkszoon Keyser and Frederick de Houtman. It represents the dolphinfish of tropical waters, not the more familiar goldfish commonly found in ponds and aquaria. The constellation has also been depicted as a swordfish.

Southern Hemisphere

The Goldfish

》 FEATURES OF INTEREST

Beta (β) Doradus 👁 🔭 A bright Cepheid variable, ranging between magnitudes 3.4 and 4.1 every 9.8 days.

R Doradus 👁 🔭 A red giant that varies somewhat erratically between 5th and 6th magnitudes every 6 months or so.

NGC 2070 (The Tarantula Nebula) 👁 🔭 🏃 🖥 The most remarkable object in the Large Magellanic Cloud, bright enough to be seen by the naked eye. At its heart, a cluster of new-born stars is visible through binoculars and small telescopes. Supernova 1987A, the first supernova visible to the naked eye for nearly 400 years, exploded near the Tarantula Nebula.

The Large Magellanic Cloud (LMC) 👁 🔭 🏃 A satellite galaxy of our own, some 170,000 light-years away. It looks at first sight like a detached part of the Milky Way, spanning about 12 full Moon diameters. Binoculars and small telescopes bring numerous star clusters and nebulous patches within it into view. Though named after the leader of the first expedition to circumnavigate the Earth, the first probable mention of the LMC is in the work of al-Sufi, a 10th-century Arab astronomer.

》 The Tarantula Nebula
Photographs show loops of gas in the Tarantula Nebula, like a spider's legs, from which the object gets its popular name.

Interior of Tarantula Nebula
This infrared view of the spidery Tarantula Nebula, seen through the Hubble Space Telescope, reveals newly formed stars that are blocked from view at optical wavelengths by intervening dust. The Tarantula Nebula is visible to the naked eye in the Large Magellanic Cloud, some 170,000 light-years away in the southern constellation Dorado.

Wheel of fire
This ring of stars, some 150,000 light-years across, resulted when a smaller galaxy, not visible on this image, passed through a normal spiral galaxy, disrupting it and firing up the formation of new stars. The ring galaxy, catalogued as AM 0644-741, lies 300 million light-years away in Volans.

Volans Volantis (Vol)

WIDTH 🖐 **DEPTH** 🤚 **SIZE RANKING** 76th **FULLY VISIBLE** 14°N–90°S

Southern Hemisphere

This small and faint constellation of the southern sky between Carina and the Large Magellanic Cloud was introduced in the late 16th century by the Dutch navigator-astronomers Pieter Dirkszoon Keyser and Frederick de Houtman, originally under the name Piscis Volans, the flying fish, which has since been shortened. It represents the tropical fish that uses its outstretched fins as wings to glide through the air, a creature that made a great impression on early European explorers. Although it lies on the edge of the Milky Way, Volans is surprisingly bereft of deep-sky objects.

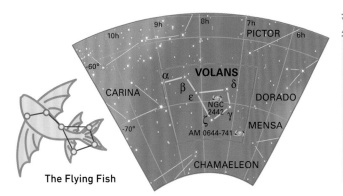

The Flying Fish

⟫ **FEATURES OF INTEREST**

Gamma (γ) Volantis 🏹 A 4th-magnitude orange star with a 6th-magnitude yellow companion visible through small telescopes, forming a beautifully coloured double.

Epsilon (ε) Volantis 🏹 Another double, for viewing through small telescopes, although not as colourful as Gamma Volantis. It consists of 4th and 7th magnitude components.

Mensa Mensae (Men)

WIDTH 🖐 **DEPTH** 🖐 **SIZE RANKING** 75th **FULLY VISIBLE** 5°N–90°S

A small constellation of the south polar region of the sky, Mensa is the faintest of all 88 constellations. Its brightest star, Alpha (α) Mensae, is only magnitude 5.1. Its main point of interest is the part of the Large Magellanic Cloud that overlaps it from neighbouring Dorado. Mensa was introduced in the 18th century by the French astronomer Nicolas Louis de Lacaille to commemorate Table Mountain, near Cape Town in South Africa, from where he observed the southern stars. The wispy appearance of the Large Magellanic Cloud reminded him of the cloud sometimes seen over the real Table Mountain.

Southern Hemisphere

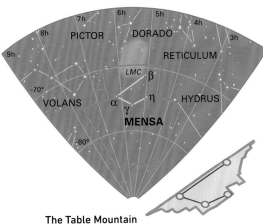

The Table Mountain

《 The Large Magellanic Cloud
Extremely rich in gas and dust, this irregular galaxy in orbit around our own has many regions of intensive star formation, including the spidery Tarantula Nebula, at upper left in this view.

Southern Hemisphere

Chamaeleon Chamaeleontis (Cha)

WIDTH 🖐🖐 **DEPTH** 🖐 **SIZE RANKING** 79th **FULLY VISIBLE** 7°N–90°S

Chamaeleon is a small, faint constellation of the south polar region of the sky, named after the lizard that can change its skin colour to match its surroundings. The constellation was introduced at the end of the 16th century by the Dutch navigator-astronomers Pieter Dirkszoon Keyser and Frederick de Houtman. Chamaeleon is placed in the sky next to Musca, the fly, another Keyser and de Houtman creation. On some early charts, the chamaeleon was depicted sticking its tongue out to catch the fly.

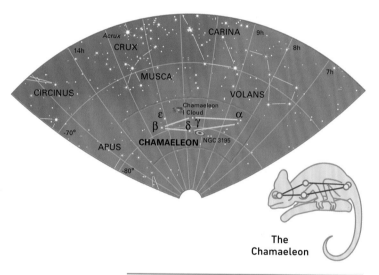

The Chamaeleon

》 FEATURES OF INTEREST

Delta (δ) Chamaeleontis 🔭 A wide pair of unrelated 4th- and 5th-magnitude stars, easily seen through binoculars.

NGC 3195 🏹 A planetary nebula of similar apparent size to Jupiter, but it is relatively faint and so requires a moderate-sized telescope to be seen.

》 Complex clouds
Clouds of gas and dust lit up by hot young stars are seen in this false-colour view captured by the European Southern Observatory in Chile. The area lies around 500 light-years away in the constellation Chamaeleon, and is one of the nearest star-forming regions to us.

Apus Apodis (Aps)

WIDTH ⁀ **DEPTH** ⁀ **SIZE RANKING** 67th **FULLY VISIBLE** 7°N–90°S

Apus, lying in the almost featureless region around the south pole of the sky, represents a bird of paradise. It is one of the figures introduced in the late 16th century by the Dutch navigator-astronomers Pieter Dirkszoon Keyser and Frederick de Houtman, but is a disappointing tribute to such an exotic family of birds. The constellation appeared on the celestial globe produced by Dutch cartographer Petrus Plancius in 1598 as "Paradysvogel Apis Indica". In works by 17th- and 18th-century astronomers, it is often referred to either as *Apis Indica* or simply *Avis Indica* (the Indian bird).

Southern Hemisphere

The Bird of Paradise

》 FEATURES OF INTEREST

Delta (δ) Apodis ⋔ A wide pair of unrelated 5th-magnitude red giants. They are easily divisible with binoculars.

Theta (θ) Apodis ⋔ A red giant that varies somewhat erratically between 5th and 7th magnitudes every four months or so.

☑ **Far southern globular**
IC 4499 is a small, faint globular cluster in Apus. Barely visible through amateur telescopes, the superior vision of the Hubble Space Telescope resolves it into a sparkling splash of stars. The cluster lies over 50,000 light-years from us.

Southern Hemisphere

Pavo Pavonis (Pav)

WIDTH ✋🖐 **DEPTH** ✋ **SIZE RANKING** 44th **FULLY VISIBLE** 15°N–90°S

Pavo is one of the far southern constellations introduced at the end of the 16th century by the Dutch navigator-astronomers Pieter Dirkszoon Keyser and Frederick de Houtman. It represents the peacock of southeast Asia, which the explorers encountered on their travels. It lies on the edge of the Milky Way next to another exotic bird, the toucan (the constellation Tucana). In Greek mythology, the peacock was the sacred bird of Hera, wife of Zeus, who travelled through the air in a chariot drawn by peacocks.

⟫ FEATURES OF INTEREST

Kappa (κ) Pavonis 👁 ⚲ One of the brighter Cepheid variables. Its variations, between magnitudes 3.9 and 4.8 every 9.1 days, are easy to follow with the naked eye and binoculars.

Xi (ξ) Pavonis ⟡ A double star with components of unequal brightness, 4th and 8th magnitudes. The fainter star is difficult to see with the smallest apertures because the brighter neighbour overwhelms it.

NGC 6744 ⟡ A large barred spiral galaxy presented virtually face-on to us, visible as an elliptical haze in telescopes of small to moderate aperture. It lies about 30 million light-years away.

NGC 6752 ⚲ ⟡ One of the largest and brightest globular clusters, just at the limit of naked-eye visibility and easily found with binoculars, covering half the apparent width of the full Moon. Telescopes with apertures of 75mm (3in) or more will resolve its brightest individual stars.

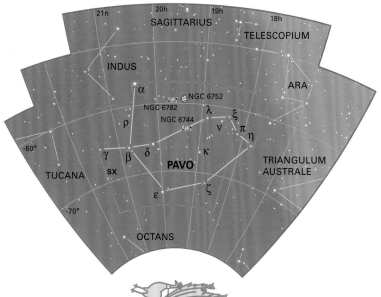

The Peacock

⟪ NGC 6752
This large and bright globular cluster, too far south to have appeared on Charles Messier's list, makes an excellent sight in all apertures.

⟫ Spiral galaxy NGC 6744
This beautiful face-on barred spiral galaxy is detectable with moderate-sized amateur telescopes. Our own galaxy, now thought to be a barred spiral, might look similar to this.

Octans Octantis (Oct)

WIDTH 🖑🖑 **DEPTH** 🖑🖑 **SIZE RANKING** 50th **FULLY VISIBLE** 0°N–90°S

Octans is the constellation at the south celestial pole, devised in the 18th century by the French astronomer Nicolas Louis de Lacaille. It represents a navigational instrument called an octant, a predecessor of the sextant. The octant was invented in 1731 by the English instrument maker John Hadley. This area of sky is quite barren, with no bright star to mark the position of the pole as Polaris does in the northern hemisphere. Because of the effect of precession (see p.151) the positions of the celestial poles are constantly changing, and the south celestial pole is moving in the direction of Chamaeleon.

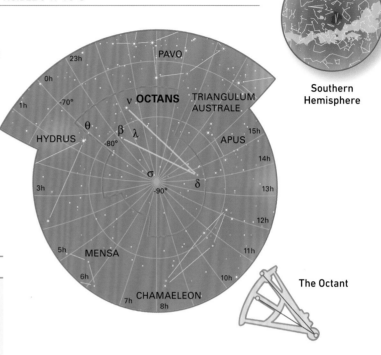

Southern Hemisphere

The Octant

⟫ FEATURES OF INTEREST

Lambda (λ) Octantis 🔭 A double star divisible with a small telescope. The components are of 5th and 7th magnitudes.

Sigma (σ) Octantis 👁 🔭 The nearest star to the south celestial pole detectable with the naked eye. A yellow-white giant, it lies 280 light-years away. However, it is only of magnitude 5.4 and hence far from prominent. Currently it is just over one degree from the celestial pole but this distance is increasing due to precession.

⌣ Southern pole star trails
Stars spin around the south celestial pole in this long-exposure photograph taken from Australia. Unlike in the north, there is no bright star near the southern pole.

Pleiades
Among the most beautiful sky
sights is the Pleiades star cluster in
Taurus, well placed for observation
from November to March. The faint
nebulosity around the stars is visible
only on photographs.

Monthly sky guide

The charts on the following pages show the stars as they appear each month around 10pm, as seen from various latitudes in the northern and southern hemispheres. Three horizons are marked on each of the charts, making them usable throughout most of the inhabited world. A bright object that is not on the charts will be one of the planets, which change position from night to night. A difference of a few degrees in latitude will have little effect on the stars you can see.

Each monthly chart is a representation of the entire sky above you, with the horizon around the rim and the zenith (the point directly overhead) in the middle. If you are facing north, hold the page so that the label NORTH on the rim of the chart is at the bottom. Similarly, if you are facing south, or any other direction, hold the page so that the label for that particular direction is at the bottom. What stars are on view in the night sky depends on your latitude on Earth.

60°N	
40°N	
20°N	
0°	
20°S	
40°S	

⏶ Find your latitude
Check your position on Earth on the map above and choose which latitude is closest to where you live. Colour-coded horizon lines on each monthly chart correspond to the latitude lines shown above. The horizons range from 60° north to 40° south.

⏵⏵ Whole sky charts
Each chart shows the sky as it appears at 10pm in mid-month. The sky will look the same at 11pm at the start of the month and 9pm at the end. Add an hour when daylight saving time is in operation.

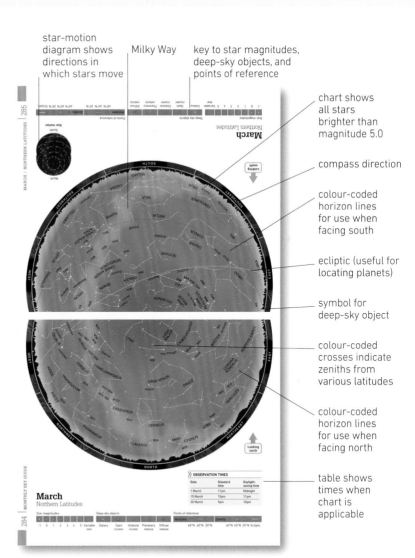

star-motion diagram shows directions in which stars move

Milky Way

key to star magnitudes, deep-sky objects, and points of reference

chart shows all stars brighter than magnitude 5.0

compass direction

colour-coded horizon lines for use when facing south

ecliptic (useful for locating planets)

symbol for deep-sky object

colour-coded crosses indicate zeniths from various latitudes

colour-coded horizon lines for use when facing north

table shows times when chart is applicable

March
Northern Latitudes

OBSERVATION TIMES		
Date	Standard time	Daylight-saving time
1 March	11pm	Midnight
15 March	10pm	11pm
30 March	9pm	10pm

January
Northern Latitudes

Star magnitudes

✦	-1
✦	0
✦	1
✦	2
·	3
·	4
·	5
✦	Variable star

Deep-sky objects

⬡	Galaxy
✦	Open cluster
⊛	Globular cluster
⬭	Planetary nebula
⬭	Diffuse nebula

Points of reference

Horizons	—— 60°N 40°N 20°N
Zeniths	+ + + 60°N 40°N 20°N Ecliptic

▶▶ OBSERVATION TIMES

Date	Standard time	Daylight-saving time
1 January	11pm	Midnight
15 January	10pm	11pm
30 January	9pm	10pm

Looking north →

Constellations and objects labelled on chart:
PEGASUS, ANDROMEDA, PISCES, LACERTA, CASSIOPEIA, TRIANGULUM, PERSEUS, CYGNUS, CEPHEUS, CAMELOPARDALIS, AURIGA, Deneb, M39, M52, M103, NGC 884, NGC 869, M31, M33, M34, Capella, LYRA, Vega, M57, DRACO, Polaris, URSA MINOR, LYNX, HERCULES, M92, M81, LEO MINOR, M13, M101, Mizar, The Plough, URSA MAJOR, CORONA BOREALIS, BOÖTES, M51, CANES VENATICI, LEO, M3, COMA BERENICES, M53, M64, M87

WEST, NORTHWEST, NORTH, NORTHEAST, EAST

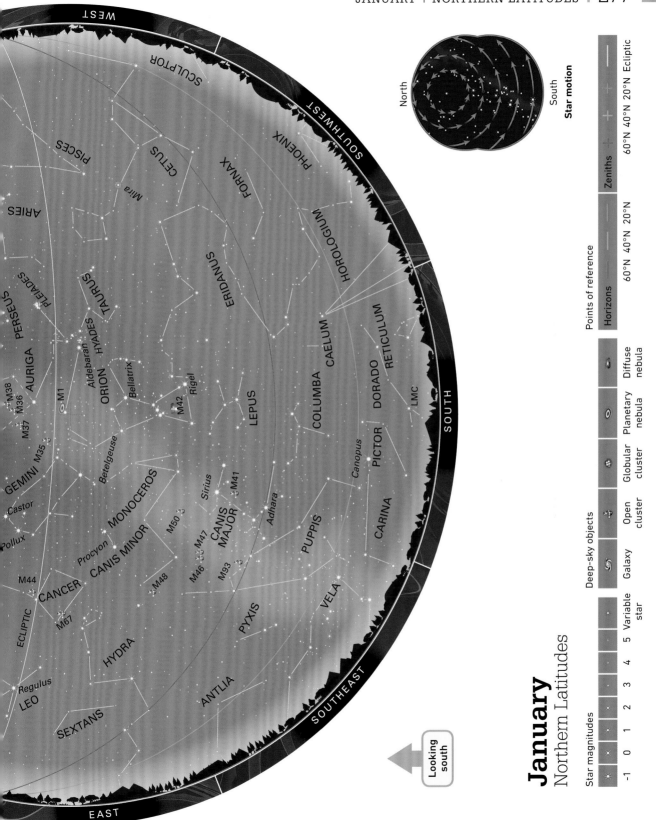

WEST

SCULPTOR

SOUTHWEST

PISCES

CETUS

PHOENIX

Mira

ARIES

FORNAX

PERSEUS

PLEIADES

HOROLOGIUM

TAURUS

ERIDANUS

HYADES

Aldebaran

AURIGA

CAELUM

M38

ORION

Bellatrix

CAELUM

COLUMBA

DORADO

RETICULUM

M36

M1

M37

Rigel

M42

LEPUS

SOUTH

M35

Betelgeuse

PICTOR

LMC

GEMINI

MONOCEROS

Canopus

Castor

Sirius

M41

CARINA

Pollux

M50

Adhara

CANIS MAJOR

M44

Procyon

M47

PUPPIS

CANCER

CANIS MINOR

M46

M93

M67

M48

VELA

ECLIPTIC

PYXIS

HYDRA

Regulus

ANTLIA

LEO

SEXTANS

SOUTHEAST

EAST

North — Star motion — South

Star motion

January
Northern Latitudes

Looking south

Points of reference

Horizons			Zeniths		
60°N	40°N	20°N	60°N	40°N	20°N Ecliptic

Deep-sky objects

Galaxy	Open cluster	Globular cluster	Planetary nebula	Diffuse nebula

Star magnitudes

-1	0	1	2	3	4	5 Variable star

January
Southern Latitudes

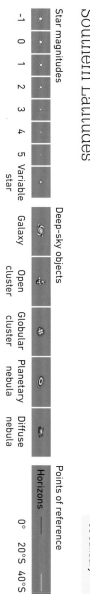

Star magnitudes

-1	0	1	2	3	4	5	Variable star

Deep-sky objects

Galaxy	Open cluster	Globular cluster	Planetary nebula	Diffuse nebula

Points of reference

Horizons				Zeniths		
0°	20°S	40°S		0°	20°S	40°S	Ecliptic

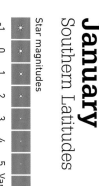

» OBSERVATION TIMES

Date	Standard time	Daylight-saving time
1 January	11pm	Midnight
15 January	10pm	11pm
30 January	9pm	10pm

Looking north →

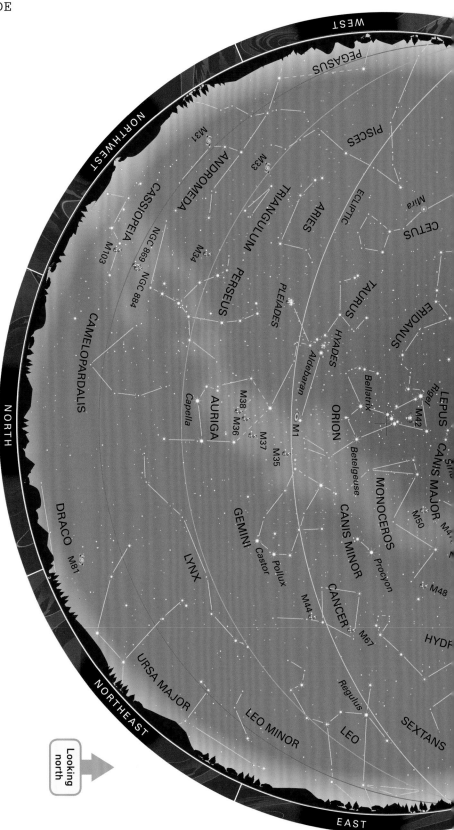

January
Southern Latitudes

Looking south

Star magnitudes

| -1 | 0 | 1 | 2 | 3 | 4 | 5 | Variable star |

Deep-sky objects

| Galaxy | Open cluster | Globular cluster | Planetary nebula | Diffuse nebula |

Points of reference

Horizons
0°S   20°S   40°S

Zeniths
0°S   20°S   40°S   Ecliptic

North

South
Star motion

February
Northern Latitudes

Star magnitudes

-1	0	1	2	3	4	5	Variable star

Deep-sky objects

Galaxy	Open cluster	Globular cluster	Planetary nebula	Diffuse nebula

Points of reference

	Horizons	Zeniths	
	60°N 40°N 20°N	60°N 40°N 20°N	Ecliptic

OBSERVATION TIMES

Date	Standard time	Daylight-saving time
1 February	11pm	Midnight
15 February	10pm	11pm
1 March	9pm	10pm

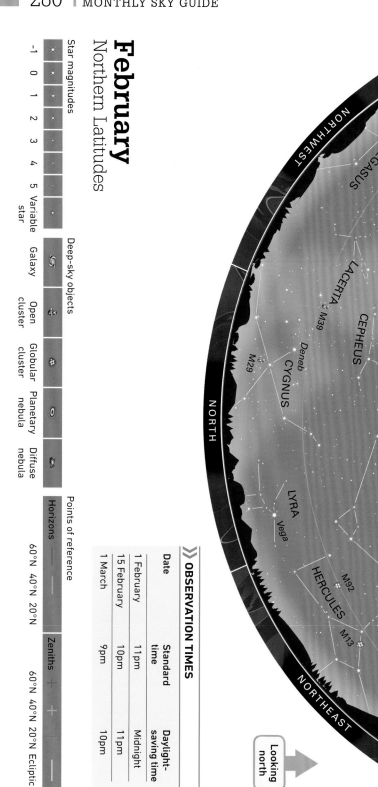

Looking north

WEST

NORTHWEST

NORTH

NORTHEAST

EAST

PISCES
ARIES
TRIANGULUM
M33
M34
PLEIADES
PEGASUS
ANDROMEDA
M31
CASSIOPEIA
M52
M103
NGC 869
NGC 884
PERSEUS
CAMELOPARDALIS
Capella
AURIGA
LACERTA
M39
CEPHEUS
Deneb
CYGNUS
M29
LYRA
Vega
Polaris
URSA MINOR
DRACO
M81
The Plough
LYNX
URSA MAJOR
LEO MINOR
M101
Mizar
M51
CANES VENATICI
HERCULES
M92
M13
M3
BOÖTES
M64
COMA BERENICES
M53
CORONA BOREALIS
Arcturus

WEST

CETUS

Mira

TAURUS

ERIDANUS

FORNAX

SOUTHWEST

HYADES

Aldebaran

M1

AURIGA

M35

Bellatrix

ORION

M42

Rigel

LEPUS

COLUMBA

CAELUM

PICTOR

DORADO

GEMINI

Castor

Pollux

CANCER

Betelgeuse

CANIS MINOR

NGC 2244

M50

Sirius

CANIS MAJOR

M41

Adhara

Canopus

North

South

Star motion

MONOCEROS

Procyon

M48

M47

M46

M93

PUPPIS

VELA

CARINA

VOLANS

SOUTH

M44

M67

ECLIPTIC

HYDRA

PYXIS

ANTLIA

LEO

Regulus

SEXTANS

CRATER

CORVUS

SOUTHEAST

M87

VIRGO

M104

EAST

Looking
south

February
Northern Latitudes

Star magnitudes

-1	0	1	2	3	4	5	Variable star

Deep-sky objects

Galaxy · Open cluster · Globular cluster · Planetary nebula · Diffuse nebula

Points of reference

Horizons · Zeniths · Ecliptic

60°N · 40°N · 20°N · 60°N · 40°N · 20°N

February
Southern Latitudes

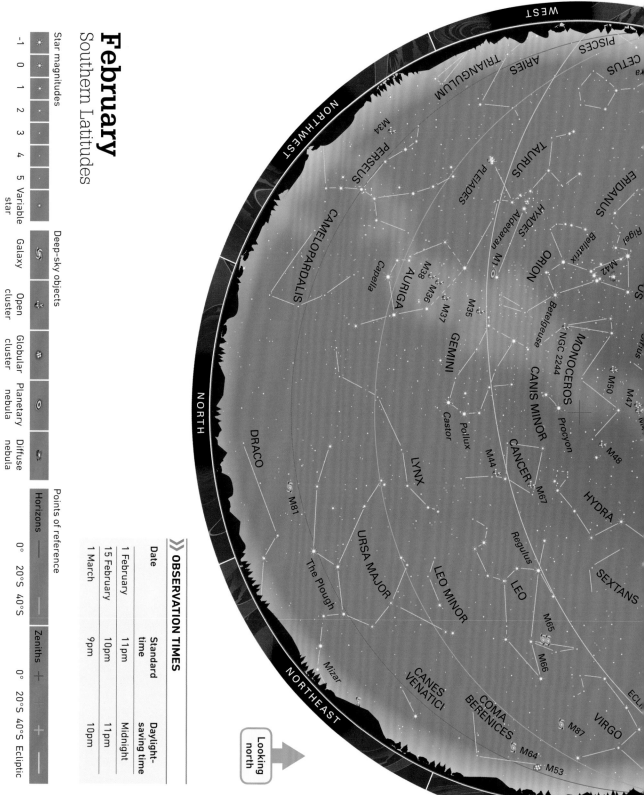

Star magnitudes

-1	0	1	2	3	4	5	Variable star

Deep-sky objects

Galaxy	Open cluster	Globular cluster	Planetary nebula	Diffuse nebula

Points of reference

Horizons			Zeniths			
0°	20°S	40°S	0°	20°S	40°S	Ecliptic

OBSERVATION TIMES

Date	Standard time	Daylight-saving time
1 February	11pm	Midnight
15 February	10pm	11pm
1 March	9pm	10pm

Looking north

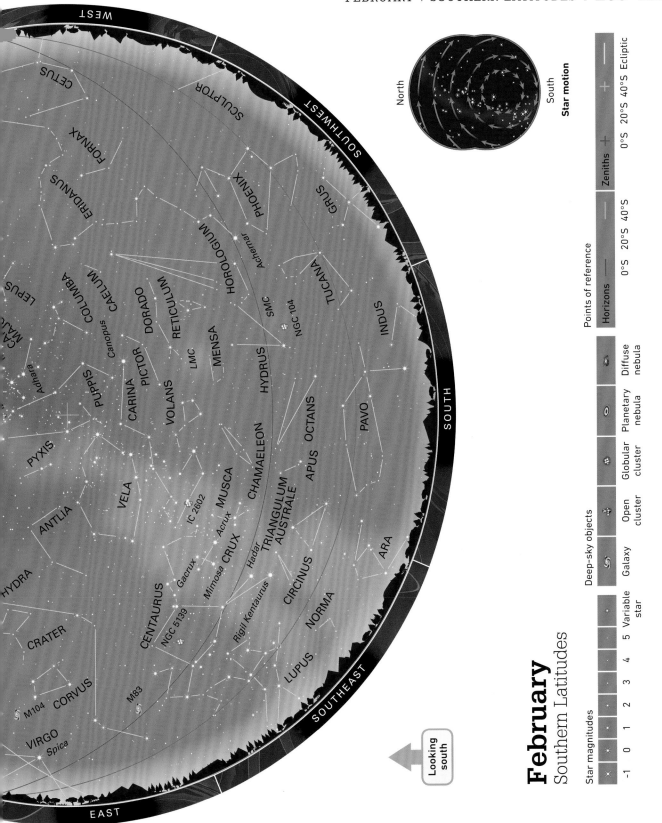

Star motion

North

South

February
Southern Latitudes

Looking south

Points of reference

Horizons			Zeniths			Ecliptic
0°S	20°S	40°S	0°S	20°S	40°S	

Deep-sky objects

Galaxy	Open cluster	Globular cluster	Planetary nebula	Diffuse nebula

Star magnitudes

-1	0	1	2	3	4	5	Variable star

Constellation and star labels

WEST
SOUTHWEST
SOUTH
SOUTHEAST
EAST

CETUS
SCULPTOR
FORNAX
ERIDANUS
PHOENIX
HOROLOGIUM
Achernar
GRUS
TUCANA
SMC
NGC 104
INDUS
RETICULUM
DORADO
CAELUM
COLUMBA
Canopus
PICTOR
CARINA
MENSA
LMC
VOLANS
HYDRUS
PAVO
OCTANS
APUS
CHAMAELEON
MUSCA
Acrux
IC 2602
PUPPIS
Adhara
PYXIS
VELA
ANTLIA
TRIANGULUM AUSTRALE
Hadar
Gacrux
Mimosa CRUX
Rigil Kentaurus
CIRCINUS
ARA
NORMA
LUPUS
CENTAURUS
NGC 5139
HYDRA
CRATER
CORVUS
M104
M83
VIRGO
Spica
LEPUS

March
Northern Latitudes

Star magnitudes

| -1 | 0 | 1 | 2 | 3 | 4 | 5 | Variable star |

Deep-sky objects

| Galaxy | Open cluster | Globular cluster | Planetary nebula | Diffuse nebula |

Points of reference

Horizons			Zeniths		
60°N	40°N	20°N	60°N	40°N	20°N Ecliptic

≫ OBSERVATION TIMES

Date	Standard time	Daylight-saving time
1 March	11pm	Midnight
15 March	10pm	11pm
30 March	9pm	10pm

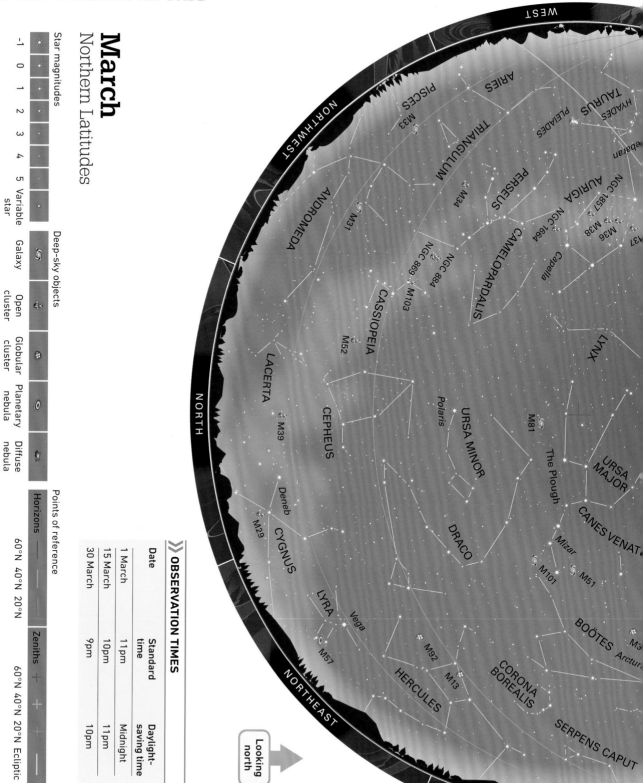

Looking north

North

South

Star motion

March
Northern Latitudes

Looking south

Points of reference

Horizons				Zeniths				Ecliptic
60°N	40°N	20°N		60°N	40°N	20°N		

Deep-sky objects

Galaxy	Open cluster	Globular cluster	Planetary nebula	Diffuse nebula

Star magnitudes

-1	0	1	2	3	4	5	Variable star

March
Southern Latitudes

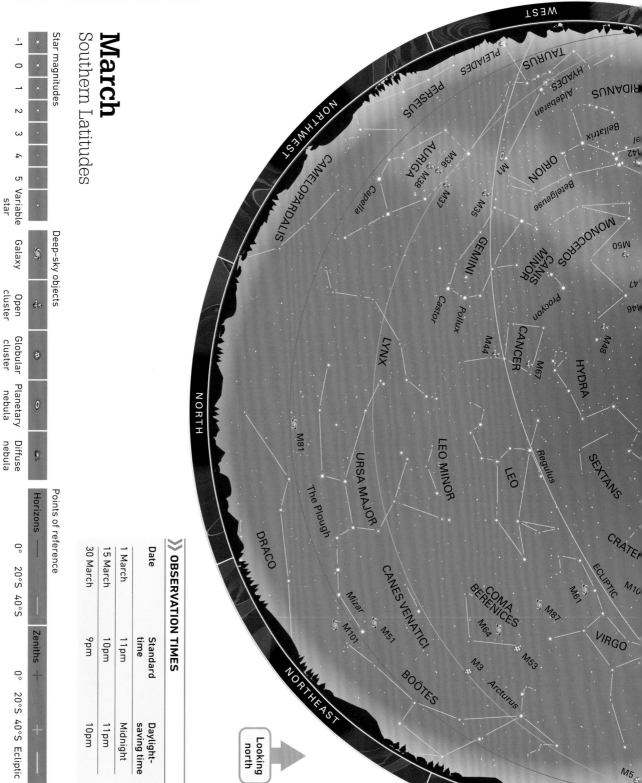

Star magnitudes

✦	-1
✦	0
✦	1
·	2
·	3
·	4
·	5
✦	Variable star

Deep-sky objects

🌀	Galaxy
❋	Open cluster
⊕	Globular cluster
◎	Planetary nebula
▨	Diffuse nebula

Points of reference

Horizons	—		
	0°	20°S	40°S
Zeniths	+	+	+
	0°	20°S	40°S Ecliptic

≫ OBSERVATION TIMES

Date	Standard time	Daylight-saving time
1 March	11pm	Midnight
15 March	10pm	11pm
30 March	9pm	10pm

Looking north →

WEST

FORNAX

LEPUS

ERIDANUS

CANIS MAJOR

Sirius

M41

Adhara

CAELUM

COLUMBA

Canopus

PICTOR

DORADO

HOROLOGIUM

SOUTHWEST

PHOENIX

Achernar

M9

PYXIS

PUPPIS

CARINA

VOLANS

LMC

RETICULUM

MENSA

HYDRUS

SMC

NGC 104

TUCANA

ANTLIA

VELA

CHAMAELEON

OCTANS

INDUS

SOUTH

HYDRA

CENTAURUS

Gacrux

Acrux

Mimosa

CRUX

NGC 4755

Hadar

MUSCA

CIRCINUS

APUS

PAVO

TRIANGULUM
AUSTRALE

NGC 5139

Rigil Kentaurus

ARA

TELESCOPIUM

CORVUS

M83

NORMA

LUPUS

Spica

LIBRA

M4

SCORPIUS

Shaula

SOUTHEAST

M80 Antares

M62

GO

EAST

**Looking
south**

North

South

Star motion

Points of reference

Horizons — 0°S 20°S 40°S   Zeniths + 0°S 20°S 40°S Ecliptic

Deep-sky objects

| Galaxy | Open cluster | Globular cluster | Planetary nebula | Diffuse nebula |

Variable star

Star magnitudes

-1 0 1 2 3 4 5

March
Southern Latitudes

April
Northern Latitudes

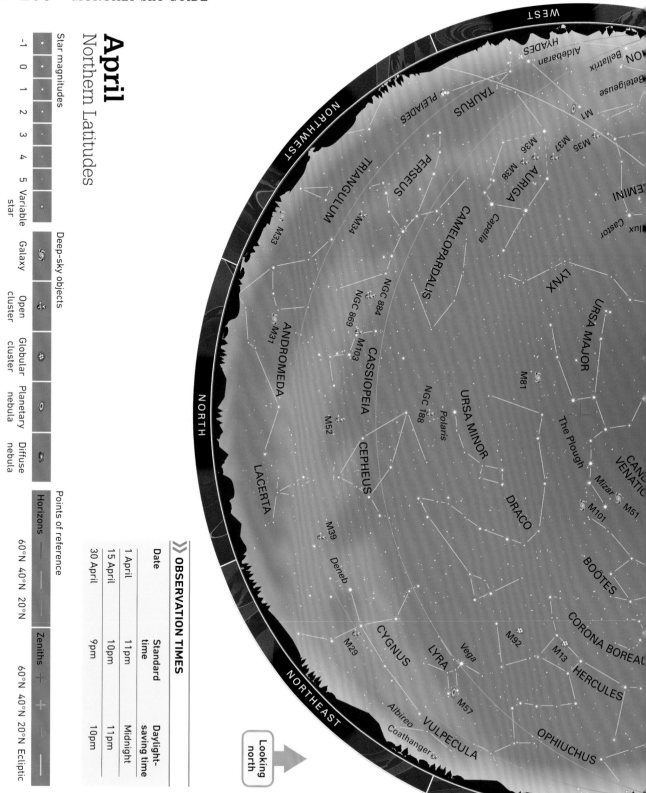

Star magnitudes

-1	0	1	2	3	4	5	Variable star

Deep-sky objects

Galaxy · Open cluster · Globular cluster · Planetary nebula · Diffuse nebula

Points of reference

Horizons — Zeniths + Ecliptic

| | 60°N 40°N 20°N | 60°N 40°N 20°N |

≫ OBSERVATION TIMES

Date	Standard time	Daylight-saving time
1 April	11pm	Midnight
15 April	10pm	11pm
30 April	9pm	10pm

Looking north

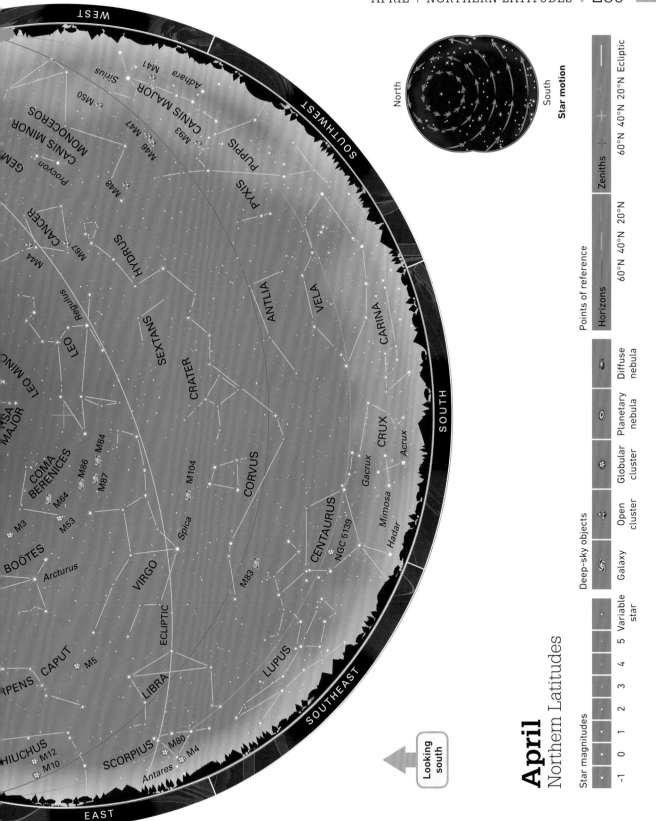

April
Northern Latitudes

Looking south

Star motion

North

South

Star motion

Points of reference

Horizons

| 60°N | 40°N | 20°N |

Zeniths

| 60°N | 40°N | 20°N | Ecliptic |

Deep-sky objects

| Galaxy | Open cluster | Globular cluster | Planetary nebula | Diffuse nebula |

Star magnitudes

| -1 | 0 | 1 | 2 | 3 | 4 | 5 | Variable star |

April
Southern Latitudes

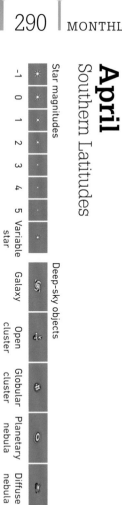

Star magnitudes

-1	✦
0	✦
1	✦
2	✦
3	∗
4	∗
5	·
Variable star	✴

Deep-sky objects

Galaxy	⬭
Open cluster	✿
Globular cluster	✺
Planetary nebula	◉
Diffuse nebula	▨

Points of reference

Horizons	Zeniths	Ecliptic
——	+	——

OBSERVATION TIMES

Date	Standard time	Daylight-saving time
1 April	11pm	Midnight
15 April	10pm	11pm
30 April	9pm	10pm

	Horizons			Zeniths		
	0°	20°S	40°S	0°	20°S	40°S

Looking north →

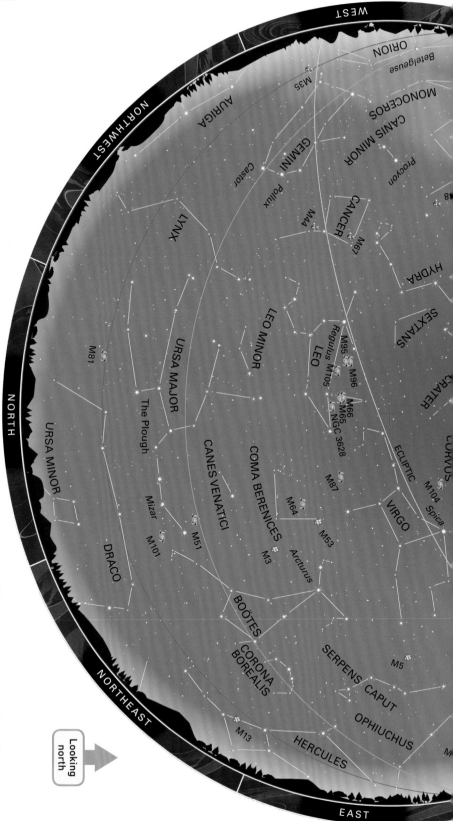

April
Southern Latitudes

North / South
Star motion

Points of reference

Horizons — 0°S — 20°S — 40°S

Zeniths + 0°S + 20°S + 40°S Ecliptic

Deep-sky objects

| Galaxy | Open cluster | Globular cluster | Planetary nebula | Diffuse nebula |

Star magnitudes

| -1 | 0 | 1 | 2 | 3 | 4 | 5 | Variable star |

Looking south

WEST

ORION
LEPUS
CANIS MAJOR
Sirius
M41
Adhara
M93
M
PUPPIS
PYXIS
ANTLIA
VELA
HYDRA
CORVUS
CENTAURUS
NGC 5139
M83
Gacrux
Mimosa
CRUX
Acrux
Hadar
MUSCA
Rigil Kentaurus
CIRCINUS
LUPUS
NORMA
TRIANGULUM AUSTRALE
LIBRA
SCORPIUS
M4
Antares
M80
M19
M62
Shaula
M6
M7
CORONA
AUSTRALIS
M69
SAGITTARIUS
M54
TELESCOPIUM
ARA
M10
OPHIUCHUS
M9
M8
M23
M21
M28
M24
M22
M54

COLUMBA
CAELUM
DORADO
PICTOR
CARINA
Canopus
VOLANS
CHAMAELEON
MENSA
LMC
APUS
OCTANS
PAVO
INDUS

SOUTHWEST
ERIDANUS
HOROLOGIUM
RETICULUM
Achernar
HYDRUS
SMC
NGC 104
PHOENIX
TUCANA

SOUTH

SOUTHEAST

EAST

May
Northern Latitudes

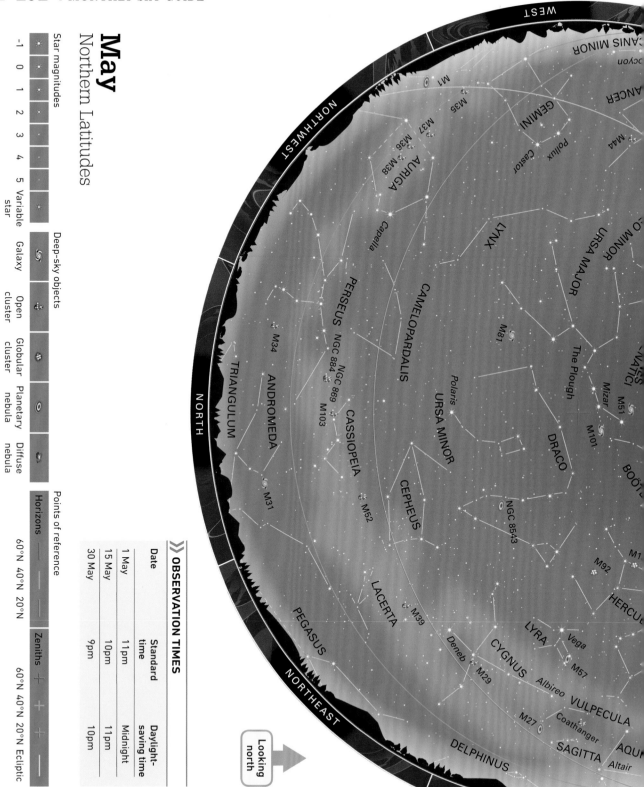

Star magnitudes

-1	0	1	2	3	4	5	Variable star

Deep-sky objects

Galaxy	Open cluster	Globular cluster	Planetary nebula	Diffuse nebula

Points of reference

Horizons				Zeniths		
60°N	40°N	20°N		60°N	40°N	20°N Ecliptic

›› OBSERVATION TIMES

Date	Standard time	Daylight-saving time
1 May	11pm	Midnight
15 May	10pm	11pm
30 May	9pm	10pm

Looking north →

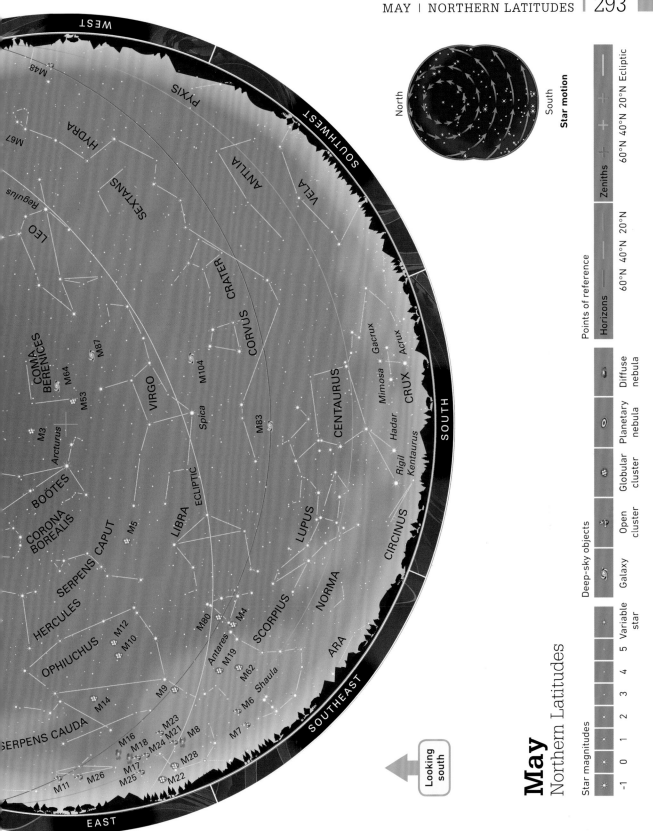

Star motion

North

South

WEST

PYXIS

M48

SOUTHWEST

ANTLIA

VELA

HYDRA

SEXTANS

M67

CRATER

Regulus

CORVUS

LEO

COMA BERENICES

M87

M104

Gacrux

M64

M53

VIRGO

Acrux

CRUX

M83

Mimosa

CENTAURUS

Hadar

M3

Spica

Arcturus

Rigil Kentaurus

BOÖTES

ECLIPTIC

SOUTH

CORONA BOREALIS

LIBRA

LUPUS

M5

SERPENS CAPUT

CIRCINUS

HERCULES

NORMA

OPHIUCHUS

M12

M10

SCORPIUS

ARA

M80

M4

Antares

M19

M9

M62

M14

M6 Shaula

SERPENS CAUDA

M7

M16

M23

M18

M21

M24

M8

M11

M17

M28

M26

M25

M22

SOUTHEAST

EAST

Looking south

May
Northern Latitudes

Star magnitudes

| -1 | 0 | 1 | 2 | 3 | 4 | 5 | Variable star |

Deep-sky objects

| Galaxy | Open cluster | Globular cluster | Planetary nebula | Diffuse nebula |

Points of reference

| Horizons | Zeniths | |
| 60°N | 40°N | 20°N | 60°N | 40°N | 20°N | Ecliptic |

May
Southern Latitudes

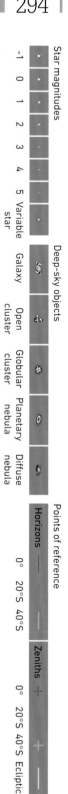

Star magnitudes

-1	★
0	★
1	★
2	★
3	•
4	•
5	·
Variable star	·

Deep-sky objects

Galaxy	Open cluster	Globular cluster	Planetary nebula	Diffuse nebula

Points of reference

Horizons			Zeniths		
0°	20°S	40°S	0°	20°S	40°S Ecliptic

OBSERVATION TIMES

Date	Standard time	Daylight-saving time
1 May	11pm	Midnight
15 May	10pm	11pm
30 May	9pm	10pm

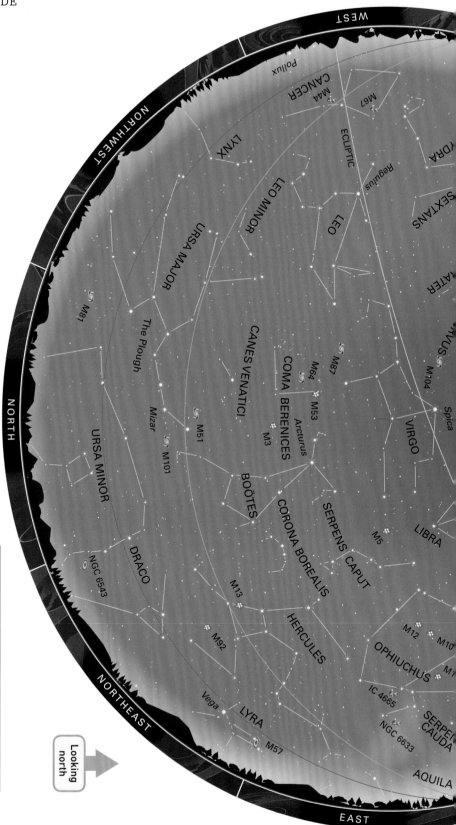

Looking north

WEST

SOUTHWEST

SOUTH

SOUTHEAST

EAST

MONOCEROS

CANIS MAJOR

M41

M47 M46 M93

Adhara

COLUMBA

PYXIS

PUPPIS

ANTLIA

VELA

Canopus

DORADO

HOROLOGIUM

RETICULUM

MENSA LMC

NGC 2516

VOLANS

CARINA

NGC 3114

NGC 3532

CRUX

Gacrux

Mimosa

Hadar

Coalsack Nebula

Acrux

MUSCA

CHAMAELEON

HYDRUS

Achernar

PHOENIX

SMC

NGC 104

TUCANA

CENTAURUS

NGC 5139

CRATER

CORVUS

HYDRA

M83

Rigil Kentaurus

LUPUS

CIRCINUS

TRIANGULUM AUSTRALE

APUS

OCTANS

NORMA

ARA

PAVO

TELESCOPIUM

INDUS

GRUS

MICROSCOPIUM

M4

Antares

M80

M19 M62

SCORPIUS

OPHIUCHUS Shaula

M9

M6

M7

CORONA AUSTRALIS

M23 M21 M8

M28

M69

M54

SAGITTARIUS

M55

M116 M17

M18 M24

M25 M22

SCUTUM

M26

111

AQUILA

M

North

South

Star motion

Looking south

May
Southern Latitudes

Points of reference

Horizons		Zeniths		Ecliptic
0°S 20°S 40°S		0°S 20°S 40°S		0°S 20°S 40°S

Deep-sky objects

Galaxy	Open cluster	Globular cluster	Planetary nebula	Diffuse nebula

Star magnitudes

-1	0	1	2	3	4	5	Variable star

June
Northern Latitudes

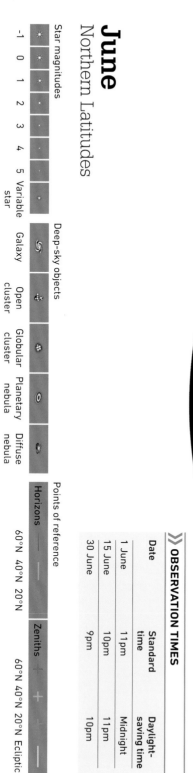

Star magnitudes

-1	
0	
1	
2	
3	
4	
5	
Variable star	

Deep-sky objects

Galaxy	Open cluster	Globular cluster	Planetary nebula	Diffuse nebula

Points of reference

Horizons			Zeniths		
60°N	40°N	20°N	60°N	40°N	20°N Ecliptic

≫ OBSERVATION TIMES

Date	Standard time	Daylight-saving time
1 June	11pm	Midnight
15 June	10pm	11pm
30 June	9pm	10pm

Looking north →

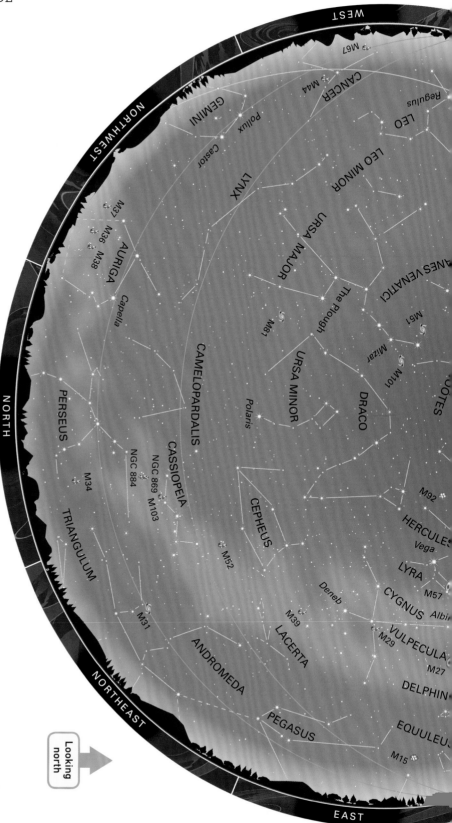

June
Northern Latitudes

Looking south

North / South

Star motion

Points of reference

Horizons
60°N 40°N 20°N

Zeniths
60°N 40°N 20°N Ecliptic

Deep-sky objects

Galaxy | Open cluster | Globular cluster | Planetary nebula | Diffuse nebula

Star magnitudes

Variable star
5 4 3 2 1 0 -1

Constellations and objects visible on the chart

WEST
SEXTANS
LEO
CRATER
HYDRA
SOUTHWEST
M104
CORVUS
VIRGO
M87
COMA BERENICES
M64
M53
Spica
M83
ECLIPTIC
NGC 5139
CENTAURUS
Gacrux
Mimosa
BOÖTES
M3
Arcturus
CORONA BOREALIS
SERPENS CAPUT
M5
LIBRA
LUPUS
Hadar
Rigil Kentaurus
CIRCINUS
SOUTH
M13
HERCULES
OPHIUCHUS
M12
M10
M80
M4
SCORPIUS
NORMA
TRIANGULUM AUSTRALE
M19 Antares
VULPECULA
M14
M9
M23
M62
Shaula
ARA
SERPENS CAUDA
SCUTUM
M16
M17
M18
M24 M21
M28 M8
M6
M7
TELESCOPIUM
SAGITTA
M11 M26
M25
M22
M69
CORONA AUSTRALIS
Altair
AQUILA
M55
M54
SAGITTARIUS
CAPRICORNUS
SOUTHEAST
EAST

June
Southern Latitudes

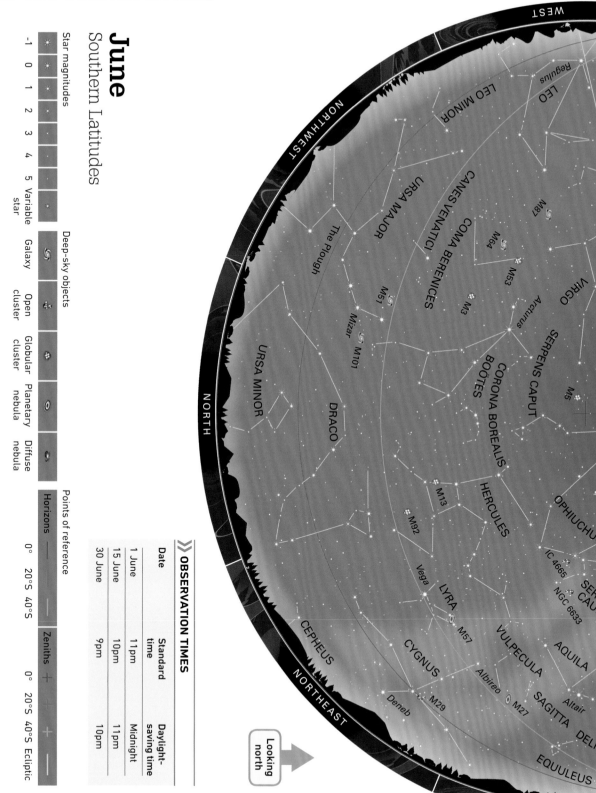

Star magnitudes

-1	0	1	2	3	4	5	Variable star	

Deep-sky objects

Galaxy	Open cluster	Globular cluster	Planetary nebula	Diffuse nebula

Points of reference

Horizons				Zeniths				Ecliptic
0°	20°S	40°S		0°	20°S	40°S		

▶ OBSERVATION TIMES

Date	Standard time	Daylight-saving time
1 June	11pm	Midnight
15 June	10pm	11pm
30 June	9pm	10pm

Looking north →

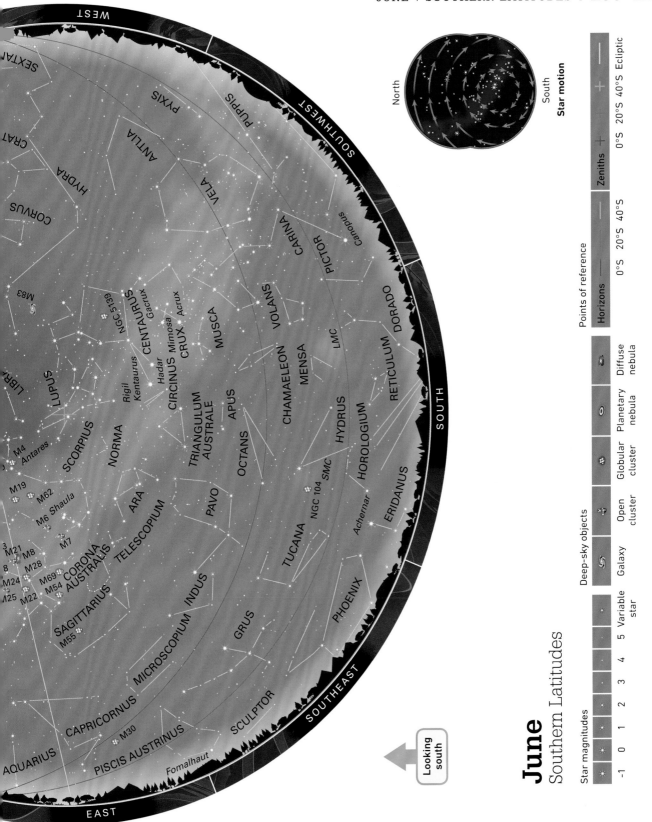

June
Southern Latitudes

Looking south

Star motion

North

South

Points of reference

Horizons — 0°S — 20°S — 40°S

Zeniths + 0°S + 20°S + 40°S Ecliptic

Deep-sky objects

Galaxy — Open cluster — Globular cluster — Planetary nebula — Diffuse nebula

Star magnitudes

-1 0 1 2 3 4 5 Variable star

WEST

SOUTHWEST

SOUTH

SOUTHEAST

EAST

SEXTANS

CRATER

HYDRA

CORVUS

LIBRA

LUPUS

SCORPIUS

NORMA

CIRCINUS

CENTAURUS

NGC 5139

Gacrux

Mimosa

Hadar

Rigil Kentaurus

CRUX

Acrux

MUSCA

TRIANGULUM AUSTRALE

APUS

OCTANS

ARA

PAVO

TELESCOPIUM

CORONA AUSTRALIS

SAGITTARIUS

INDUS

MICROSCOPIUM

CAPRICORNUS

AQUARIUS

PISCIS AUSTRINUS

Fomalhaut

SCULPTOR

GRUS

TUCANA

NGC 104 SMC

PHOENIX

HYDRUS

HOROLOGIUM

ERIDANUS

Achernar

RETICULUM

DORADO

LMC

MENSA

CHAMAELEON

VOLANS

CARINA

Canopus

PICTOR

ANTLIA

PYXIS

PUPPIS

VELA

M83

Antares

M4

M19

M62

M6 Shaula

M7

M21 M8

M28

M69 M54

M22 M24

M25

M30

M55

M55

July
Northern Latitudes

Star magnitudes

-1	✦
0	✦
1	✦
2	✦
3	✦
4	⋅
5	·
Variable star	⊙

Deep-sky objects

Galaxy	𝖘
Open cluster	✲
Globular cluster	⬤
Planetary nebula	⊘
Diffuse nebula	▨

Points of reference

	60°N	40°N	20°N	
Horizons	—	—	—	
Zeniths	+	+	+	
	60°N	40°N	20°N	Ecliptic

≫ OBSERVATION TIMES

Date	Standard time	Daylight-saving time
1 July	11pm	Midnight
15 July	10pm	11pm
30 July	9pm	10pm

Looking north →

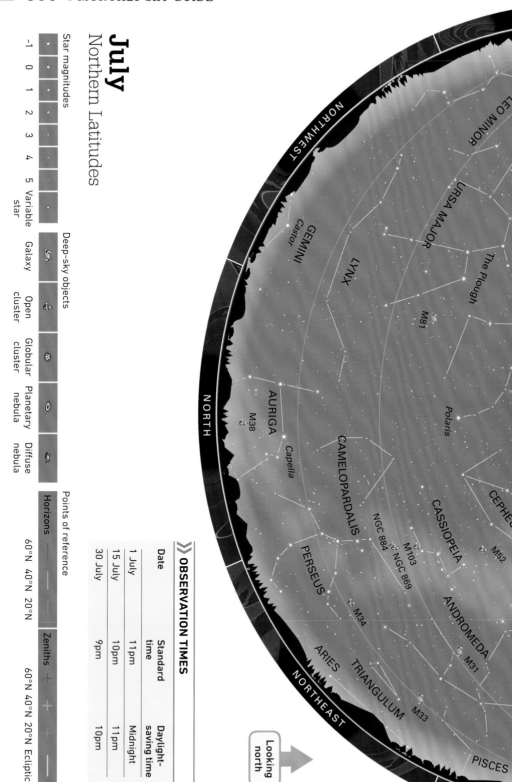

Star motion

North

South

Star motion

Points of reference

Horizons   Zeniths

60°N 40°N 20°N   60°N 40°N 20°N Ecliptic

Deep-sky objects

| Galaxy | Open cluster | Globular cluster | Planetary nebula | Diffuse nebula |

Star magnitudes

| -1 | 0 | 1 | 2 | 3 | 4 | 5 | Variable star |

July
Northern Latitudes

Looking south

July
Southern Latitudes

Star magnitudes

-1	0	1	2	3	4	5	Variable star

Deep-sky objects

Galaxy	Open cluster	Globular cluster	Planetary nebula	Diffuse nebula

Points of reference

Horizons			Zeniths			Ecliptic
0°	20°S	40°S	0°	20°S	40°S	

⟫ OBSERVATION TIMES

Date	Standard time	Daylight-saving time
1 July	11pm	Midnight
15 July	10pm	11pm
30 July	9pm	10 pm

Looking north →

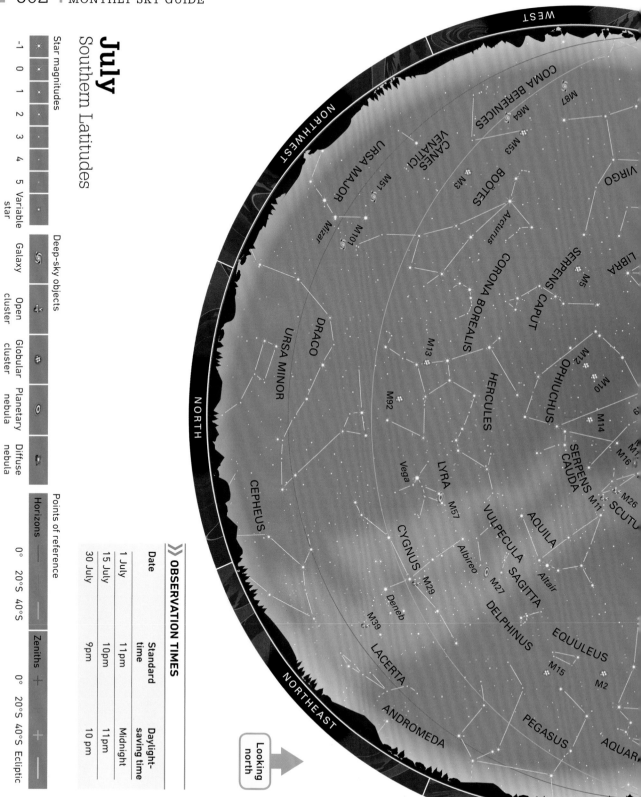

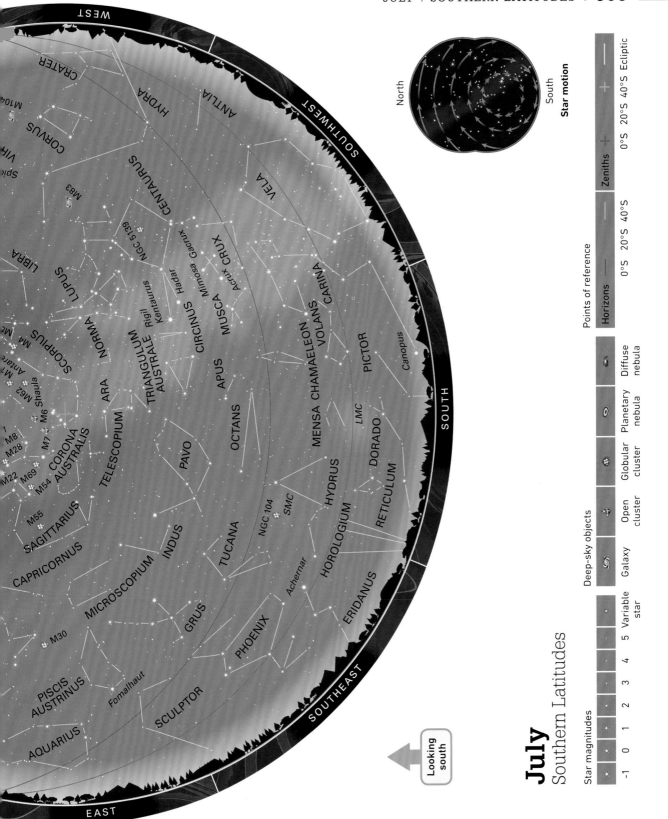

WEST

CRATER

M104

HYDRA

ANTLIA

CORVUS

Spica

VIR

SOUTHWEST

CENTAURUS

NGC 5139

VELA

LIBRA

Hadar

Mimosa Gacrux

LUPUS

Rigil Kentaurus

CIRCINUS

Acrux CRUX

MUSCA

CARINA

CHAMAELEON

VOLANS

NORMA

TRIANGULUM AUSTRALE

APUS

PICTOR

SCORPIUS

ARA

Canopus

M4 M

Antare

Mt Shaula

M62 M6

TELESCOPIUM

OCTANS

MENSA

M9

CORONA AUSTRALIS

PAVO

DORADO

SOUTH

M7

LMC

HYDRUS

RETICULUM

1

M8

M28

TUCANA

M22 M69

SMC

M54

NGC 104

SAGITTARIUS

INDUS

HOROLOGIUM

M55

Achernar

CAPRICORNUS

MICROSCOPIUM

GRUS

ERIDANUS

M30

PHOENIX

PISCIS AUSTRINUS

Fomalhaut

SCULPTOR

SOUTHEAST

AQUARIUS

EAST

North

South

Star motion

July
Southern Latitudes

Looking south

Points of reference

Horizons ——— ——— ——— Zeniths + + + Ecliptic

0°S 20°S 40°S 0°S 20°S 40°S

Deep-sky objects

Galaxy	Open cluster	Globular cluster	Planetary nebula	Diffuse nebula

Star magnitudes

-1   0   1   2   3   4   5   Variable star

August
Northern Latitudes

Star magnitudes

✦	-1
✦	0
✦	1
✦	2
✦	3
✦	4
✦	5
✦	Variable star

Deep-sky objects

Galaxy	
Open cluster	
Globular cluster	
Planetary nebula	
Diffuse nebula	

Points of reference

Horizons	
	60°N 40°N 20°N Zeniths
	60°N 40°N 20°N Ecliptic

≫ OBSERVATION TIMES

Date	Standard time	Daylight-saving time
1 August	11pm	Midnight
15 August	10pm	11pm
30 August	9pm	10pm

Looking north →

The sky chart shows constellations including: COMA BERENICES, M64, M53, Arcturus, M3, BOÖTES, CORONA BOREALIS, CANES VENATICI, M51, M101, Mizar, HERCULES, M13, M92, LYRA, Vega, URSA MAJOR, The Plough, URSA MINOR, DRACO, LYNX, M81, CYGNUS, Deneb, Polaris, CEPHEUS, M39, LACERTA, CAMELOPARDALIS, CASSIOPEIA, M52, ANDROMEDA, NGC 869, NGC 884, M103, AURIGA, Capella, PERSEUS, M31, PEGASUS, M34, M37, M38, M36, TRIANGULUM, M33, PISCES, TAURUS, PLEIADES, ARIES

Compass points: WEST, NORTHWEST, NORTH, NORTHEAST, EAST

WEST

VIRGO

LIBRA

SERPENS CAPUT

M5

LUPUS

SOUTHWEST

OPHIUCHUS

M12

M10

M14

SCORPIUS

M4

M80

M19 M62

Antares

Shaula

NORMA

M9

HERCULES

SERPENS CAUDA

M23

M16

M17

M21

M24

M8

M6

ARA

SCUTUM

M18

M25

M28

M7

M11

M26

M22

M69

M20

CORONA AUSTRALIS

VULPECULA

M57

Albireo

SAGITTA

AQUILA

M54

SAGITTARIUS

TELESCOPIUM

M27

LYRA

CYGNUS

M55

DELPHINUS

Altair

ECLIPTIC

PAVO

SOUTH

DELPHINUS

M15

EQUULEUS

M2

CAPRICORNUS

MICROSCOPIUM

INDUS

M30

PEGASUS

PISCIS AUSTRINUS

M55

GRUS

AQUARIUS

Fomalhaut

SCULPTOR

PISCES

PHOENIX

SOUTHEAST

CETUS

EAST

Looking south

North

South

Star motion

August
Northern Latitudes

Star magnitudes

| -1 | 0 | 1 | 2 | 3 | 4 | 5 | Variable star |

Deep-sky objects

| Galaxy | Open cluster | Globular cluster | Planetary nebula | Diffuse nebula |

Points of reference

Horizons

60°N 40°N 20°N

Zeniths

60°N 40°N 20°N Ecliptic

August
Southern Latitudes

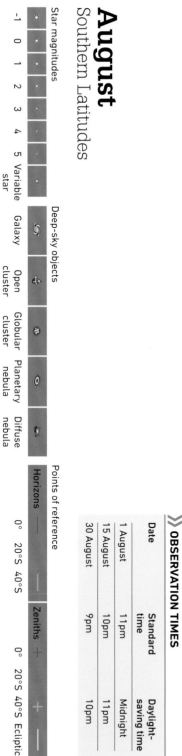

Star magnitudes

-1	0	1	2	3	4	5	Variable star

Deep-sky objects

Galaxy | Open cluster | Globular cluster | Planetary nebula | Diffuse nebula

Points of reference

Horizons — 0° 20°S 40°S

Zeniths + 0° 20°S 40°S

Ecliptic —

» OBSERVATION TIMES

Date	Standard time	Daylight-saving time
1 August	11pm	Midnight
15 August	10pm	11pm
30 August	9pm	10pm

Looking north →

August
Southern Latitudes

Looking south

Star motion

North

South

Points of reference

Horizons ——— 0°S 20°S 40°S

Zeniths ——+—— 0°S 20°S 40°S Ecliptic

Deep-sky objects

Galaxy	Open cluster	Globular cluster	Planetary nebula	Diffuse nebula

Star magnitudes

-1	0	1	2	3	4	5	Variable star

September
Northern Latitudes

Star magnitudes

-1	0	1	2	3	4	5	Variable star

Deep-sky objects

Galaxy	Open cluster	Globular cluster	Planetary nebula	Diffuse nebula

Points of reference

Horizons —	Zeniths +	Ecliptic —
60°N 40°N 20°N	60°N 40°N 20°N	

OBSERVATION TIMES

Date	Standard time	Daylight-saving time
1 September	11pm	Midnight
15 September	10pm	11pm
30 September	9pm	10pm

Looking north

September
Northern Latitudes

Looking south

Star motion

North

South

Points of reference

Horizons

60°N 40°N 20°N

Zeniths

60°N 40°N 20°N

Ecliptic

Deep-sky objects

| Galaxy | Open cluster | Globular cluster | Planetary nebula | Diffuse nebula |

Star magnitudes

| -1 | 0 | 1 | 2 | 3 | 4 | 5 | Variable star |

September
Southern Latitudes

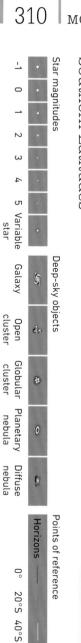

Star magnitudes

-1	0	1	2	3	4	5	Variable star

Deep-sky objects

Galaxy	Open cluster	Globular cluster	Planetary nebula	Diffuse nebula

Points of reference

Horizons			Zeniths			Ecliptic
0°	20°S	40°S	0°	20°S	40°S	

》 OBSERVATION TIMES

Date	Standard time	Daylight-saving time
1 September	11pm	Midnight
15 September	10pm	11pm
30 September	9pm	10pm

Looking north →

WEST

SERPENS CAPUT

M13

HERCULES

OPHIUCHUS

M14 M10 M12

NORTHWEST

M92

SERPENS CAUDA

SCUTUM

M11 M26

DRACO

LYRA

Vega

M57 Albireo

CYGNUS

M29

VULPECULA

SAGITTA

M27

DELPHINUS

AQUILA

Altair

CAPRICORNUS

Deneb M39

EQUULEUS

M15

NORTH

CEPHEUS

LACERTA

M2

M52

ANDROMEDA

PEGASUS

AQUARIUS

M103

CASSIOPEIA

NGC 869

NGC 884

M31

ECLIPTIC

PISCES

PERSEUS

M33

TRIANGULUM

ARIES

Mira

CETUS

M34

NORTHEAST

EAST

WEST

OPHIUCHU

LIBRA

M80
M4
Antares
M19
Shaula M62
M9
M23
M6
M8
M7
M21
M24
M28
M25
M22
M54
M69
M55

SCORPIUS

LUPUS

NORMA

SAGITTARIUS

CORONA AUSTRALIS

TELESCOPIUM

ARA

TRIANGULUM AUSTRALE

CIRCINUS

Hadar

CENTAURUS

NGC 5139

SOUTHWEST

MICROSCOPIUM

INDUS

PAVO

APUS

MUSCA

Rigil Kentaurus
Acrux
Mimosa
Gacrux

CRUX

CAPRICORNUS

M30

CHAMAELEON

CARINA

SOUTH

GRUS

TUCANA

OCTANS

AQUARIUS

PISCIS AUSTRINUS

Fomalhaut

SMC

MENSA

LMC

VOLANS

VELA

NGC 104

HYDRUS

RETICULUM

PHOENIX

Achernar

SCULPTOR

DORADO

PICTOR

ERIDANUS

HOROLOGIUM

CAELUM

Canopus

PUPPIS

CETUS

FORNAX

COLUMBA

SOUTHEAST

EAST

Looking south

September
Southern Latitudes

North · South
Star motion

Points of reference

Horizons		Zeniths	
0°S	20°S 40°S		0°S 20°S 40°S Ecliptic

Deep-sky objects

Galaxy	Open cluster	Globular cluster	Planetary nebula	Diffuse nebula

Star magnitudes

-1	0	1	2	3	4	5 Variable star

October
Northern Latitudes

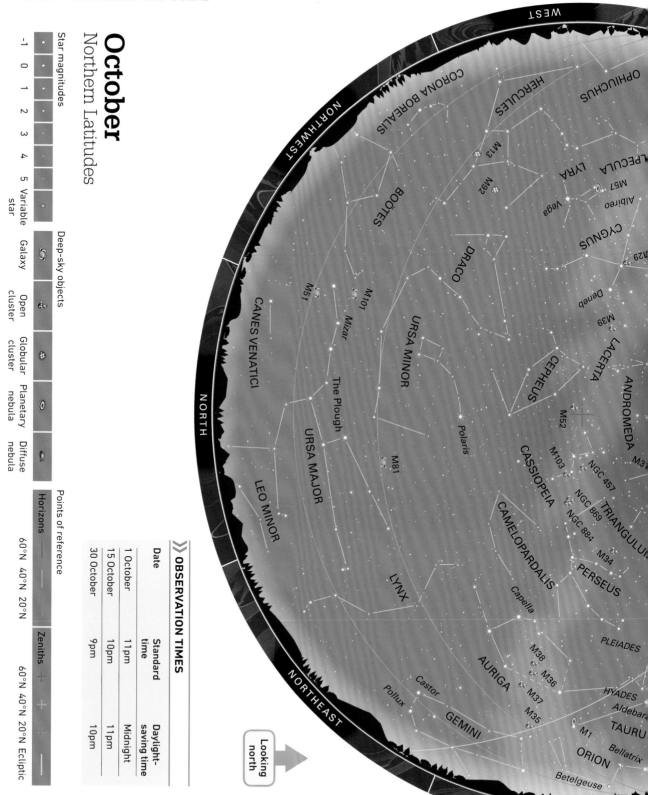

Star magnitudes

| | -1 | 0 | 1 | 2 | 3 | 4 | 5 | Variable star |

Deep-sky objects

Galaxy · Open cluster · Globular cluster · Planetary nebula · Diffuse nebula

Points of reference

Horizons — 60°N 40°N 20°N

Zeniths + + + 60°N 40°N 20°N Ecliptic

» OBSERVATION TIMES

Date	Standard time	Daylight-saving time
1 October	11pm	Midnight
15 October	10pm	11pm
30 October	9pm	10pm

Looking north →

WEST · NORTHWEST · NORTH · NORTHEAST · EAST

CORONA BOREALIS · HERCULES · OPHIUCHUS · BOÖTES · LYRA · VULPECULA · M13 · M92 · Vega · Albireo · CYGNUS · M29 · Deneb · M39 · LACERTA · ANDROMEDA · M31 · DRACO · M51 · M101 · Mizar · URSA MINOR · CEPHEUS · M52 · CASSIOPEIA · M103 · NGC 457 · TRIANGULUM · The Plough · Polaris · CANES VENATICI · NGC 869 · NGC 884 · M34 · CAMELOPARDALIS · PERSEUS · URSA MAJOR · M81 · LEO MINOR · Capella · LYNX · PLEIADES · AURIGA · M38 · M36 · M37 · HYADES · Castor · M35 · Aldebaran · Pollux · GEMINI · M1 · TAURUS · ORION · Bellatrix · Betelgeuse

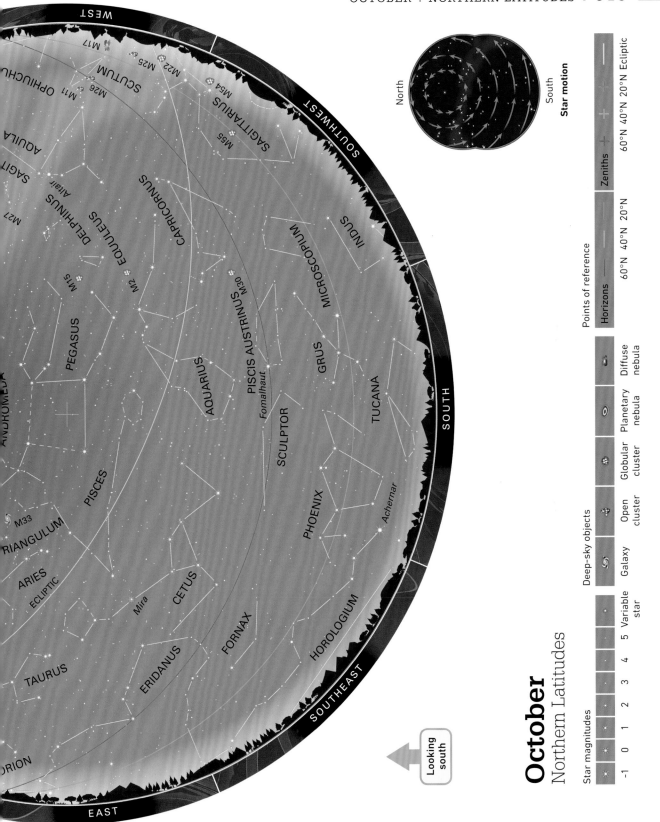

WEST

OPHIUCHUS

M17

M25 M22 M26 SCUTUM

M11

AQUILA

SAGITTA

M54 SAGITTARIUS

SOUTHWEST

M55

DELPHINUS

Altair

EQUULEUS M2 CAPRICORNUS

M27

M15

PEGASUS

INDUS

MICROSCOPIUM

PISCIS AUSTRINUS M30

AQUARIUS

Fomalhaut

GRUS

TUCANA

ANDROMEDA

SCULPTOR

PISCES

PHOENIX

Achernar

M33

TRIANGULUM

ARIES

ECLIPTIC

CETUS

Mira

FORNAX

SOUTH

HOROLOGIUM

TAURUS

ERIDANUS

ORION

SOUTHEAST

EAST

Looking
south

North

South

Star motion

October
Northern Latitudes

Points of reference

Horizons			Zeniths	Ecliptic
60°N	40°N	20°N		
60°N	40°N	20°N		20°N

Deep-sky objects

Galaxy	Open cluster	Globular cluster	Planetary nebula	Diffuse nebula

Star magnitudes

-1	0	1	2	3	4	5	Variable star

October
Southern Latitudes

Star magnitudes

-1	0	1	2	3	4	5	Variable star

Deep-sky objects

Galaxy	Open cluster	Globular cluster	Planetary nebula	Diffuse nebula

Points of reference

Horizons	Zeniths
0° 20°S 40°S	0° 20°S 40°S Ecliptic

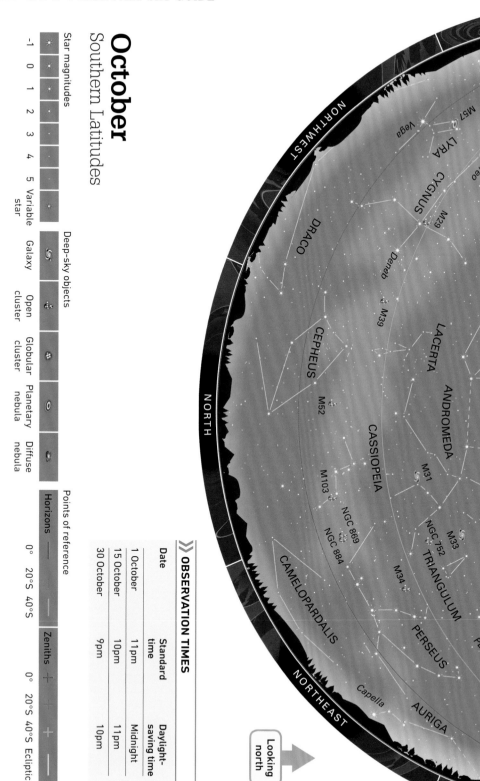

OBSERVATION TIMES

Date	Standard time	Daylight-saving time
1 October	11pm	Midnight
15 October	10pm	11pm
30 October	9pm	10pm

Looking north →

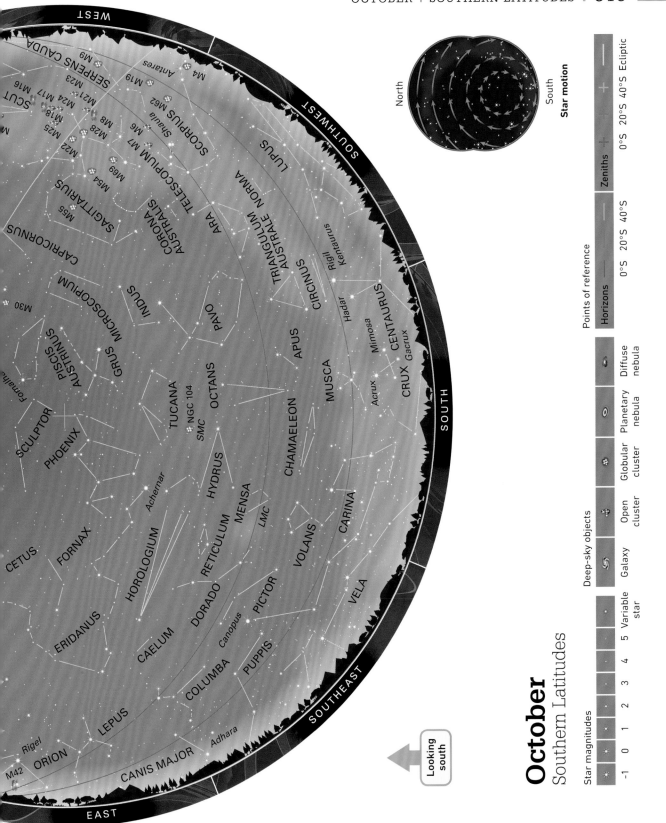

October
Southern Latitudes

Looking south

North
South
Star motion

Points of reference

Horizons
0°S 20°S 40°S

Zeniths
0°S 20°S 40°S Ecliptic

Deep-sky objects

Galaxy | Open cluster | Globular cluster | Planetary nebula | Diffuse nebula

Star magnitudes

-1 0 1 2 3 4 5 Variable star

November
Northern Latitudes

Star magnitudes

-1	✦
0	✦
1	✦
2	✦
3	•
4	•
5	·
Variable star	⊙

Deep-sky objects

Galaxy	⬭
Open cluster	⊛
Globular cluster	⊙
Planetary nebula	◎
Diffuse nebula	⬭

Points of reference

	Horizons	60°N	40°N	20°N	Zeniths	60°N	40°N	20°N	Ecliptic

» OBSERVATION TIMES

Date	Standard time	Daylight-saving time
1 November	11pm	Midnight
15 November	10pm	11pm
30 November	9pm	10pm

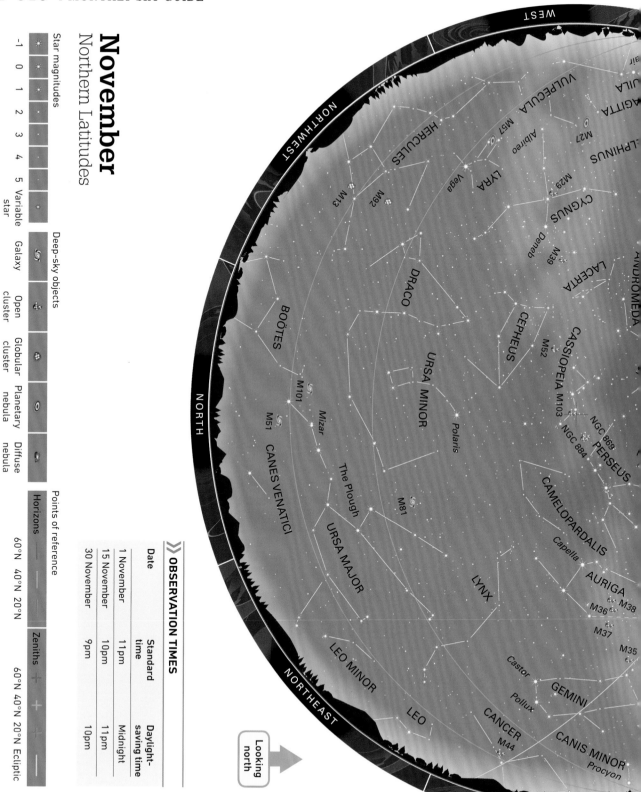

Looking north

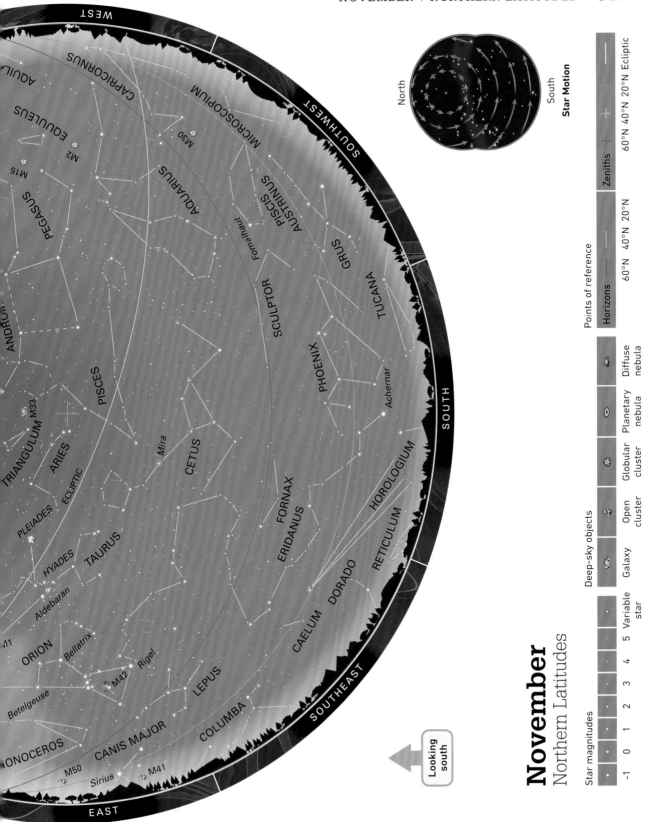

Star Motion

North

South

Points of reference

Horizons

60°N 40°N 20°N

Zeniths

60°N 40°N 20°N

Ecliptic

Deep-sky objects

Galaxy

Open cluster

Globular cluster

Planetary nebula

Diffuse nebula

Star magnitudes

-1 0 1 2 3 4 5 Variable star

November
Northern Latitudes

Looking south

November
Southern Latitudes

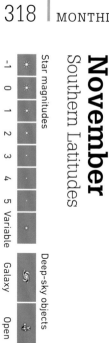

Star magnitudes

-1	
0	
1	
2	
3	
4	
5	
Variable star	

Deep-sky objects

Galaxy	
Open cluster	
Globular cluster	
Planetary nebula	
Diffuse nebula	

Points of reference

Horizons		
0°	20°S	40°S

Zeniths		
0°	20°S	40°S Ecliptic

❯❯ OBSERVATION TIMES

Date	Standard time	Daylight-saving time
1 November	11pm	Midnight
15 November	10pm	11pm
30 November	9pm	10pm

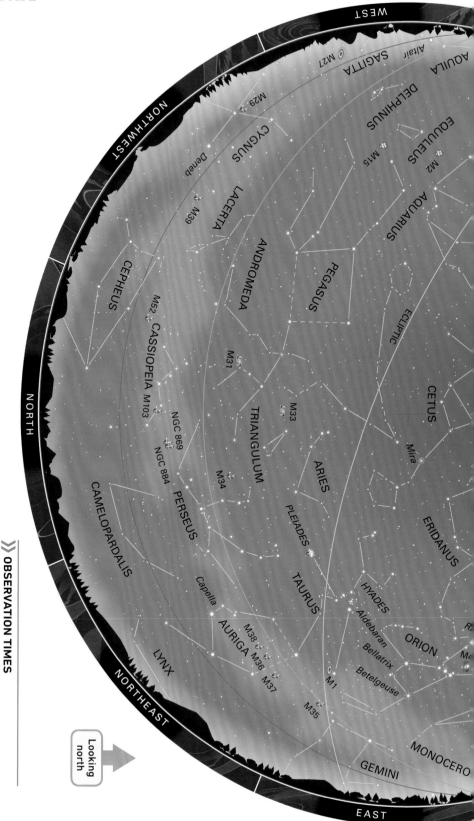

Looking north →

North

South

Star motion

Points of reference

Horizons — 0°S 20°S 40°S

Zeniths + 0°S 20°S 40°S

Ecliptic ---- 40°S 20°S 0°S

Deep-sky objects

Galaxy | Open cluster | Globular cluster | Planetary nebula | Diffuse nebula

Variable star

Star magnitudes

-1 0 1 2 3 4 5

November
Southern Latitudes

Looking south

WEST

SOUTHWEST

SOUTH

SOUTHEAST

EAST

CAPRICORNUS

M22

M69 M7

SAGITTARIUS

CORONA M54

AUSTRALIS

M55

Shaula

SCORPIUS

MICROSCOPIUM

TELESCOPIUM

ARA

NORMA

PISCIS

AUSTRINUS

M30

AQUARIUS

Fomalhaut

INDUS

GRUS

PAVO

TRIANGULUM

AUSTRALE

CIRCINUS

SCULPTOR

PHOENIX

TUCANA

NGC 104

SMC

OCTANS

APUS

Rigil

Kentaurus

Hadar

CENTAURUS

Achernar

CHAMAELEON

MUSCA

Mimosa

Acrux

CRUX

Gacrux

FORNAX

ERIDANUS

HOROLOGIUM

HYDRUS

RETICULUM

MENSA

CAELUM

DORADO

LMC

VOLANS

PICTOR

CARINA

VELA

COLUMBA

Canopus

PUPPIS

LEPUS

M79

M41

Adhara

Sirius

M50

MONOCEROS

CANIS

MAJOR

NGC 2362

M47 M93

M46

PYXIS

December
Northern Latitudes

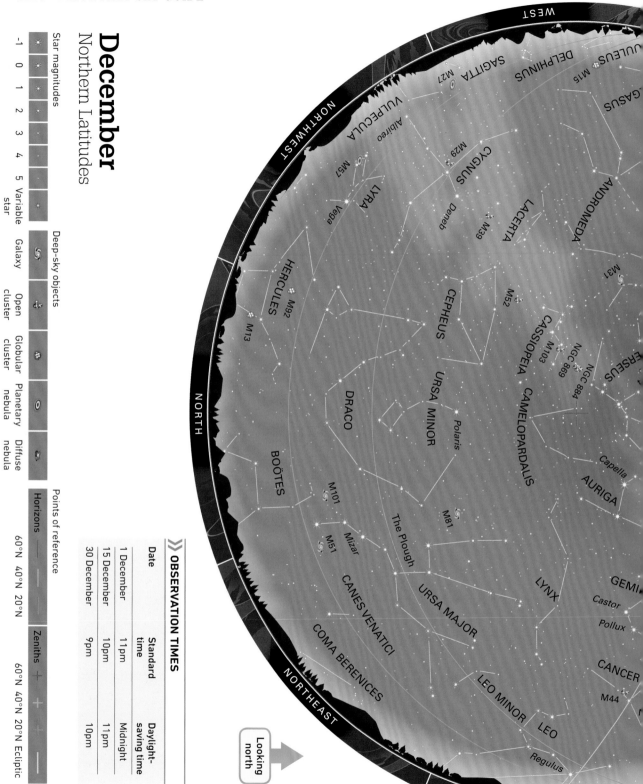

Star magnitudes

| -1 | 0 | 1 | 2 | 3 | 4 | 5 | Variable star |

Deep-sky objects

| Galaxy | Open cluster | Globular cluster | Planetary nebula | Diffuse nebula |

Points of reference

	60°N	40°N	20°N		60°N	40°N	20°N	
Horizons				Zeniths				Ecliptic

≫ OBSERVATION TIMES

Date	Standard time	Daylight-saving time
1 December	11pm	Midnight
15 December	10pm	11pm
30 December	9pm	10pm

Looking north →

WEST

PISCIS AUSTRINUS

Fomalhaut

AQUARIUS

SOUTHWEST

PEGASUS

ECLIPTIC

SCULPTOR

ANDROMEDA

PISCES

CETUS

PHOENIX

Mira

TRIANGULUM

ARIES

FORNAX

HYDRUS

Achernar

PERSEUS

PLEIADES

TAURUS

ERIDANUS

HOROLOGIUM

RETICULUM

AURIGA

HYADES

Aldebaran

CAELUM

SOUTH

38
M36
M37 M1
M35

ORION

Bellatrix

Rigel

DORADO

PICTOR

M42

Betelgeuse

LEPUS

Canopus

GEMINI

CANIS MAJOR

COLUMBA

CANIS MINOR

MONOCEROS

Sirius

M41

Adhara

Procyon

M50

M47

M93

PUPPIS

ANCER

M46

SOUTHEAST

HYDRA

M48

EAST

Looking south

North

South

Star motion

December
Northern Latitudes

Star magnitudes

-1	0	1	2	3	4	5	Variable star

Deep-sky objects

Galaxy	Open cluster	Globular cluster	Planetary nebula	Diffuse nebula

Points of reference

Horizons
60°N 40°N 20°N

Zeniths
60°N 40°N 20°N

Ecliptic

December
Southern Latitudes

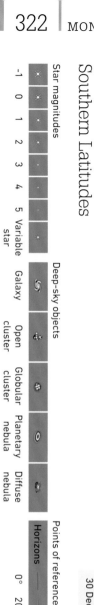

Star magnitudes

| -1 | 0 | 1 | 2 | 3 | 4 | 5 | Variable star |

Deep-sky objects

| Galaxy | Open cluster | Globular cluster | Planetary nebula | Diffuse nebula |

Points of reference

Horizons			Zeniths		
0°	20°S	40°S	0°	20°S	40°S Ecliptic

Date	Standard time	Daylight-saving time
1 December	11pm	Midnight
15 December	10pm	11pm
30 December	9pm	10pm

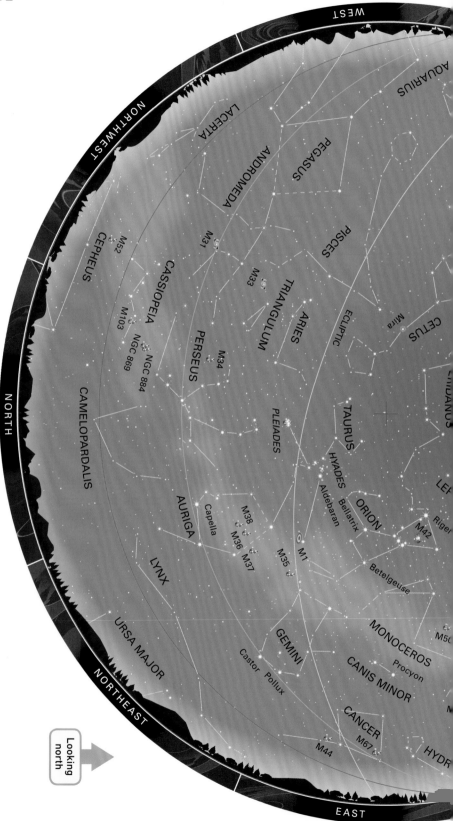

Looking north

December
Southern Latitudes

Looking south

Star motion

North

South

Points of reference

Horizons
0°S   20°S   40°S

Zeniths
0°S   20°S   40°S   Ecliptic

Deep-sky objects

Galaxy | Open cluster | Globular cluster | Planetary nebula | Diffuse nebula

Star magnitudes

-1   0   1   2   3   4   5   Variable star

Diamond ring
Total solar eclipses are the
most spectacular sky sights
in nature. Just before or after
totality, a sliver of sunlight
creates a "diamond ring" effect.

Almanac

This section is essentially a calendar giving the dates of the new and full Moon, eclipses, transits, and the principal aspects of the naked-eye planets for a period of 14 years (2018–31). Some of these events can obviously be seen from any part of the Earth. Others, such as the transits of Venus and Mercury and the eclipses of the Sun and Moon, are visible only from certain continents or regions. In these cases, the Almanac indicates from which parts of the world the eclipses can be seen.

Eclipses of the Sun occur when the Moon passes in front of the Sun and blocks off its light. Total eclipses are visible only from a narrow band within which the Moon covers the entire Sun, but partial eclipses are visible over a much wider area. When the Moon is near the most distant point in its orbit from Earth, it may not cover the Sun completely, leaving a ring, or annulus, of light surrounding it at mid-eclipse. This is known as an annular eclipse. Rarely, an eclipse can be total at the middle of the eclipse track, but annular at either end. Total or annular eclipses last only a few minutes, but partial ones can last up to three hours. Never look directly at the Sun; see p.93 for guidelines on methods of observing the Sun safely.

Eclipses of the Moon occur when the Moon enters the Earth's shadow and can last up to four hours. Even when the Moon is totally eclipsed, it does not disappear entirely, as some sunlight reaches it via the Earth's atmosphere.

On rare occasions, the inner planets Mercury and Venus pass in front of the Sun as seen from Earth, an event known as a transit. Mercury or Venus can then be seen as a black dot crossing the face of the Sun over a period of hours. In the period covered here, there is one transit of Mercury, in November 2019.

Planetary aspects

The aspects given in the Almanac indicate good times to view certain planets. For the inner planets, Mercury and Venus, this is at the point of greatest elongation – the maximum separation of the planet from the Sun, either in the morning sky (rising before the Sun) or evening sky (setting after the Sun). For the outer planets, the Almanac gives dates of opposition – when a planet lies in the opposite direction from the Sun as seen from Earth. At these times the planet is passing closest to Earth, and appears at its largest and brightest. The planet is also then visible all night. Mars comes to opposition every two years two months, Jupiter every 13 months, and Saturn every year and two weeks. Not all oppositions are equally favourable, because the planets' orbits are elliptical, and their distance from the Earth varies from one opposition to the next. The orbit of Mars is particularly elliptical, and its closest oppositions occur every 15 or 17 years.

Planets in the sky
Close pairings of planets in the sky present an attractive sight. Here Venus and Jupiter are seen in the morning twilight. Venus is brighter and lower.

2018

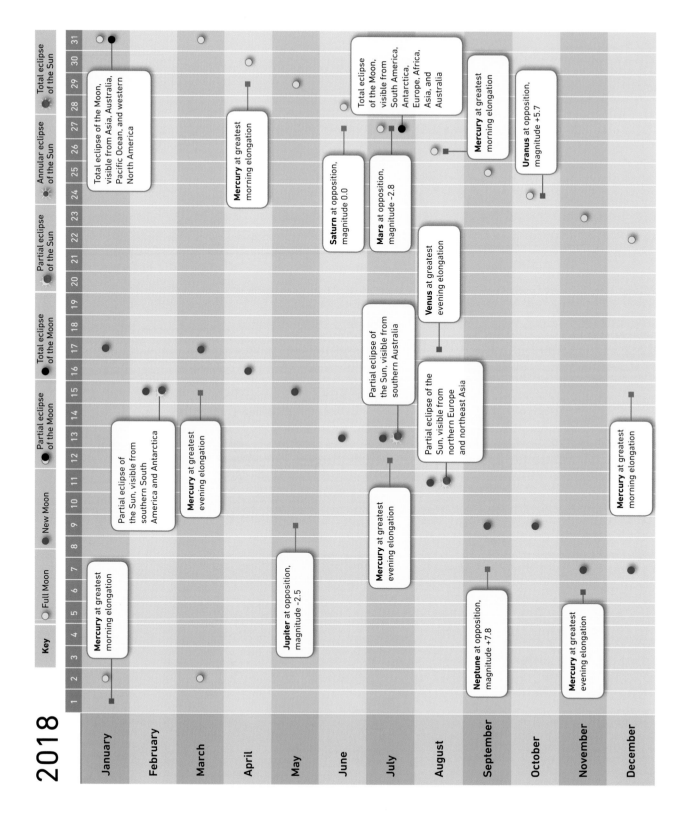

Key

○ Full Moon
● New Moon
◐ Partial eclipse of the Moon
● Total eclipse of the Moon
☀ Partial eclipse of the Sun
☀ Annual eclipse of the Sun
● Total eclipse of the Sun

January

Mercury at greatest morning elongation

Total eclipse of the Moon, visible from Asia, Australia, Pacific Ocean, and western North America

February

Partial eclipse of the Sun, visible from southern South America and Antarctica

March

Mercury at greatest evening elongation

April

Mercury at greatest morning elongation

May

Jupiter at opposition, magnitude -2.5

June

Saturn at opposition, magnitude 0.0

July

Mercury at greatest evening elongation

Partial eclipse of the Sun, visible from southern Australia

Mars at opposition, magnitude -2.8

Total eclipse of the Moon, visible from South America, Antarctica, Europe, Africa, Asia, and Australia

August

Partial eclipse of the Sun, visible from northern Europe and northeast Asia

Venus at greatest evening elongation

September

Neptune at opposition, magnitude +7.8

Mercury at greatest morning elongation

October

Uranus at opposition, magnitude +5.7

November

Mercury at greatest evening elongation

Mercury at greatest morning elongation

December

Mercury at greatest evening elongation

2019

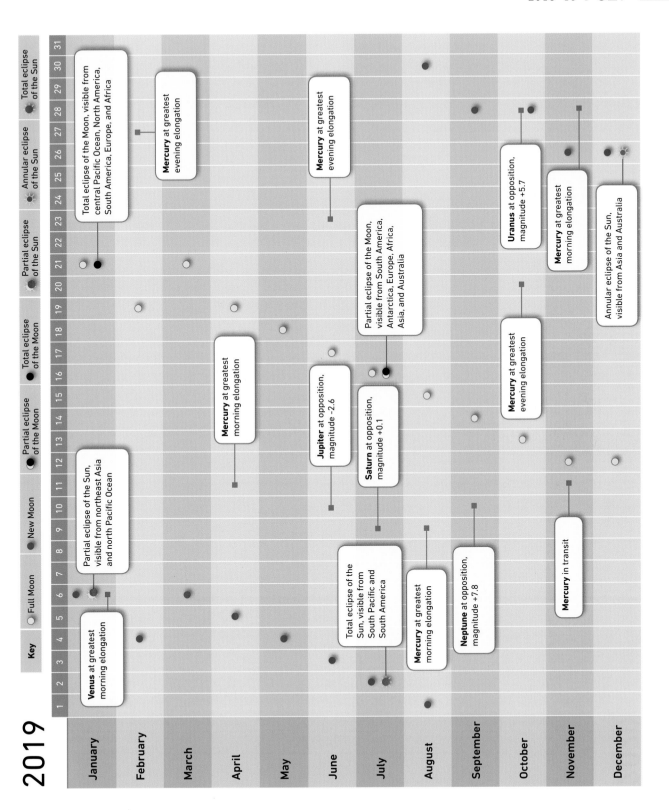

Key

○ Full Moon ● New Moon ◐ Partial eclipse of the Moon ● Total eclipse of the Moon ◑ Partial eclipse of the Sun ◉ Annular eclipse of the Sun ● Total eclipse of the Sun

| | 1 | 2 | 3 | 4 | 5 | 6 | 7 | 8 | 9 | 10 | 11 | 12 | 13 | 14 | 15 | 16 | 17 | 18 | 19 | 20 | 21 | 22 | 23 | 24 | 25 | 26 | 27 | 28 | 29 | 30 | 31 |

January

Venus at greatest morning elongation

Partial eclipse of the Sun, visible from northeast Asia and north Pacific Ocean

Total eclipse of the Moon, visible from central Pacific Ocean, North America, South America, Europe, and Africa

Mercury at greatest evening elongation

February

March

April

Mercury at greatest morning elongation

May

June

Jupiter at opposition, magnitude −2.6

Mercury at greatest evening elongation

July

Total eclipse of the Sun, visible from South Pacific and South America

Mercury at greatest morning elongation

Saturn at opposition, magnitude +0.1

Partial eclipse of the Moon, visible from South America, Antarctica, Europe, Africa, Asia, and Australia

August

Neptune at opposition, magnitude +7.8

Mercury at greatest morning elongation

Mercury at greatest evening elongation

September

October

Uranus at opposition, magnitude +5.7

November

Mercury in transit

Mercury at greatest morning elongation

December

Annular eclipse of the Sun, visible from Asia and Australia

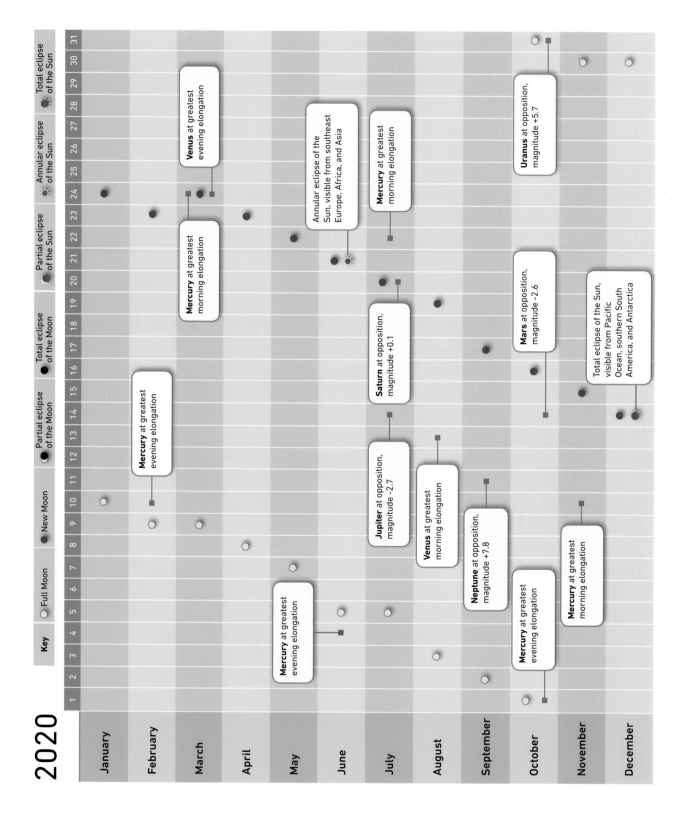

2020

Key

○ Full Moon   ● New Moon   ◐ Partial eclipse of the Moon   ● Total eclipse of the Moon   ☀ Partial eclipse of the Sun   ☀ Annular eclipse of the Sun   ☀ Total eclipse of the Sun

Mercury at greatest evening elongation

Mercury at greatest morning elongation

Venus at greatest evening elongation

Mercury at greatest evening elongation

Jupiter at opposition, magnitude −2.7

Saturn at opposition, magnitude +0.1

Annular eclipse of the Sun, visible from southeast Europe, Africa, and Asia

Mercury at greatest morning elongation

Venus at greatest morning elongation

Neptune at opposition, magnitude +7.8

Mars at opposition, magnitude −2.6

Uranus at opposition, magnitude +5.7

Mercury at greatest evening elongation

Mercury at greatest morning elongation

Total eclipse of the Sun, visible from Pacific Ocean, southern South America, and Antarctica

January / February / March / April / May / June / July / August / September / October / November / December

1 2 3 4 5 6 7 8 9 10 11 12 13 14 15 16 17 18 19 20 21 22 23 24 25 26 27 28 29 30 31

2021

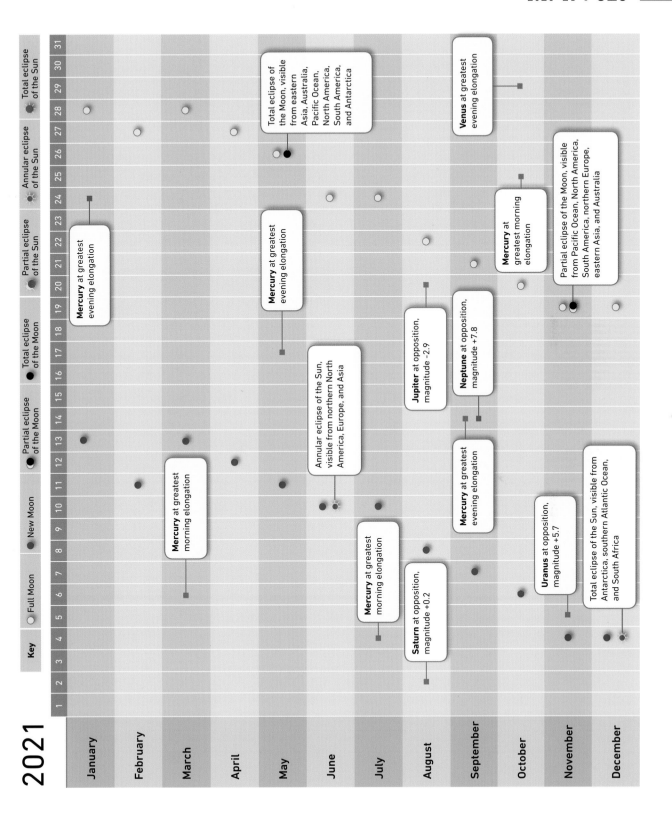

Key
- ○ Full Moon
- ● New Moon
- ◐ Partial eclipse of the Moon
- ● Total eclipse of the Moon
- ◔ Partial eclipse of the Sun
- ◑ Annular eclipse of the Sun
- ● Total eclipse of the Sun

January — Mercury at greatest evening elongation

February

March — Mercury at greatest morning elongation

April

May — Mercury at greatest evening elongation; Total eclipse of the Moon, visible from eastern Asia, Australia, Pacific Ocean, North America, South America, and Antarctica

June — Annular eclipse of the Sun, visible from northern North America, Europe, and Asia

July — Mercury at greatest morning elongation

August — Saturn at opposition, magnitude +0.2; Jupiter at opposition, magnitude -2.9

September — Mercury at greatest evening elongation; Neptune at opposition, magnitude +7.8

October — Mercury at greatest morning elongation

November — Uranus at opposition, magnitude +5.7; Partial eclipse of the Moon, visible from Pacific Ocean, North America, South America, northern Europe, eastern Asia, and Australia; Venus at greatest evening elongation

December — Total eclipse of the Sun, visible from Antarctica, southern Atlantic Ocean, and South Africa

2022

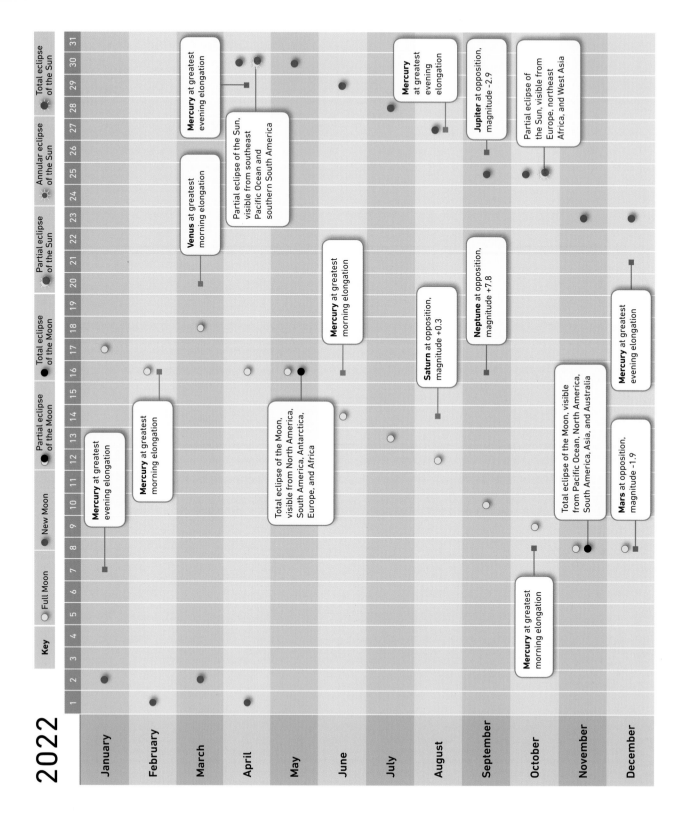

Key

○ Full Moon ● New Moon ◐ Partial eclipse of the Moon ● Total eclipse of the Moon ☀ Partial eclipse of the Sun ✦ Annular eclipse of the Sun ● Total eclipse of the Sun

January

February

March

April

May

June

July

August

September

October

November

December

Mercury at greatest evening elongation

Mercury at greatest morning elongation

Mercury at greatest evening elongation

Venus at greatest morning elongation

Partial eclipse of the Sun, visible from southeast Pacific Ocean and southern South America

Total eclipse of the Moon, visible from North America, South America, Antarctica, Europe, and Africa

Mercury at greatest morning elongation

Saturn at opposition, magnitude +0.3

Mercury at greatest evening elongation

Neptune at opposition, magnitude +7.8

Jupiter at opposition, magnitude −2.9

Partial eclipse of the Sun, visible from Europe, northeast Africa, and West Asia

Mercury at greatest morning elongation

Total eclipse of the Moon, visible from Pacific Ocean, North America, South America, Asia, and Australia

Mars at opposition, magnitude −1.9

Mercury at greatest evening elongation

2023

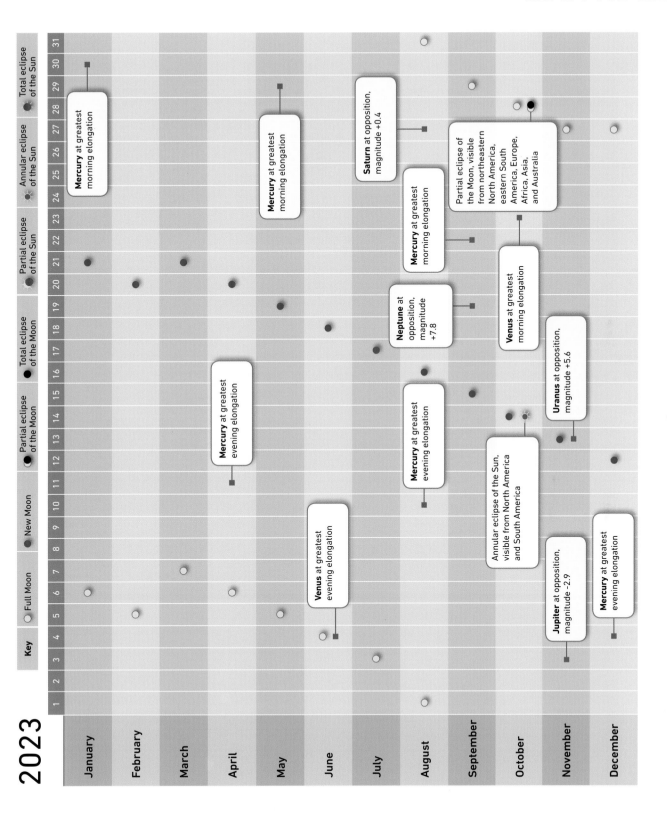

Key

- ○ Full Moon
- ● New Moon
- ◑ Partial eclipse of the Moon
- ● Total eclipse of the Moon
- ◉ Partial eclipse of the Sun
- ☀ Annular eclipse of the Sun
- ● Total eclipse of the Sun

January — **Mercury** at greatest morning elongation

April — **Mercury** at greatest evening elongation

May — **Mercury** at greatest morning elongation

June — **Venus** at greatest evening elongation

July — **Neptune** at opposition, magnitude +7.8; **Mercury** at greatest evening elongation

August — **Saturn** at opposition, magnitude +0.4; **Mercury** at greatest morning elongation; Partial eclipse of the Moon, visible from northeastern North America, eastern South America, Europe, Africa, Asia, and Australia

September — **Venus** at greatest morning elongation

October — Annular eclipse of the Sun, visible from North America and South America

November — **Jupiter** at opposition, magnitude −2.9; **Uranus** at opposition, magnitude +5.6

December — **Mercury** at greatest evening elongation

2024

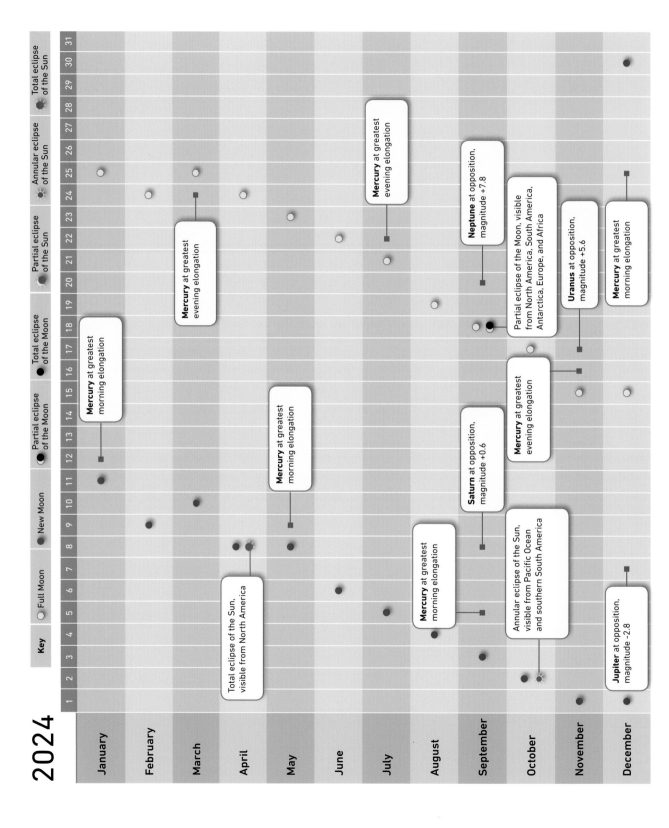

Key

- ○ Full Moon
- ● New Moon
- ◉ Partial eclipse of the Moon
- ● Total eclipse of the Moon
- ◉ Partial eclipse of the Sun
- ☀ Annular eclipse of the Sun
- ● Total eclipse of the Sun

January — **Mercury** at greatest morning elongation

March — **Mercury** at greatest evening elongation

Total eclipse of the Sun, visible from North America

May — **Mercury** at greatest morning elongation

June — **Mercury** at greatest morning elongation

July — **Mercury** at greatest evening elongation

September — **Saturn** at opposition, magnitude +0.6

Neptune at opposition, magnitude +7.8

October — **Mercury** at greatest evening elongation

Annular eclipse of the Sun, visible from Pacific Ocean and southern South America

Partial eclipse of the Moon, visible from North America, South America, Antarctica, Europe, and Africa

November — **Uranus** at opposition, magnitude +5.6

December — **Jupiter** at opposition, magnitude −2.8

Mercury at greatest morning elongation

January, February, March, April, May, June, July, August, September, October, November, December

2025

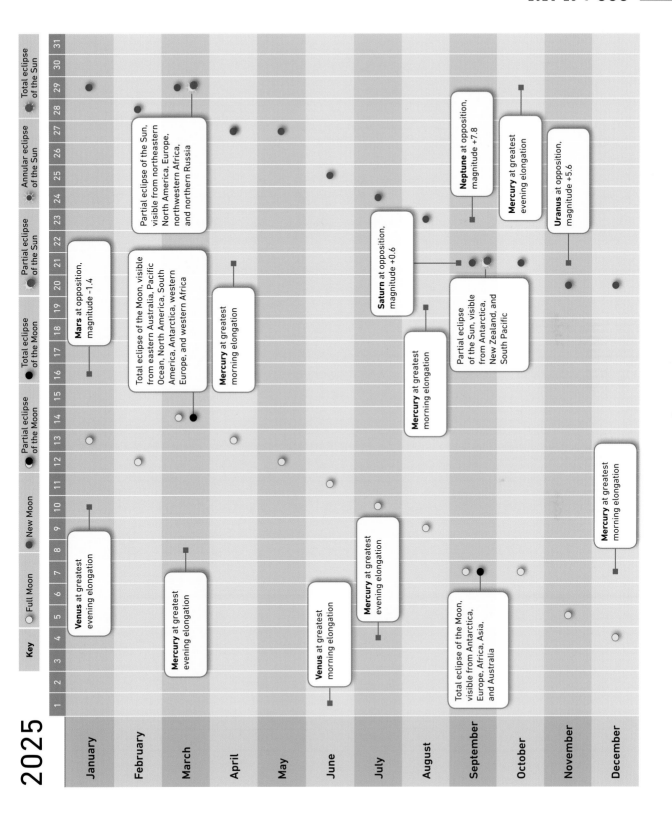

Key

○ Full Moon   ● New Moon   ◐ Partial eclipse of the Moon   ● Total eclipse of the Moon   ◑ Partial eclipse of the Sun   ◑ Annular eclipse of the Sun   ● Total eclipse of the Sun

January — **Venus** at greatest evening elongation

February — **Mars** at opposition, magnitude –1.4

March — **Mercury** at greatest evening elongation

March — Total eclipse of the Moon, visible from eastern Australia, Pacific Ocean, North America, South America, Antarctica, western Europe, and western Africa

March — Partial eclipse of the Sun, visible from northeastern North America, Europe, northwestern Africa, and northern Russia

April — **Mercury** at greatest morning elongation

June — **Venus** at greatest morning elongation

July — **Mercury** at greatest evening elongation

August — Total eclipse of the Moon, visible from Antarctica, Europe, Africa, Asia, and Australia

August — **Mercury** at greatest morning elongation

September — **Saturn** at opposition, magnitude +0.6

September — Partial eclipse of the Sun, visible from Antarctica, New Zealand, and South Pacific

September — **Neptune** at opposition, magnitude +7.8

October — **Mercury** at greatest evening elongation

November — **Uranus** at opposition, magnitude +5.6

December — **Mercury** at greatest morning elongation

2026

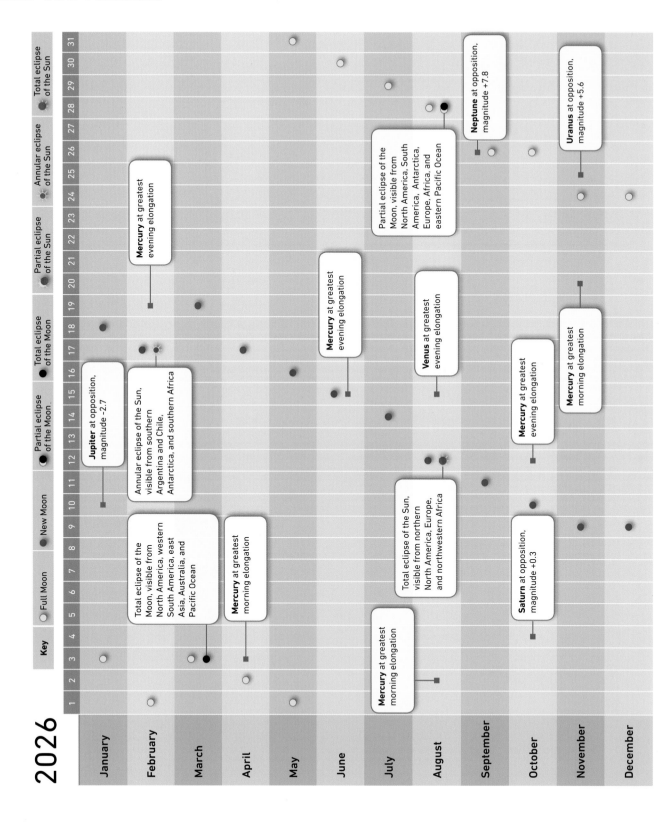

Key

○ Full Moon ● New Moon ◐ Partial eclipse of the Moon ● Total eclipse of the Moon ◑ Partial eclipse of the Sun ◔ Annular eclipse of the Sun ◕ Total eclipse of the Sun

Months: January, February, March, April, May, June, July, August, September, October, November, December

Days: 1–31

January

Jupiter at opposition, magnitude –2.7

February

Total eclipse of the Moon, visible from North America, western South America, east Asia, Australia, and Pacific Ocean

Annular eclipse of the Sun, visible from southern Argentina and Chile, Antarctica, and southern Africa

Mercury at greatest evening elongation

March

Mercury at greatest morning elongation

April

May

June

Mercury at greatest evening elongation

July

Mercury at greatest morning elongation

Total eclipse of the Sun, visible from northern North America, Europe, and northwestern Africa

August

Venus at greatest evening elongation

Partial eclipse of the Moon, visible from North America, South America, Antarctica, Europe, Africa, and eastern Pacific Ocean

September

Saturn at opposition, magnitude +0.3

Neptune at opposition, magnitude +7.8

October

Mercury at greatest evening elongation

November

Mercury at greatest morning elongation

Uranus at opposition, magnitude +5.6

December

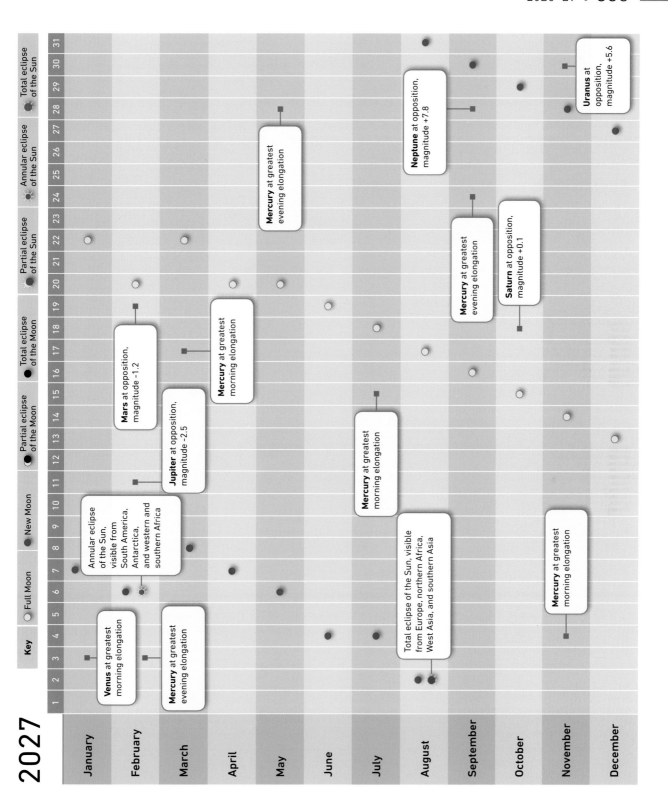

2027

Key

○ Full Moon
● New Moon
◐ Partial eclipse of the Moon
● Total eclipse of the Moon
● Partial eclipse of the Sun
Annular eclipse of the Sun
● Total eclipse of the Sun

Venus at greatest morning elongation

Mercury at greatest evening elongation

Annular eclipse of the Sun, visible from South America, Antarctica, and western and southern Africa

Jupiter at opposition, magnitude −2.5

Mars at opposition, magnitude −1.2

Mercury at greatest morning elongation

Mercury at greatest evening elongation

Mercury at greatest morning elongation

Total eclipse of the Sun, visible from Europe, northern Africa, West Asia, and southern Asia

Neptune at opposition, magnitude +7.8

Mercury at greatest evening elongation

Saturn at opposition, magnitude +0.1

Uranus at opposition, magnitude +5.6

Mercury at greatest morning elongation

January
February
March
April
May
June
July
August
September
October
November
December

2028

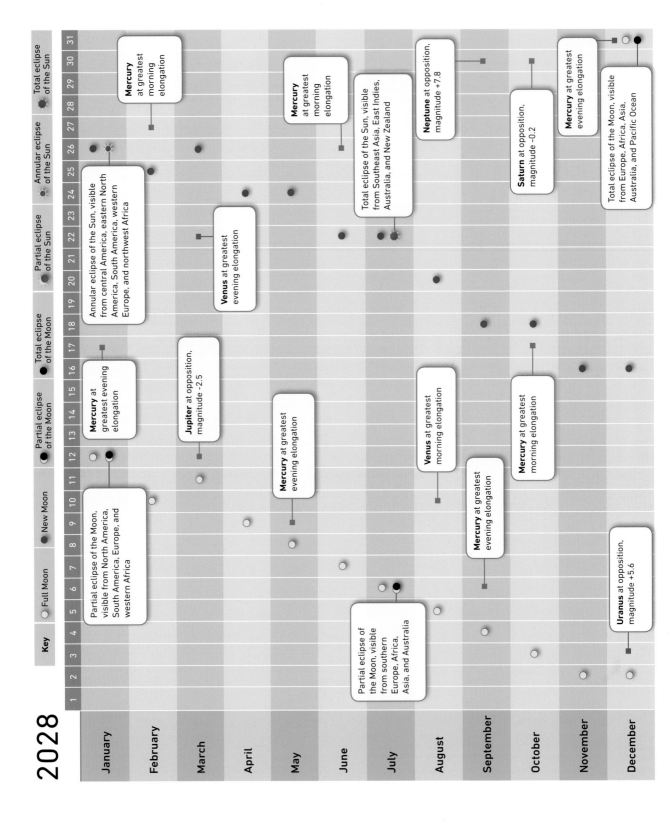

Key

- ○ Full Moon
- ● New Moon
- ◑ Partial eclipse of the Moon
- ● Total eclipse of the Moon
- ☀ Partial eclipse of the Sun
- ☀ Annular eclipse of the Sun
- ● Total eclipse of the Sun

January
- Partial eclipse of the Moon, visible from North America, South America, Europe, and western Africa
- **Mercury** at greatest evening elongation

February
- Annular eclipse of the Sun, visible from central America, eastern North America, South America, western Europe, and northwest Africa
- **Mercury** at greatest morning elongation

March
- **Jupiter** at opposition, magnitude −2.5

May
- **Venus** at greatest evening elongation
- **Mercury** at greatest evening elongation

June
- **Mercury** at greatest morning elongation

July
- Partial eclipse of the Moon, visible from southern Europe, Africa, Asia, and Australia
- Total eclipse of the Sun, visible from Southeast Asia, East Indies, Australia, and New Zealand

August
- **Venus** at greatest morning elongation

September
- **Mercury** at greatest evening elongation
- **Neptune** at opposition, magnitude +7.8

October
- **Mercury** at greatest morning elongation
- **Saturn** at opposition, magnitude −0.2

November
- **Uranus** at opposition, magnitude +5.6
- **Mercury** at greatest evening elongation

December
- Total eclipse of the Moon, visible from Europe, Africa, Asia, Australia, and Pacific Ocean

2029

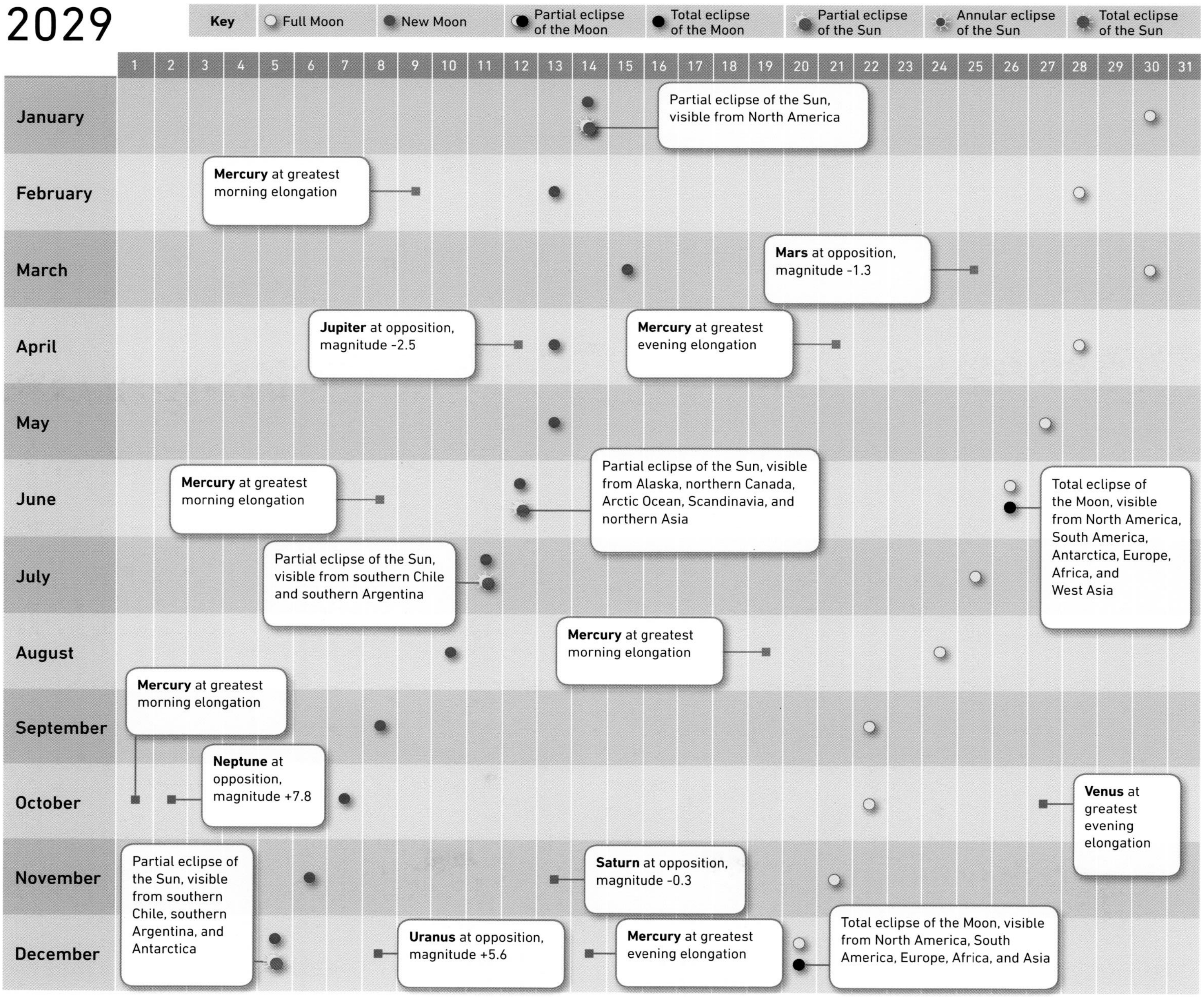

Key

- Full Moon
- New Moon
- Partial eclipse of the Moon
- Total eclipse of the Moon
- Partial eclipse of the Sun
- Total eclipse of the Moon
- Partial eclipse of the Sun
- Annular eclipse of the Sun
- Total eclipse of the Sun

January — Partial eclipse of the Sun, visible from North America

February — Mercury at greatest morning elongation

March — Mars at opposition, magnitude -1.3

April — Jupiter at opposition, magnitude -2.5; Mercury at greatest evening elongation

May — Mercury at greatest morning elongation

June — Partial eclipse of the Sun, visible from Alaska, northern Canada, Arctic Ocean, Scandinavia, and northern Asia; Total eclipse of the Moon, visible from North America, South America, Antarctica, Europe, Africa, and West Asia

July — Partial eclipse of the Sun, visible from southern Chile and southern Argentina; Mercury at greatest morning elongation

August — Mercury at greatest morning elongation

September — Mercury at greatest morning elongation; Neptune at opposition, magnitude +7.8

October — Partial eclipse of the Sun, visible from southern Chile, southern Argentina, and Antarctica; Saturn at opposition, magnitude -0.3

November — Uranus at opposition, magnitude +5.6; Mercury at greatest evening elongation; Venus at greatest evening elongation

December — Total eclipse of the Moon, visible from North America, Europe, Africa, and Asia

2030

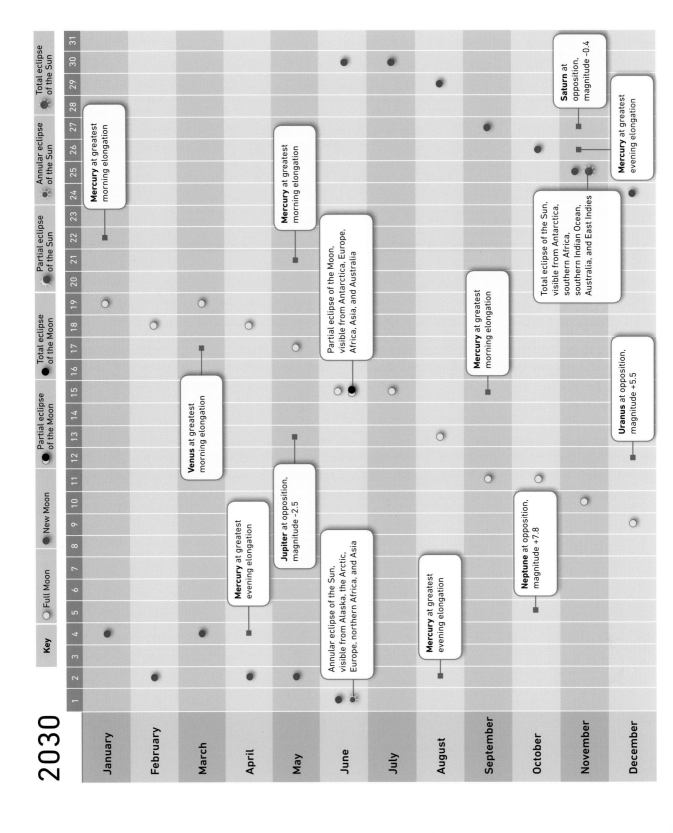

Key
- ○ Full Moon
- ● New Moon
- ◑ Partial eclipse of the Moon
- ● Total eclipse of the Moon
- ☀ Partial eclipse of the Sun
- ☀ Annular eclipse of the Sun
- ● Total eclipse of the Sun

January

February

March

April

May

June

July

August

September

October

November

December

Mercury at greatest morning elongation

Mercury at greatest evening elongation

Venus at greatest morning elongation

Annular eclipse of the Sun, visible from Alaska, the Arctic, Europe, northern Africa, and Asia

Jupiter at opposition, magnitude −2.5

Mercury at greatest morning elongation

Partial eclipse of the Moon, visible from Antarctica, Europe, Africa, Asia, and Australia

Mercury at greatest evening elongation

Mercury at greatest morning elongation

Neptune at opposition, magnitude +7.8

Total eclipse of the Sun, visible from Antarctica, southern Africa, southern Indian Ocean, Australia, and East Indies

Saturn at opposition, magnitude −0.4

Mercury at greatest evening elongation

Uranus at opposition, magnitude +5.5

2031

Key: ○ Full Moon | ● New Moon | ◑ Partial eclipse of the Moon | ● Total eclipse of the Moon | ☀ Partial eclipse of the Sun | ☀ Annular eclipse of the Sun | ● Total eclipse of the Sun

	1	2	3	4	5	6	7	8	9	10	11	12	13	14	15	16	17	18	19	20	21	22	23	24	25	26	27	28	29	30	31
January				■				○															●								
February							○														●		●								
March									○									■					●								
April							○														●										
May																					● ☀										
June					○										■			●													
July			○																●												
August		○																●													
September	○															●													■	○	
October										■				●		◐												○		○	
November							■										■			■								○			
December														●																	

Mercury at greatest morning elongation (January)

Mercury at greatest morning elongation (March)

Mars at opposition, magnitude −1.8

Mercury at greatest evening elongation (March–April)

Venus at greatest evening elongation

Jupiter at opposition, magnitude −2.6

Mercury at greatest evening elongation

Annular eclipse of the Sun, visible from southern Africa, southern Asia, East Indies, and Australia

Mercury at greatest morning elongation (August)

Venus at greatest morning elongation

Neptune at opposition, magnitude +7.8

Uranus at opposition, magnitude +5.5

Mercury at greatest evening elongation (October)

Saturn at opposition, magnitude −0.5

Mercury at greatest morning elongation (December)

GLOSSARY

absolute magnitude
A figure that indicates the true light output, or luminosity, of a star. It is the magnitude that the star would appear if it were placed at a standard distance, chosen as 10 parsecs (32.6 light-years).

albedo
The amount of light reflected from the surface of a planet, moon, asteroid, or other body. A high-albedo object is light in colour, a low-albedo one is dark.

altazimuth mount
A simple form of mounting in which the telescope can pivot freely up and down (in altitude) and from side to side (in azimuth).

altitude
The angular distance of an object above the horizon, in degrees.

aperture
The width of a telescope's main lens or mirror, or the opening at the top of a telescope's tube.

aphelion
The point in an object's orbit at which it is furthest from the Sun.

apparent magnitude
The brightness of a celestial object as seen from Earth. The further away the object, the fainter it appears.

asterism
A pattern formed by stars that are part of one or more constellations, such as the Plough or Big Dipper, which is part of Ursa Major.

asteroid
A solid body smaller than a planet; also known as a minor planet. Most asteroids orbit the Sun in the Asteroid Belt between Mars and Jupiter.

astronomical unit
The average distance between the Earth and the Sun, 149,597,870km (92,900,277 miles).

aurora
A glow in the Earth's upper atmosphere caused by interactions with particles from the Sun.

axis
The imaginary line through the centre of an object around which it rotates. The axis joins the poles.

azimuth
The angle of an object around the horizon measured in degrees from north via east and back to north again.

barred spiral galaxy
A type of spiral galaxy in which the central regions are elongated rather than rounded. Spiral arms emerge from each end of the central bar.

Big Bang
The explosive event that marked the origin of the Universe about 13.8 billion years ago.

binary star
A pair of stars linked by gravity, orbiting around their common centre of mass. *See also* eclipsing binary; spectroscopic binary.

black hole
A volume of space in which gravity is so great that nothing can escape, not even light. Black holes are thought to be formed when massive stars die.

brown dwarf
A gaseous object midway in size between a planet and a star that has insufficient mass to spark hydrogen fusion reactions at its core.

Cassegrain telescope
A type of reflecting telescope in which a secondary mirror reflects the light back through a hole in the centre of the main mirror, where the eyepiece or other detectors are placed.

catadioptric
A type of reflecting telescope with a thin lens placed across its aperture that gives the telescope a wide field of view combined with a short focal length.

CCD
see charge-coupled device.

celestial equator
An imaginary circle on the celestial sphere midway between the celestial poles. It divides the celestial sphere into two equal halves, one half north of the equator and the other half south.

celestial poles
The two points on the celestial sphere directly above the Earth's north and south poles. The celestial sphere appears to turn around an axis joining the celestial poles each day.

celestial sphere
An imaginary sphere surrounding the Earth, upon which celestial objects appear to lie.

Cepheid variable
A type of variable star that changes regularly in brightness every few days or weeks as it pulsates in size. Cepheids are named after their prototype, Delta Cephei.

charge-coupled device (CCD)
A light-sensitive electronic detector used to record images and spectra in place of film. CCDs consist of millions of tiny picture elements known as pixels.

chromosphere
A layer of gas above the Sun's visible surface, or photosphere. The chromosphere can be seen only when the brighter photosphere is blocked out.

circumpolar

A term referring to celestial objects that circle the pole without setting, as seen from a particular location.

comet

A small body consisting of ice and dust. When far from the Sun, a comet is frozen solid. Closer to the Sun, it warms up and releases dust and gas to form a large glowing head (the coma) and sometimes a tail.

conjunction

An occasion when two bodies in the Solar System, such as a planet and the Sun, line up as seen from Earth. *See also* inferior conjunction; superior conjunction.

constellation

Originally a star pattern but now an area of sky within boundaries laid down by the International Astronomical Union. There are 88 constellations.

corona

The Sun's tenuous outermost layer, visible only when the Sun is obscured at a total eclipse.

declination

The angular distance of an object north or south of the celestial equator, measured in degrees. It is the equivalent of latitude on Earth.

deep-sky object

An object outside the Solar System, such as a star cluster, nebula, or galaxy.

diffuse nebula

A bright cloud of gas, illuminated by stars within it. The Orion Nebula is a famous example.

Dobsonian mount

A simple form of altazimuth mounting, commonly used for Newtonian reflectors.

double star

A pair of stars that appear close together as seen from Earth. In most cases, the stars are related, forming a binary star.

But in some cases, the stars lie at different distances, and are termed an optical double.

dwarf planet

A celestial body in the Solar System that orbits the Sun and is massive enough for it to be spherical but, unlike the classical planets, has not cleared the region around its orbit of other bodies.

eclipsing binary

A pair of stars in orbit around each other in which one star periodically passes in front of the other as seen from Earth, cutting off its light.

ecliptic

The path followed by the Sun on the celestial sphere during the year, which is actually due to the Earth's orbital movement. The planets always appear close to the ecliptic because their orbits all lie in a similar plane to that of the Earth.

elongation

The angle between a planet and the Sun, or a moon and a planet. When Mercury and Venus are at their maximum angular separation from the Sun, they are said to be at greatest elongation, either east of the Sun (in the evening sky) or west of it (in the morning sky).

equatorial mount

A type of telescope mounting in which one axis, termed the polar axis, is aligned parallel to the Earth's axis. By turning this axis, the telescope can be kept aimed at a celestial object as the Earth spins.

equinox

The occasion when the Sun lies on the celestial equator. This occurs twice a year, on 20 March (the spring or vernal equinox) and 22 or 23 September (the autumnal equinox). At an equinox, day and night are roughly equal in length everywhere on Earth.

extragalactic

Any object outside our own galaxy.

extrasolar planet

A planet that orbits a star other than the Sun.

eyepiece

A lens (or, in practice, a combination of two or more lenses) used to magnify the image produced by a telescope.

finder

A small telescope or sighting device to help aim a larger telescope at a target.

galaxy

A mass of stars held together by gravity. Galaxies are of two main types: spirals, which have arms; and ellipticals, which do not. Diameters of galaxies range from about a thousand light-years to hundreds of thousands of light-years.

giant star

A star that has become bigger and brighter towards the end of its life. Stars more than about ten times the mass of the Sun become supergiants.

globular cluster

A dense, ball-shaped cluster containing tens or hundreds of thousands of stars. Globular clusters contain some of the oldest stars known.

Hubble constant

A measure of the rate at which the Universe is expanding, found by plotting the distance of galaxies against their redshifts.

inferior conjunction

The occasion when either Mercury or Venus lies between the Sun and the Earth.

Kuiper Belt

A swarm of icy asteroids beyond the orbit of Neptune.

light-year

The distance covered by a beam of light in a calendar year, 9,460 billion km (5,875 billion miles).

Local Group
The cluster of about 50 galaxies that includes our own galaxy. The largest is the Andromeda Galaxy. Our galaxy is the second-largest.

Magellanic Clouds
Two small galaxies that accompany our own Milky Way galaxy.

magnitude
A scale for measuring the brightness of celestial objects. The brightest objects are given small or even negative (–) numbers whereas faint objects have large positive (+) numbers.

main sequence
A stage in the life of a star when it creates energy by converting hydrogen into helium by nuclear reactions at its centre.

mare
Any of the dark lowland areas on the Moon.

meridian
An imaginary line in the sky running from north to south through the celestial poles and the observer's zenith. An object on the meridian is at its highest point above the horizon.

Messier Catalogue
A list of over 100 deep-sky objects that could be mistaken for comets, produced by the French astronomer Charles Messier (1730–1817).

meteor
A streak of light, also called a shooting star, caused by a speck of dust burning up in the atmosphere.

meteorite
A chunk of rock or metal from space that lands on the surface of the Earth or another Solar System body.

Milky Way
The faint, hazy band of light that can be seen crossing the sky on dark nights, composed of distant stars within our own galaxy. The name is also used for our galaxy as a whole.

Mira variable
A red giant or supergiant star that varies in brightness over a period of months or years due to pulsations in its size. It is named after the star Mira, the first of the type to be discovered.

moon
A natural satellite of a planet. A capital letter is used when referring to our own Moon.

nebula
A cloud of gas and dust, usually found in the spiral arms of a galaxy. Some nebulae are bright, being lit up by stars within them, while others are dark. *See also* planetary nebula.

neutron star
A small, highly dense star consisting of the atomic particles known as neutrons and thought to be created when a massive star dies in a supernova.

New General Catalogue (NGC)
A listing of nearly 8,000 deep-sky objects compiled by the Danish astronomer J.L.E. Dreyer (1852–1926).

Newtonian
A design of reflecting telescope in which the eyepiece is positioned at the side of the telescope tube.

nova
A star that erupts temporarily, becoming thousands of times brighter for a few weeks or months.

occultation
An event in which one celestial body passes in front of another, usually when the Moon passes in front of a star or planet.

Oort Cloud
A spherical swarm of comets surrounding the Solar System, extending halfway to the nearest star.

open cluster
An irregularly shaped group of dozens or hundreds of stars, usually found in the spiral arms of a galaxy. The stars in open clusters are relatively young.

opposition
The occasion when a body in the Solar System lies in the opposite direction to the Sun, as seen from Earth.

parallax
The change in position of an object when seen from two different locations. Nearby stars show a slight parallax shift as the Earth orbits the Sun, from which their distances can be calculated.

parsec
A unit of distance used by astronomers. It is the distance at which a star would have a parallax of one second of arc. One parsec is equal to 30,900 billion km (19,200 billion miles), or 3.2616 light-years.

perihelion
The point in an object's orbit at which it is closest to the Sun.

phase
The fraction of the disc of a planet or moon that is illuminated by the Sun, as seen from Earth.

photosphere
The visible surface of the Sun or another star.

planet
A celestial body that orbits the Sun, has sufficient mass for its gravity to create a nearly round shape, and has eliminated any body capable of moving in a neighbouring orbit. *See also* dwarf planet, extrasolar planet.

planetary nebula
A shell of gas thrown off by a star towards the end of its life.

precession
A slow wobble of the Earth in space, which causes its axis to describe a circle on the celestial sphere every 25,800 years. Because of precession, the coordinates of stars are continually changing.

proper motion
The movement of a star relative to the Sun. As a result of stars' proper motions, the shapes of the constellations change over hundreds of thousands of years.

pulsar
A neutron star that emits pulses of radio waves and other radiation as it spins.

quasar
The highly luminous core of a distant galaxy, thought to be caused by ultra-hot gas circulating around a massive black hole at the galaxy's centre.

radial velocity
The speed at which a star, planet, or other body is moving away from or approaching an observer on Earth.

radiant
The point in the sky from which the members of a meteor shower appear to diverge.

red dwarf
A star that is less massive, cooler, and dimmer than the Sun.

red giant
A star that has become larger and cooler as it nears the end of its life.

redshift
A shift in the lines in a spectrum towards longer wavelengths, caused by the movement of the emitting object away from us.

resolution
The ability of an optical instrument to distinguish fine detail, such as the individual stars in a close double star or markings on the planets.

retrograde motion
Motion from east to west, opposite to the normal direction of motion in the Solar System.

right ascension
A coordinate on the celestial sphere that is the equivalent of longitude on Earth.

It is measured in hours (1 hr = 15°), and starts at the point where the Sun crosses the celestial equator every March, known as the spring (or vernal) equinox.

satellite
Any body that orbits another, usually a moon of a planet.

Schmidt–Cassegrain
A design of telescope that incorporates a thin correcting lens across the front of the telescope tube to increase the field of view. The eyepiece is in a hole in the centre of the main mirror.

seeing
A term used to describe the steadiness of the atmosphere, which affects the quality of the image seen through a telescope. Good seeing means that the air is steady and fine detail can be distinguished.

Solar System
The family of planets and dwarf planets, their moons, and objects such as asteroids and comets that orbit the Sun.

solar wind
A stream of atomic particles from the Sun, mostly protons and electrons, which flows outwards through the Solar System.

solstice
The occasion when the Sun reaches its furthest point north or south of the celestial equator (around 21 June in the northern hemisphere, and 22 December in the southern hemisphere).

spectroscopic binary
A pair of stars so close together that they cannot be seen separately through any telescope. The binary nature of a star is revealed only when its light is examined through a spectroscope.

star
A sphere of gas that produces energy at its centre by nuclear reactions.

sunspot
A cooler patch on the Sun's surface that appears darker by contrast with its surroundings.

supergiant star
The largest and most luminous type of star. Stars at least 10 times as massive as the Sun swell into supergiants at the ends of their lives.

superior conjunction
The occasion when Mercury or Venus lies on the far side of the Sun from Earth.

supernova
A star that explodes at the end of its life, brightening by millions of times for a few weeks or months.

Universe
Everything that exists, including all matter, space, and time. The Universe is thought to have begun in a Big Bang about 13.8 billion years ago.

variable star
Any star that appears to change in brightness.

white dwarf
A small, dense star with a mass similar to that of the Sun, but only about the diameter of the Earth. White dwarfs are the shrunken remains of stars such as the Sun that have burnt out.

zenith
The point in the sky directly above an observer.

zodiac
The band of sky either side of the ecliptic through which the Sun and planets move.

INDEX

ACKNOWLEDGMENTS

First Edition

DK UK
Senior Editor Peter Frances
Senior Art Editor Peter Laws
Managing Editor Liz Wheeler
Managing Art Editor Phil Ormerod
Art Director Bryn Walls
Publisher Jonathan Metcalf
DTP Designer John Goldsmid
Production Controller Joanna Bull

Produced for Dorling Kindersley by Cooling Brown Ltd

Creative Director Arthur Brown
Project Editor Fiona Wild
Designers Peter Cooling, Murdo Culver, Elaine Hewson,
Ted Kinsey, Tish Jones
Editors Ferdie McDonald, Amanda Lebentz
Picture Researcher Louise Thomas

Cooling Brown would like to thank the following people for their help in the preparation of this book: Hilary Bird for indexing, Constance Novis for proofreading. Special thanks go to Peter Gallon and Ted Harrison at Telescope House for their technical expertise and loan of equipment. You can visit their astronomy shop online at www.telescopehouse.co.uk. Thanks too to binoculars and telescopes photographer Dave King, and model Vicky Brown.

Second Edition

DK Delhi would like to thank:
Janashree Singha and Shambhavi Thatte for editorial assistance; Pooja Pipil for design assistance; and Deepak Negi for image research.

Picture credits
The publisher would like to thank the following for their kind permission to reproduce their photographs:

Picture Key: a-above; b-below/bottom; c-centre; f-far; l-left; r-right; t-top

2–3 Seán Doran: NASA / SwRI / MSSS / Gerald Eichstädt. **4 akg-images:** Austrian National Library / Erich Lessing. **5 NASA:** Johnson Space Center. **6–7 NASA:** ESA and The Hubble Heritage Team (STScI/AURA) (t). **8–9 NASA:** Composition: Mattias Malmer, Image Data: Cassini Imaging Team. **10 DK Images:** The British Museum, London (bc). **11 Laurie Hatch Photography. 12–13 Corbis:** Reuters / Ian Waldle (b). **13 Science Photo Library:** John Chumack (tl). **14–15 Alamy Images:** Visual Arts Library (London). **16 Ancient Art & Architecture Collection:** (tl). w ww.bridgeman.co.uk: British Library, London (Ms Add 24189 fol.15) (bl). **17 Alamy Images:** David Ball (cra). **www.bridgeman.co.uk:** University Library, Istanbul (bl). **18 Corbis:** Bettmann (br). **DK Images:** Australian National Maritime Museum, Sydney (tr). **Science & Society Picture Library:** Science Museum, London (bc). **19 akg-images:** Collection Schloß Ambras / Erich Lessing (crb). **The Picture Desk:** Art Archive / Maritiem Museum Prins Hendrik Rotterdam / Dagli Orti (t). **20 Corbis:** Gianni Dagli Orti (cra). **DK Images:** Courtesy of The Science Museum, London / Clive Streeter (tl). **NOAO/AURA/NSF:** Lowell Observatory (b). **21 Corbis:** Richard T. Nowitz (bl); Jim Sugar (clb). **DK Images:** Courtesy of The Science Museum, London / Dave King (ca). **Royal Astronomical Society:** (crb). **22–23 Bridgeman Images:** Index **24 DK Images:** National Maritime Museum, London / Tina Chambers (ca). **NOAO/AURA/NSF:** Adam Block (b). **Royal Astronomical Society:** (fclb). **25 American Institute of Physics, Emilio Segre Visual Archives:** (bl). **Corbis:** Bettmann (br). **Galaxy Picture Library:** (tr). **26 Corbis:** Roger Ressmeyer (tl). **Getty Images:** Time Life Pictures / Mansell (b). **Royal Astronomical Society:** (cr). **27 Corbis:** Bettmann (tr). **NASA:** DMR, COBE, Two-Year Sky Map (crb). **Science Photo Library:** Hale Observatories (cl). **28 Alamy Images:** Popperfoto (b). **NASA:** MSFC (tl). **29 Corbis:** Bettmann (br). **DK Images:** (tr). **Science Photo Library:** Novosti (cla); Detlev Van Ravenswaay (bl). **30 Alamy Images:** Popperfoto (tl). **NASA:** Langley Research Center (bl). **Science Photo Library:** (cr). **31 NASA:** Johnson Space Center (cla); Saturn Apollo Program (cl); Saturn Apollo Program (r). **32 NASA:** Johnson Space Center (tl, cra, bl). **33 Getty Images:** NASA / Handout (b). **NASA:** (t). **34–35 NASA:** Johnson Space Center. **36 NASA:** (tl, br). **NSSDC/ GSFC/ NASA:** (cra, bl, br). **37 Getty Images:** Hulton Archive / Santi Visalli Inc. (cra). **NASA:** Courtesy JPL-Caltech (bl); Courtesy JPL-Caltech / Space Science Institute (br); Courtesy JPL-Caltech / Mars Exploration Rover (t). **38 Science & Society Picture Library:** Science Museum, London (cr). **39 Alamy Images:** Homer Sykes (b). **NASA:** Jeff Hester and Paul Scowen (Arizona State University) (tr). **Sudbury Neutrino Observatory:** Photo courtesy of Ernest Orlando Lawrence Berkeley National Laboratory (tl). **40 Denise Applewhite, Princeton University:** (c). **40–41 Corbis:** Roger Ressmeyer (b). **41 NASA:** ESA, P. Kalas and J. Graham (University of California, Berkeley), and M. Clampin (NASA's Goddard Space Flight Center) (c); M. Brown (Caltech), C. Trujillo (Gemini), D. Rabinowitz (Yale), NSF (cr). **42 Corbis:** Hulton-Deutsch Collection / Raymond S. Kleboe (cr). **43 Courtesy of the NAIC - Arecibo Observatory, a facility of the NSF:** (b). **Image Courtesy NRAO / AUI / NSF:** R. Perley, C. Carilli & J. Dreher (t). **44–45 NASA:** ESA, STScI, J. Hester and P. Scowen (Arizona State University). **46 NASA:** Courtesy JPL-Caltech/STScI/ CXC/SAO. **47 NASA:** HST. **48 Courtesy of Daisuke Kawata and Brad K. Gibson, Swinburne University of Technology:** (tl). **NASA:** GSFC. Image by Reto Stöckli, enhancements by Robert Simmon (cb). **Science Photo Library:** Detlev van Ravenswaay (cr). **49 Chandra X-Ray Observatory:** Kitt Peak (t). **Science Photo Library:** David A. Hardy. Futures: 50 Years in Space (cl). **50 Corbis:** Bettmann (ca). **51 Volker Wendel & Bernd Flach-Wilken (www.spiegelteam.de):** (cb). **53 Image Courtesy NRAO / AUI / NSF:** D. S. Adler, D. J. Westpfahl (br). **54–55 Andrey Kravtsov:** Simulations were performed at the National Center for Supercomputing Applications (Urbana-Champaign, Illinois) by Andrey Kravtsov (The Univ. of Chicago) and Anatoly Klypin (New Mexico State Univ.). Visualisations by Andrey Kravtsov (b). **NASA:** ESA, S. Beckwith (STScI) and the HUDF Team (t). **55 NASA:** Michael Corbin (CSC/STScI) (cl). **Science Photo Library:** NASA (clb, cb); NASA / ESA / STScI (cr); The Virgo collaboration / CCO 1.0 / SCIENCE PHOTO LIBRARY (tr). **56–57 NASA and The Hubble Heritage Team (AURA/STScI):** ESA, H. Teplitz and M. Rafelski (IPAC / Caltech), A. Koekemoer (STScI), R. Windhorst (Arizona State University), and Z. Levay (STScI). **58 European Southern Observatory:** MPG / ESO 2.2m + WFI (br). **59 Science Photo Library:** Harvard College Observatory (tr). **60 NASA:** ESA, and The Hubble Heritage Team (STScI/AURA) (bl). **62 NASA:** ESA, HEIC, and The Hubble Heritage Team (STScI/AURA). **63 NASA:** H. Ford (JHU), G. Illingworth (UCSC/LO), M.Clampin (STScI), G. Hartig (STScI), the ACS Science Team, and ESA. **64 Corbis:** Bettmann (br). **NASA:** The Hubble Heritage Team (AURA/STScI/NASA) (cl). **67 Science Photo Library:** National Optical Astronomy Observatories (br). **69 David Malin Images:** Anglo-Australian Observatory / Photo by David Malin (bl). **NASA:** ESA, C.R. O'Dell (Vanderbilt University), and M. Meixner, P. McCullough (cla); Andrew Fruchter and the ERO Team [Sylvia Baggett (STScI), Richard Hook (ST-ECF), Zoltan Levay (STScI)] (tr). **70 European Southern Observatory:** VLT Kueyen + FORS2 (bl). **Galaxy Picture Library:** STScI (tl). **NASA:** Jeff Hester (Arizona State University) (br). **71 Science Photo Library:** NASA. **72–73 NASA:** The Hubble Heritage Team (AURA/STScI). **74 Getty Images:** Stocktrek Images (clb). **75 www.bridgeman.co.uk:** Private Collection (bl). **Science Photo Library:** Mount Stromlo and Siding Spring Observatories (tl). **77 NASA:** Andrea Dupree (Harvard-Smithsonian CfA), Ronald Gilliland (STScI) and ESA (br). **NOAO/AURA/NSF:** (tr). **The Picture Desk:** Art Archive / Royal Astronomical Society / Eileen Tweedy (tl). **78 NASA:** JPL-Caltech (crb). **Science Photo Library:** Chris Butler (bl). **79 David Malin Images:** Akira Fujii (r). **Volker Wendel & Bernd Flach-Wilken (www.spiegelteam.de):** (c). **Chandra X-Ray Observatory:** NASA / CXC / MIT / F.K. Baganoff et al. (bc). **Loke Tan:** (clb). **80 Sven Kohle, AlltheSky.com:** (br). **NASA:** ESA, and The Hubble Heritage Team (STScI/AURA) (bl). **NOAO/AURA/NSF:** Adam Block (tr). **Science Photo Library:** MPIA-HD, Birkle, Slawik (tl). **81 Chandra X-Ray Observatory:** X-ray (NASA/CXC/M. Karovska et al.); Radio 21-cm image (NRAO/VLA/J.Van Gorkom/Schminovich et al.); Radio continuum image (NRAO/VLA/J. Condon et al.); Optical (Digitized Sky Survey U.K. Schmidt Image/STScI). **82 Courtesy of the Archives, California Institute of Technology:** (br). **NASA:** N. Benitez (JHU), T. Broadhurst (The Hebrew University), H. Ford (JHU), M. Clampin (STScI), G. Hartig (STScI), G. Illingworth (UCO/Lick Observatory), the ACS Science Team and ESA (bl). **83 NASA:** JPL-Caltech (cb). **European Southern Observatory:** (cla); VLT / NACO (tr). **84 NASA:** John Spencer (Lowell Observatory). **85 NASA:** SDO (bc). **86 NASA:** (l). **88 DK Images:** NASA (t). **European Space Agency:** DLR/FU Berlin (G. Neukum) (br). **89 DK Images:** NASA (bc). **Science Photo Library:** Pekka Parviainen (br). **Seán Doran:** NASA / SwRI / MSSS / Gerald Eichstädt / Seán Doran (cr). **NASA:** JPL / Space Science Institute (cl). **90 Science Photo Library:** National Optical Astronomy Observatories (bc). **91 DK Images:** NASA. **Science Photo Library:** John Chumack (bl, br). **Courtesy of SOHO / EIT Consortium. SOHO is a project of international cooperation between ESA and NASA:** (tl). **92 NASA:** (clb). **Corbis:** Francesco Muntada (br). **Science Photo Library:** Jerry Lodriguss (cra). **93 Alamy Images:** John Prior Images (cl, c, cr, fcr). **DK Images:** Andy Crawford (bl). **94–95 NASA:** Image courtesy of the Lockheed Martin team of NASA's TRACE Mission. **96 NASA:** Johns Hopkins University Applied Physics Laboratory / Carnegie Institution of Washington (br). **97 NASA:** JHU / APL (cr). **Science Photo Library:** John Sanford (br). **98 NASA:** Courtesy JPL / Caltech (b, clb). **98–99 planetary.org:** Mattias Malmer, from NASA / JPL data (c). **99 DK Images:** Courtesy of ESA / James Stevenson (cb). **Galaxy Picture Library:** Robin Scagell (br). **100 Corbis:** Layne Kennedy (clb); Kevin Schafer (b). **101 European Space Agency:** Denmann production (cr). **Getty Images:** Photographer's Choice / Tom Walker (crb). **102–103 NASA:** Earth Observatory. **104 Robert Gendler:** (tl). **105 DK Images:** NASA (bl, br). **NASA:** Johnson Space Center (bc). **106 DK Images:** Bruce Forster (cr). **108–109 NASA. 107 Corbis:** Reuters / Juan Carlos Ulate (bc). **109 European Space Agency:** (c). **110 DK Images:** Alistair Duncan (clb). **Galaxy Picture Library:** Robin Scagell (bl, bc). **NASA:** Courtesy JPL / Caltech / U.S. Geological Survey (br). **111 Galaxy Picture Library:** ESO (tr); Thierry Legault (crb). **NASA:** Courtesy JPL / Caltech / U.S.Geological Survey (cra); NSSDC (cr). **112–113 NASA. 114 NASA:** Courtesy JPL / Caltech / Cornell (tl). **114–115 NASA:** JPL-Caltech / MSSS (b). **115 NASA:** Courtesy JPL / Caltech (crb). **116 European Space Agency:** DLR/FU Berlin (G. Neukum) (fcla, br). **NASA:** Courtesy JPL / Caltech / Cornell (tl); JPL / Viking Project (fcl). **117 NASA:** JPL-Caltech / MSSS (br). **Galaxy Picture Library:** Damian Peach (br); Robin Scagell (tr, cl, c). **118–119 Corbis. 120–121 ESA / Hubble:** NASA, ESA and A. Simon (GSFC) (c). **121 NASA:** JPL (clb); Courtesy JPL / Caltech (crb). **122 NASA:** Courtesy JPL / Caltech / PIRL / University of Arizona (br); Courtesy JPL / Caltech / DLR and Brown University (cla); Courtesy JPL / Caltech / DLR (cra); Courtesy JPL / Caltech / University of Arizona / LPL (fclb, clb, crb, bc); Courtesy JPL / Caltech / Arizona State University (fcrb). **123 Galaxy Picture Library:** Damian Peach (br); Robin Scagell (bl, bc). **NASA:** Courtesy JPL / Caltech / DLR and Brown University (ca); Courtesy JPL / Caltech (cra). **ESA / Hubble:** NASA, ESA (cla). **124–125 Seán Doran:** NASA / SwRI / MSSS / Gerald Eichstädt. **127 NASA:** Courtesy JPL / Caltech / University of Colorado (br); Courtesy JPL / Caltech / Space Science Institute (crb). **128 NASA:** Courtesy JPL / Caltech (tl); Courtesy JPL / Caltech / Space Science Institute (c, cr, br); JPL-Caltech / Space Science Institute (bl); JPL / University of Arizona (clb). **129 NASA:** JPL-Caltech / Space Science Institute (cl). **akg-images:** Huygens Museum Hofwijck / Nimatallah (cr). **Galaxy Picture Library:** Robin Scagell (bl, bc); Dave Tyler (br). **NASA:** Courtesy JPL / Caltech (cla) **130–131 NASA:** JPL-Caltech. **132 W.M. Keck Observatory:** Courtesy Lawrence Sromovsky, UW-Madison Space Science and Engineering Center (bl). **133 Galaxy Picture Library:** Ed Grafton (bc). **NASA:** Courtesy JPL / Caltech (tl, fcr, crb). **134 NASA:** Courtesy JPL / Caltech (clb). **135 Galaxy Picture Library:** Maurice Gavin (br). **NASA:** Courtesy JPL / Caltech (clb); Courtesy JPL / Caltech / U.S. Geological Survey (crb). **136–137 NASA:** JHUAPL / SwRI. **136 NASA. 137 NASA:** JHUAPL / SwRI (cr). **138 NASA:** Courtesy JPL / Caltech (cl). **Science Photo Library:** Walter Pacholka, Astropics (b). **139 Galaxy Picture Library:** Juan Carlos Casado (bl). **Alamy Images:** Visual&Written SL (cr). **ESA:** J. Huart (tr); Rosetta / MPS for OSIRIS Team MPS / UPD / LAM / IAA / SSO / INTA / UPM / DASP / IDA (cla). **140–141 ESA. 143 Corbis:**

Steve Kaufman (b). **DK Images:** (crb) NASA (t); Courtesy of the Natural History Museum, London / Colin Keates (fcrb (stony), fcrb (iron)). **NASA:** JPL-Caltech / UCLA / MPS / DLR / IDA (tr). **144–145 Science Photo Library:** Pekka Parviainen. **146 Science Photo Library:** Frank Zullo. **147 Image courtesy of Celestron. 149 Courtesy of Peter Wienerroither. 150–151 Courtesy of Peter Wienerroither:** (b). **151 Galaxy Picture Library:** Robin Scagell (cla). **153 Science Photo Library:** John Foster (cr). **154 Dave King:** (b). **155 156 Dave King:** (bl, bc). **157 Dave King:** (br). **158 Galaxy Picture Library:** Chris Picking (cb, crb). **Dave King:** (cra). **160 DK Images:** Andy Crawford (cla). **Galaxy Picture Library:** Optical Vision (tr). **Dave King:** (tl, bl). **161 Dave King. 162 Corbis:** Roger Ressmeyer (br). **Dave King:** (tl, bl). **163 Galaxy Picture Library:** Philip Perkins (tl). **Dave King:** (cr, bl). **164–165 Thanakrit Santikunaporn. 166 DK Images:** Till Credner www.allthesky.com. **167 DK Images:** Courtesy of the National Maritime Museum, London / Tina Chambers. **168 123RF.com:** Alzam (clb). **170 Alamy Stock Photo:** Alan Dyer / Stocktrek Images (cla). **ESO:** ESO / C. Malin (clb). **173 DK Images:** Till Credner www.allthesky.com (tr). **NOAO/AURA/NSF:** Joe and Gail Metcalf / Adam Block (bc). **174 Galaxy Picture Library:** Robin Scagell. **175 Volker Wendel & Bernd Flach-Wilken (www. spiegelteam.de). 176 Matt BenDaniel. 177 NOAO/AURA/NSF:** Hillary Mathis, N.A. Sharp. **178 Galaxy Picture Library:** Robin Scagell. **180 NOAO/AURA/NSF. 181 DK Images:** Till Credner (www. allthesky.com) (bl). **William McLaughlin:** (t). **182 NOAO/AURA/NSF:** Bill Uminski and Cyndi Kristopeit / Adam Block (bl); Jon and Bryan Rolfe / Adam Block (cra). **183 Galaxy Picture Library:** Damian Peach. **184 NOAO/AURA/NSF:** Tom Bash and John Fox / Adam Block. **185 Galaxy Picture Library:** Nik Szymanek. **186 DK Images:** Till Credner (www.allthesky.com) (bl). **William McLaughlin:** (br). **187 Volker Wendel & Bernd Flach-Wilken (www.spiegelteam.de). 188 William McLaughlin. 190–191 Science Photo Library:** Tony & Daphne Hallas. **192 NOAO/AURA/NSF:** N.A. Sharp. **193 DK Images:** Till Credner (www.allthesky.com) (cr). **Galaxy Picture Library:** Robin Scagell (fcr). **194 Galaxy Picture Library:** Robin Scagell (br). **David Malin Images:** Pasachoff / Caltech (bl). **195 Galaxy Picture Library:** Michael Stecker. **196 Galaxy Picture Library:** Robin Scagell. **197 NOAO/AURA/NSF:** Tom Bash and John Fox / Adam Block. **198 ESA / Hubble:** NASA & ESA / A. Riess (STScI) (b). **199 NASA:** (b). **200 David Malin Images:** Akira Fujii. **201 NOAO/AURA/NSF:** Morris Wade / Adam Block (br); Adam Block (cr). **202 ESA / Hubble:** NASA, ESA, and E. Perlman (Florida Institute of Technology) / Judy Schmidt (t). **203 Till Credner / AlltheSky.com:** (b). **204 David Malin Images:** Anglo-Australian Observatory / Photo by David Malin. **205 NOAO/AURA/NSF:** Jay Ballauer / Adam Block. **206 NASA:** ESA, STScI and P. Dobbie (University of Tasmania) (t). **207 NASA:** ESA, and the Hubble Heritage Team (STScI / AURA) (b). **208 Galaxy Picture Library:** Robin Scagell. **209 Galaxy Picture Library:** Robin Scagell (bl). **NOAO/AURA/NSF:** Joe and Gail Metcalf / Adam Block (br). **210 NASA:** ESA, Hubble Space Telescope (r). **NOAO/AURA/NSF:** Joe and Gail Metcalf / Adam Block (br). **211 Till Credner / AlltheSky.com. 212 NOAO/AURA/NSF:** Adam Block. **213 William McLaughlin. 214 Galaxy Picture Library:** Robin Scagell. **215 NOAO/AURA/NSF:** François and Shelley Pelletier / Adam Block. **216 DK Images:** Till Credner (www.allthesky.com). **NOAO/AURA/NSF:** Ryan Steinberg and Family / Adam Block (cl). **217 ESA / Hubble:** John Corban & the ESA / ESO / NASA Photoshop FITS Liberator (t). **219 NOAO/AURA/NSF:** Michael Petrasko and Muir Eveden / Adam Block.

220–221 NASA: STScI. **222 Galaxy Picture Library:** Robin Scagell. **223 Volker Wendel & Bernd Flach-Wilken (www.spiegelteam.de):** (br). **NOAO/AURA/NSF:** Adam Block (bl). **224 Daniel Verschatse - Observatorio Antilhue - Chile. 225 NASA:** X-ray / CXC / Univ. of Alabama / K.Wong et al / Optical / ESO / VLT (b). **226 NASA:** RX J1131 (b). **227 NOAO/AURA/NSF:** Bob and Bill Twardy / Adam Block. **228 NOAO/AURA/NSF:** Adam Block. **229 Galaxy Picture Library:** 2MASS (cra); Gordon Garradd (b). **230 Volker Wendel & Bernd Flach-Wilken (www.spiegelteam.de):** (bl). **DK Images:** Till Credner (www.allthesky.com) (br). **231 Volker Wendel & Bernd Flach-Wilken (www.spiegelteam.de):** (t). **232–233 Science Photo Library:** Allan Morton / Dennis Milon. **234 NOAO/AURA/NSF:** Allan Cook / Adam Block (bl); N.A.Sharp, Mark Hanna, REU program (br). **235 Alamy Stock Photo:** NASA Image Collection (b). **236 ESA / Hubble:** NASA, The Hubble Heritage Team (STScI / AURA)-ESA / Hubble Collaboration and A. Evans (University of Virginia, Charlottesville / NRAO / Stony Brook University) (t). **237 NASA:** ESA, and R. Sharples (University of Durham) (b). **238 ESA / Hubble:** NASA / D. Calzetti (University of Massachusetts) and the LEGUS Team (b). **239 NASA:** ESA / Hubble (b). **240 ESO:** (b). **241 NOAO/AURA/NSF:** Nicole Bies and Esidro Hernandez / Adam Block. **242 Galaxy Picture Library:** DSS (bl). **NOAO/AURA/NSF:** Adam Block (br). **243 NASA:** JPL / Caltech / SSC (b). **244 NASA and The Hubble Heritage Team (AURA/STScI):** ESA, and the Hubble Heritage Team (STScI / AURA) (b). **245 NOAO/AURA/NSF. 246 Galaxy Picture Library:** Chris Picking (fbl). **Daniel Verschatse - Observatorio Antilhue - Chile:** (b). **247 Volker Wendel & Bernd Flach-Wilken (www.spiegelteam.de). 248–249 NASA:** ESA, and the Hubble Heritage Project (STScI / AURA). **250 Galaxy Picture Library:** Yoji Hirose. **251 Daniel Verschatse - Observatorio Antilhue - Chile. 252 NASA and The Hubble Heritage Team (AURA/STScI):** Andrew S. Wilson (University of Maryland); Patrick L. Shopbell (Caltech); Chris Simpson (Subaru Telescope); Thaisa Storchi-Bergmann and F. K. B. Barbosa (UFRGS, Brazil); and Martin J. Ward (University of Leicester, U.K.) (b). **253 ESO:** (b). **254 ESA / Hubble:** ESA, the Hubble Heritage Team (STScI / AURA)-ESA / Hubble Collaboration and A. Evans (University of Virginia, Charlottesville / NRAO / Stony Brook University) (t). **255 Alamy Stock Photo:** Robert Gendler / Stocktrek Images (b). **256 NASA:** CXC / JPL-Caltech / CfA (b). **257 ESA / Hubble:** J. Barrington (t). **258 NASA:** ESA / Hubble / R. Tugral (b). **259 ESO:** (bc). **260 NASA:** X-ray / CXC / MIT / M. McDonald et al.; Optical: NASA / STScI (b). **261 Volker Wendel & Bernd Flach-Wilken (www.spiegelteam.de):** (cra). **Daniel Verschatse - Observatorio Antilhue - Chile:** (br). **262 NASA:** ESA and the Hubble Heritage Team (STScI / AURA) (b). **263 ESO:** (clb). **NASA:** Space Telescope Science Institute (br). **265 Volker Wendel & Bernd Flach-Wilken (www.spiegelteam.de). 266–267 NASA and The Hubble Heritage Team (AURA/STScI):** ESA, and E. Sabbi (STScI). **268 NASA:** Hubble Heritage Team (Aura / Stsci), J. Higdon (Cornell) Esa (t). **269 Volker Wendel & Bernd Flach-Wilken (www.spiegelteam.de). 270 ESO:** (r). **271 ESA / Hubble:** NASA (b). **272 Volker Wendel & Bernd Flach-Wilken (www.spiegelteam.de):** (br). **Galaxy Picture Library:** Gordon Garradd (crb). **273 Alamy Images:** Adam van Bunnens. **274 Matt BenDaniel. 324 Corbis:** Roger Ressmeyer. **325 Science Photo Library:** Rev. Ronald Royer.

All other images © Dorling Kindersley
For further information see: www.dkimages.com